THE NEXT WAVE IN COMPUTING, OPTIMIZATION, AND DECISION TECHNOLOGIES

OPERATIONS RESEARCH/COMPUTER SCIENCE INTERFACES SERIES

Other published titles in the series:

Greenberg /*A Computer-Assisted Analysis System for Mathematical Programming Models and Solutions: A User's Guide for ANALYZE*
Greenberg / *Modeling by Object-Driven Linear Elemental Relations: A Users Guide for MODLER*
Brown & Scherer / *Intelligent Scheduling Systems*
Nash & Sofer / *The Impact of Emerging Technologies on Computer Science & Operations Research*
Barth / *Logic-Based 0-1 Constraint Programming*
Jones / *Visualization and Optimization*
Barr, Helgason & Kennington / *Interfaces in Computer Science & Operations Research: Advances in Metaheuristics, Optimization, & Stochastic Modeling Technologies*
Ellacott, Mason & Anderson / *Mathematics of Neural Networks: Models, Algorithms & Applications*
Woodruff / *Advances in Computational & Stochastic Optimization, Logic Programming, and Heuristic Search*
Klein / *Scheduling of Resource-Constrained Projects*
Bierwirth / *Adaptive Search and the Management of Logistics Systems*
Laguna & González-Velarde / *Computing Tools for Modeling, Optimization and Simulation*
Stilman / *Linguistic Geometry: From Search to Construction*
Sakawa / *Genetic Algorithms and Fuzzy Multiobjective Optimization*
Ribeiro & Hansen / *Essays and Surveys in Metaheuristics*
Holsapple, Jacob & Rao / *Business Modelling: Multidisciplinary Approaches — Economics, Operational and Information Systems Perspectives*
Sleezer, Wentling & Cude/*Human Resource Development And Information Technology: Making Global Connections*
Voß & Woodruff / *Optimization Software Class Libraries*
Upadhyaya et al / *Mobile Computing: Implementing Pervasive Information and Communications Technologies*
Reeves & Rowe / *Genetic Algorithms—Principles and Perspectives: A Guide to GA Theory*
Bhargava & Ye / *Computational Modeling And Problem Solving In The Networked World: Interfaces in Computer Science & Operations Research*
Woodruff / *Network Interdiction And Stochastic Integer Programming*
Anandalingam & Raghavan / *Telecommunications Network Design And Management*
Laguna & Martí / *Scatter Search: Methodology And Implementations In C*
Gosavi/ *Simulation-Based Optimization: Parametric Optimization Techniques and Reinforcement Learning*
Koutsoukis & Mitra / *Decision Modelling And Information Systems: The Information Value Chain*
Milano / *Constraint And Integer Programming: Toward a Unified Methodology*
Wilson & Nuzzolo / *Schedule-Based Dynamic Transit Modeling: Theory and Applications*

THE NEXT WAVE IN COMPUTING, OPTIMIZATION, AND DECISION TECHNOLOGIES

Edited by

BRUCE GOLDEN
University of Maryland

S. RAGHAVAN
University of Maryland

EDWARD WASIL
American University

Bruce Golden	S. Raghavan	Edward Wasil
University of Maryland	University of Maryland	American University
USA	USA	USA

Library of Congress Cataloging-in-Publication Data

The next wave in computing, optimization, and decision technologies / edited by
Bruce Golden, S. Raghavan, Edward Wasil.
p. cm. – (Operations research/computer science interfaces series ; ORCS 29)
Includes bibliographical references.
ISBN 1-4614-9851-1 (hc.: alk. paper) ISBN 0-387-23529-9 (eBook)
1. Computer science. 2. Computers. I. Golden, Bruce L., 1950- II. Raghavan, S. (Subramanian) III. Wasil, Edward A. IV. Series.

QA76.16.N49 2005
004—dc22

2004059048

9 8 7 6 5 4 3 2 1 SPIN 11054771

springeronline.com

Contents

Preface

This book is the companion volume to the Ninth INFORMS Computing Society Conference (ICS 2005), held in Annapolis, Maryland, from January 5 to 7, 2005. It contains 25 high-quality research articles that focus on the interface between operations research/management science (OR/MS) and computer science (CS).

The articles in this book were each carefully reviewed and revised accordingly. We thank the authors and a small group of academics and practitioners for serving as referees. The book is divided into six sections. The first section contains two papers on network models. The second section focuses on integer and mixed integer programming. The third section contains papers in which heuristic search and metaheuristics are applied. Three papers using stochastic modeling comprise the fourth section. In the fifth section, the unifying theme is software and modeling. The sixth section contains four papers on classification, clustering, and ranking.

Taken collectively, these articles are indicative of the state-of-the-art in the interface between OR/MS and CS and of the high-caliber research being conducted by members of the INFORMS Computing Society.

We thank the University of Maryland, American University, and George Mason University for sponsoring ICS 2005. In addition, we thank the authors for their hard work and professionalism and Stacy Calo for her invaluable help in producing this book. Finally, we note, with great pride, that two of us (BG and EW) have attended each and every one of the nine ICS conferences. The three of us hope to attend many more.

BRUCE GOLDEN, S. RAGHAVAN, AND EDWARD WASIL

I

NETWORKS

ON THE COMPLEXITY OF DELAYING AN ADVERSARY'S PROJECT

Gerald G. Brown,[1] W. Matthew Carlyle, [2] Johannes O. Royset,[3] and R. Kevin Wood[4]

Naval Postgraduate School
Monterey, CA 93943

[1] gbrown@nps.edu

[2] mcarlyle@nps.edu

[3] joroyset@nps.edu

[4] kwood@nps.edu

Abstract A "project manager" wishes to complete a project (e.g., a weapons-development program) as quickly as possible. Using a limited interdiction budget, an "interdictor" wishes to delay the project's overall completion time by interdicting and thereby delaying some of the project's component tasks. We explore a variety of PERT-based interdiction models for such problems and show that the resulting problem complexities run the gamut: polynomially solvable, weakly NP-complete, strongly NP-complete or NP-hard. We suggest methods for solving the problems that are easier than worst-case complexity implies.

Keywords: Interdiction, PERT, NP-complete

1. Introduction

Brown et al. (2004) (see also Reed 1994 and Skroch 2004) model the completion of an adversarial nation's nuclear-weapons program using general techniques of PERT. (See PERT 1958 and Malcolm et al. 1959 for the original descriptions of PERT, and see Moder et al. 1983 for a comprehensive review.) Brown et al. (2004) ask the question: How do we most effectively employ limited interdiction resources, e.g., military strikes or embargoes on key materials, to delay the project's component tasks, and thereby delay its overall completion time? They answer the question by describing an interdiction model that maximizes minimum project-completion time. This model is a Stackelberg game (von Stackelberg 1952), formulated as a bilevel integer-linear program (Moore and Bard 1990).

Brown et al. (2004) consider a highly general model for project networks. Specifically, they (*i*) allow the interdictor to employ various interdiction resources, (*ii*) allow the project manager to "crash" the project to speed project completion by applying various, constrained, task-expediting resources, and (*iii*) allow the project manager to employ alternative technologies to complete the project. The authors successfully test an algorithm that solves a realistic example of the resulting interdiction problem, but we shall see that the most general problem is NP-hard. Thus, other large, general problems could be extremely difficult to solve.

This paper therefore asks: How hard is the "project interdiction problem" when full modeling generality is unnecessary? Can we assure analysts that their version of the problem is not too difficult if modeling a single interdiction resource suffices; or crashing is impossible; or only a single technology, or modest number of technologies, need be modeled?

We show that these less general problems are, in fact, easier to solve, and go on to describe special solution techniques for them. All of these techniques are simpler than the decomposition algorithm described by Brown et al. (2004), which requires that an alternating sequence of two integer-linear programs be solved. Thus, simpler, more accessible and more efficient solution methods may be employed for these problem restrictions.

Before beginning mathematical developments, we note that we have chosen the activity-on-arc (AOA) model of a project network rather than the interchangeable activity-on-node (AON) model. The AON model is the more common of the two nowadays; however, the mathematics in this paper prove easier to describe using the AOA model, so we adopt that model from the outset.

The next section provides basic definitions for our project interdiction problems. Section 3 describes the most general model, which includes multiple technologies and project crashing. Subsequent sections discuss restricted model variants and solution techniques for them.

2. Basic Definitions

Let $G = (N, A)$ denote a *directed acyclic graph* with *node set* N and *arc set* $A \subset N \times N$. Since G is acyclic, there exists a *topological ordering*, or labeling, $1, 2, \ldots, |N|$ of the nodes $i, j \in N$ such that $i < j$ for each arc $k = (i, j) \in A$. For graphs of interest in this paper, the first node a in any such ordering is unique, as is the last node b. The *forward star* of node i, $FS(i) \subseteq A$, is the set of all arcs of the form $k = (i, j)$; the *reverse star* of node i, $RS(i) \subseteq A$, is the set of all arcs of the form $k = (j, i)$.

G represents the *activity-on-arc* diagram used in a PERT model of a project, controlled by a project manager (e.g., Elmaghraby 1977). Each arc $k \in A$ corresponds to a *task* which must be completed in order to finish the project. For each node $i \in N$, all tasks $k \in RS(i)$ must be completed before any task $k \in FS(i)$ can begin. Every node $i \in N$ represents a *milestone event* that occurs when all predecessor tasks, i.e., all $k \in RS(i)$ are complete. A milestone event might be something important like "completion of the weapon delivery system," or might simply correspond to the completion of a group of simpler tasks along the course of the project. The latter situation may occur frequently in AOA representations of projects which often have many dummy nodes (and arcs). Node b is the *project-completion event* and, because event i may also be viewed as the start of follow-on tasks $k \in FS(i)$, node a is the *project-start event.*

Each activity k has associated with it a nominal *task completion time* $t_k \geq 0$ and a variable e_k, $0 \leq e_k \leq \bar{e}_k \leq t_k$, which denotes the reduction in the activity's completion time achieved by applying *expediting resources.* No matter how much expediting resource is a applied, however, task k cannot be completed any faster than the crashed duration, $t_k - \bar{e}_k$. For simplicity in writing models, but without loss of generality, we assume that only a single expediting resource exists (e.g., money); the unit cost of expediting task k is m_k; and a total *expediting budget* of m_0 monetary units is available to the project manager. We assume that the project manager schedules tasks in order to minimize the project's completion time. It is well known that the shortest completion time, for fixed expediting decisions, corresponds to a longest a-b path in G.

An interdictor who wishes to disrupt the project possesses a set of *interdiction resources* with which to effect this disruption. Interdiction of arc k consumes $c_{rk} \in Z^+$ units of each interdiction-resource type $r \in R$, and results in adding a *delay*, $d_k \in Z^+$, to the completion time of task k. The total *interdiction budget* for resource r is $c_{r0} \in Z^+$.

If we assume that no expediting will occur, the project-interdiction model looks much like the shortest-path interdiction model of Israeli and Wood (2002). There, the interdictor attacks a road network using limited interdiction resources, and the "network user," analogous to the project manager, moves along a post-interdiction shortest path in the network. In that model, an interdiction plan is evaluated by solving a shortest-path in a general network. Our simplest model can evaluate an interdiction plan by solving a longest-path problem in an acyclic network. However, this evaluation will require the solution of a more general linear or integer-linear program if the project manager can crash his project or can employ multiple technologies, as described below. Thus, project interdiction is truly a "system-interdiction problem" (Israeli and Wood 2002), not a network-interdiction problem.

Crowston and Thompson (1967) describe an extension of project management models in which the project manager can complete a project using alternative technologies. Brown et al. (2004) use this extension to model different means of uranium enrichment. Crowston and Thompson create graphical constructs to represent alternative technologies in their AON model, but they boil down to this in the mathematical model: Using binary variables to represent whether or not a particular technology is used, certain precedence relationships will be enforced and certain others will be relaxed.

Brown et al. (2004) also include in their model several different types of precedence relationships between tasks (Elmaghraby 1977). We do not specify details, but all models in this paper can be easily adjusted for these more general precedence relationships. A fixed "lag time" may also be interjected between any pair of tasks, if required.

3. Project Interdiction Model

Here we define the general project interdiction model, **MAXMIN0**. We assume the unit of time is "(one) week" and that each interdiction resource r is measured in "r-dollars:"

MAXMIN0

Indices and Index Sets
$i, j \in N$ generic milestone events
$a, b \in N$ project start event and project completion event, respectively
$k \in A$ tasks and precedence relationships ($k = (i, j) \in A$)

Data [units]
t_k task duration [weeks]
m_k per-unit expediting cost of task k [dollars/week]
m_0 total expediting budget [dollars]
$\bar{e}_k$ maximum expediting of task k [weeks]
d_k interdiction delay of task k [weeks]
c_{rk} interdiction cost for task k, resource r [r-dollars/week]
c_{r0} total amount of interdiction resource r available [r-dollars]
M a sufficiently large constant, e.g., $M = \sum_{k \in A}(t_k + d_k)$ [weeks] (used to relax precedence constraints)

Decision Variables [units]

s_i completion time of event i [weeks]
e_k amount that task k is expedited [weeks]
w_i 1 if technology at node i is used, else 0
x_k 1 if task k is interdicted, else 0

Formulation (**MAXMIN0**)

$$z_0^* \equiv \max_{\mathbf{x}\in\mathbf{X}} \min_{\mathbf{s},\mathbf{e},\mathbf{w}} s_b \quad (1)$$

$$\text{s.t. } s_j - s_i + e_k + M(1-w_i) \geq t_k + d_k x_k \ \forall\, k=(i,j) \quad (2)$$

$$e_k \leq \bar{e}_k \ \forall\, k \in A \quad (3)$$

$$\sum_{k\in A} m_k e_k \leq m_0 \quad (4)$$

$$\mathbf{w} \in \mathbf{W} \quad (5)$$

$$w_i = 1 \ \forall\, i \in N - N_T \quad (6)$$

$$s_a = 0 \quad (7)$$

$$s_i \geq 0 \ \forall\, i \in N \quad (8)$$

$$e_k \geq 0 \ \forall\, k \in A \quad (9)$$

$$w_i \in \{0,1\} \ \forall\, i \in N. \quad (10)$$

where

$$\mathbf{X} \equiv \left\{ \mathbf{x} \in \{0,1\}^{|A|} \,\middle|\, \sum_{k\in A} c_{rk} x_k \leq c_{r0} \ \forall\, r \in R \right\}, \quad (11)$$

and where the set $\mathbf{W} \subset \{0,1\}^{|N|}$ represents all feasible combinations of alternative technologies.

For a fixed interdiction plan $\mathbf{x} = \hat{\mathbf{x}}$, the inner minimization in **MAXMIN0** is the project manager's problem: Compute the earliest project-completion time through the objective (1), subject to standard precedence constraints (2). Assuming all $w_i = 1$ so that all terms $M(1-w_i) = 0$, these constraints state that if activity $k = (i,j)$ exists between events i and j, then event j can occur no sooner than $s_i + t_k - e_k + d_k\hat{x}_k$. For $i \in N_T$, the term $M(1-w_i)$ simply relaxes all constraints for $k \in FS(i)$ when the alternative technology associated with i is not used, i.e., if $w_i = 0$. Constraint (4) reflects the project manager's limited budget for expediting tasks.

The interdictor controls the vector $\mathbf{x}$, and will use his limited interdiction resources (constraints 11) to maximize the project manager's minimum time to project completion. This is represented by the outer maximization in **MAXMIN0**.

The formulation **MAXMIN0** clarifies the opposing forces in our "Stackelberg interdiction game." The key features of this game are: (*i*) A "leader," i.e., the interdictor, first takes his actions, (*ii*) the "follower," i.e., the project manager, sees these actions and responds optimally, and (*iii*) the game finishes. Randomized strategies, as in two-person zero-sum games, are irrelevant here because the leader has complete information regarding the follower's behavior, and the follower will not act until after obtaining complete information about the leader's actions.

If we view **MAXMIN0** as the interdictor's optimization problem

$$z_0^* \equiv \max_{\mathbf{x}\in\mathbf{X}} z(\mathbf{x}), \tag{12}$$

where $z(\mathbf{x})$ defines the value of the resulting minimization problem for any value of $\mathbf{x}$, it is easy to see that the problem may be unusually difficult: Just to evaluate a potential interdiction plan $\hat{\mathbf{x}}$, i.e., just to compute $z(\hat{\mathbf{x}})$, requires the solution of an integer-linear program (ILP). If that ILP corresponds to an NP-hard problem, then **MAXMIN0** is NP-hard. In the following, we consider some special cases that are not quite that difficult.

4. One Technology, No Expediting

Suppose that a fixed set of technologies will be used, so $N_T = \emptyset$, and $w_i = 1$ for all $i \in N$. Further, assume that expediting is impossible, i.e., $e_k = 0$ for all $k \in A$. Then, **MAXMIN0** simplifies to:

MAXMIN1

$$z_1^* \equiv \max_{\mathbf{x}\in\mathbf{X}} \min_{\mathbf{s}} s_b \tag{13}$$

$$\text{s.t. } s_j - s_i \geq t_k + d_k x_k \;\; \forall\, k = (i,j) \in A \tag{14}$$

$$s_a = 0 \tag{15}$$

$$s_i \geq 0 \;\; \forall\, i \in N \tag{16}$$

For the time being, we will also assume that only a single interdiction resource (e.g., dollars) need be modeled, so that $\mathbf{X}$ is replaced by

$$\mathbf{X}_1 \equiv \left\{ \mathbf{x} \in \{0,1\}^{|A|} \;\middle|\; \sum_{k\in A} c_k x_k \leq c_0 \right\} \tag{17}$$

For fixed $\mathbf{x} = \hat{\mathbf{x}}$, the inner minimization of **MAXMIN1** is a linear program (LP) with a corresponding dual. In fact, the inner minimization in **MAXMIN1** is the well-known "earliest project completion time" problem with the longest-path problem as its dual (e.g., Ahuja et al. 1993 pp. 732-737). Hence, fixing $\mathbf{x}$ temporarily, manipulating **MAXMIN1** slightly, taking the dual of the inner minimization, and releasing $\mathbf{x}$ leads to the following useful model:

$$z_1^* = \max_{\mathbf{x}\in\mathbf{X}_1} \max_{\mathbf{y}} \sum_k (t_k + x_k d_k) y_k \tag{18}$$

$$\text{s.t.} \sum_{k\in FS(i)} y_k - \sum_{k\in RS(i)} y_k = \begin{cases} 1 & \text{if } i = a \\ 0 & \text{if } i \in N - \{a,b\} \\ -1 & \text{if } i = b \end{cases} \tag{19}$$

$$y_k \geq 0 \ \forall\, k \in A \tag{20}$$

A max-max problem is a "simple" maximization, but the nonlinear, nonconcave objective function (18) is problematic. This model linearizes easily, however: Replace each arc k with a pair of arcs, k and k', with fixed lengths $t_k + d_k$ and t_k, respectively, and let x_k control which arc is part of the project manager's model:

MAXMAX1

$$z_1^* = \max_{\mathbf{x},\mathbf{y},\mathbf{y}'} \sum_{k\in A} (t_k + d_k) y_k + \sum_{k\in A} t_k y_k' \tag{21}$$

$$\text{s.t.} \sum_{k\in FS(i)} (y_k + y_k') - \sum_{k\in RS(i)} (y_k + y_k') = \begin{cases} 1 & \text{if } i = a \\ 0 & \text{if } i \in N - \{a,b\} \\ -1 & \text{if } i = b \end{cases} \tag{22}$$

$$y_k - x_k \leq 0 \ \forall\, k \in A \tag{23}$$

$$y_k, y_k' \geq 0 \ \forall\, k \in A \tag{24}$$

$$\mathbf{x} \in \mathbf{X}_1 \tag{25}$$

THEOREM 1 **MAXMAX1** *is solvable in* $O(c_0|A|)$ *time, i.e., in pseudo-polynomial time.*

Proof: **MAXMAX1** represents a singly-constrained longest-path problem in which traversal of arc k consumes c_k units of interdiction resource and traversal of arc k' consumes none. Thus, **MAXMAX1** may be solved through the following dynamic-programming recursion in $O(c_0|A|)$ time:

$$f(1,c) = 0 \text{ for } c = 0,\ldots,c_0 \tag{26}$$

$$f(i,c) = -\infty \ \forall\, i \in N,\ c < 0 \tag{27}$$

$$f(j,c) = \max \left\{ \begin{array}{l} f(j,c-1), \\ \max\limits_{k=(i,j)\in RS(j)} (f(i,c-c_k)+t_k+d_k), \\ \max\limits_{k=(i,j)\in RS(j)} (f(i,c)+t_k) \end{array} \right\}. \quad (28)$$

∎

Our next task is to show that **MAXMIN1** is weakly NP-complete. Later in the paper we will require the formality of decision problems to show NP-completeness, but here the reader should have no difficulty in seeing the equivalence of certain optimization problems and how that equivalence implies NP-completeness.

THEOREM 2 **MAXMIN1** *is weakly NP-complete.*
Proof: Define the *binary knapsack problem* (**BKP**) as

$$\max_{\mathbf{x}} \sum_{k\in A} d_k x_k \quad (29)$$

$$\text{s.t.} \sum_{k\in A} c_k x_k \leq c_0 \quad (30)$$

$$x_k \in \{0,1\} \ \forall\, k \in A. \quad (31)$$

BKP is known to be NP-complete (e.g., Garey and Johnson 1979, p. 247) and can be modeled as an instance of **MAXMAX1** as follows:

1 Let $t_k = 0$ for all $k \in A$.

2 Let each item k in the knapsack correspond to an arc k with length $t_k + d_k$ and with "traversal cost" c_k.

3 Place all arcs in series.

4 In parallel with each arc k place an arc k' with length $t_k = 0$ and no traversal cost.

This transformation shows that **MAXMAX1** is NP-hard. But Theorem describes a pseudo-polynomial solution procedure for **MAXMAX1**, so it must, in fact, be weakly NP-complete. Since **MAXMAX1** is equivalent to **MAXMIN1**, the result follows. ∎

If the interdictor is only limited by a specific number of interdictions, **MAXMIN1** becomes even easier:

COROLLARY 3 **MAXMIN1** *is solvable in* $O(|N||A|)$ *time when* $c_k = 1$ *for all* $k \in A$.

Proof: In this case, any value of $c_0 > |N| - 1$ is equivalent to $c_0 = N - 1$ since no *a*-*b* path in G can have more than $|N| - 1$ arcs. The complexity result in Theorem 2, plus equivalence of models, then yields the result. ∎

Being able to solve these problems by dynamic programming means that fairly large problems can be solved quite effectively. However, dynamic programming can, in fact, bog down and we suggest using the constrained-shortest-path algorithm of Carlyle and Wood (2003), which converts directly to longest paths in directed acyclic paths. These authors show orders of magnitude speedups over previously known methods, including standard dynamic-programming formulations. (See Handler and Zang 1980 for a basic reference on this topic.)

5. One Technology With Expediting

Suppose the project manager can expedite certain tasks, but still, only a single set of technologies exists. **MAXMIN0** simplifies to:

MAXMIN2

$$z_2^* \equiv \max_{\mathbf{x}\in\mathbf{X}} \min_{\mathbf{s},\mathbf{e}} s_b \tag{32}$$

$$\text{s.t. } s_j - s_i + e_k \geq t_k + d_k x_k \;\; \forall\, k = (i,j) \in A \tag{33}$$

$$e_k \leq \bar{e}_k \;\forall\, k \in A \tag{34}$$

$$\sum_{k\in A} m_k e_k \leq m_0 \tag{35}$$

$$s_a = 0 \tag{36}$$

$$s_i \geq 0 \;\; \forall\, i \in N \tag{37}$$

Similar to **MAXMIN1**, for fixed **x**, the inner minimization in **MAXMIN2** is an LP and we may thus take its dual. Doing that and manipulating the resulting model slightly leads to the following ILP:

MAXMAX2

$$z_2^* = \max_{\mathbf{x},\mathbf{y},\mathbf{y}',\pi_0} \sum_{k\in A}(t_k + d_k)y_k \; + \; \sum_{k\in A} t_k y_k' \\ -(m_0 - \sum_{k\in A} m_k \bar{e}_k)\pi_0 \tag{38}$$

$$\text{s.t. } \sum_{k\in FS(i)} (y_k + y_k')$$

$$-\sum_{k\in RS(i)} (y_k + y'_k) = \begin{cases} 1 & \text{if } i = a \\ 0 & \text{if } i \in N - \{a,b\} \\ -1 & \text{if } i = b \end{cases} \quad (39)$$

$$y_k - x_k \leq 0 \;\forall\, k \in A \quad (40)$$

$$y_k + y'_k - m_k\pi_0 \geq 0 \;\forall\, k \in A \quad (41)$$

$$y_k, y'_k \geq 0 \;\forall\, k \in A \quad (42)$$

$$\pi_0 \geq 0 \quad (43)$$

$$\mathbf{x} \in \mathbf{X} \quad (44)$$

We will next prove that **MAXMIN2** is NP-complete, or rather, that its associated decision problem, **MAXMIN2d**, is NP-complete. We need the formality of decision problems now, and the definition of **MAXMIN2d** is:

DEFINITION 4 **MAXMIN2d**. *Given: Data for* **MAXMIN2** *and threshold* $\underline{z}$. *Question: Does there exist an interdiction plan* $\mathbf{x}^*$ *such that the optimally expedited project (optimal for the project manager) has length at least* $\underline{z}$*?* ∎

And, we will use a transformation from **SETCOVERd** in the proof:

DEFINITION 5 **SETCOVERd**. *Given:* $N_2 \equiv \{m+1, m+2, \ldots, m+n\}$*, the "ground set" to be covered; subsets* $N_i \subseteq N_2$*, for* $i \in N_1 \equiv \{1, \ldots, m\}$*, and threshold* $\bar{n} \in Z^+$*. Question: Does there exist a set* $N'_1 \subset N_1$*, with* $|N'_1| \leq \bar{n}$*, such that* $\cup_{i\in N'_1} N_i = N_2$*?* ∎

For our purposes, it is easier to use **SETCOVERd** defined through the bipartite graph $G' \equiv (N_1, N_2, A')$, where $A' \equiv \{(i,j) | i \in N_1, j \in N_i \text{ for some } i\}$:
Does there exist a set $N'_1 \subset N_1$ *with* $|N'_1| \leq \bar{n}$ *such that* $\cup_{i\in N'_1} N_i = N_2$*?*

THEOREM 6 **MAXMIN2** *is strongly NP-complete.*
Proof: Since the decision version of **MAXMIN1** is NP-complete and it is a special case of **MAXMIN2d**, **MAXMIN2d** must be NP-hard. Because we can formulate an ILP to represent the optimization problem, **MAXMIN2d** must, in fact, be NP-complete. The only open questions is whether **MAXMIN2d** is NP-complete in the strong or weak sense. We will show that a standard set-covering problem, **SETCOVERd**, well-known to be strongly NP-complete, can be transformed into an instance of **MAXMIN2d**. The transformation will obviously not require an exponential increase in the size of this instance's data, so it will follow that **MAXMIN2d** is strongly NP-complete.

We are given an instance of **SETCOVERd**, defined as in Definition 5 through the bipartite graph $G' = (N_1, N_2, A)$ and the threshold parameter $\bar{n}$. Next, we form a corresponding instance of **MAXMIN2d** Create the directed, acyclic project network $G \equiv (N, A)$ from G' by adding two nodes, a and b, and two sets of arcs so that $N \equiv N_1 \cup N_2 \cup \{a, b\}$ and $A \equiv A' \cup A_1 \cup A_2$ where $A_1 \equiv \{(a, i) | i \in N_1\}$ and $A_2 \equiv \{(j, b) | j \in N_2\}$. Let $t_k = 1$ for all $k \in A$; let $d_k = 1$ for all arcs $k \in A_1$, and $d_k = 0$, otherwise; assume each arc $k \in A_2$ can be expedited by e_k, $0 \leq e_k \leq 1$; let the unit cost of expediting be 1; and assume a total of $|N_2| - 1$ units of expediting resource are available. The number $\bar{n} \in Z^+$ carries over directly from above.

So, we have created a directed acyclic network with three echelons of arcs, but only those in the first echelon may be interdicted (with any effect), and only those in the last may be expedited. The instance of **MAXMIN2d** is defined as: Does there exist a set of $\bar{n}$ or fewer interdictions of arcs in A_1 such that the longest path in G, with optimal expediting has length strictly greater than 3? The answer to this problem is "yes," if and only if the answer is "yes" to the original set-covering problem.

To see this, suppose that every collection of $\bar{n}$ subsets N_i leaves at least one element of N_2 uncovered. The corresponding interdiction plan interdicts arc (s, i) for each subset N_i. Because at least one node $j \in N_2$ is left uncovered in the set-covering problem, at least one arc (j, t) is not on an interdicted path. This means there is at least one path of length 3 in the network. Furthermore, the $N_2 - 1$ units of expediting resource suffice to reduce the length of all arcs in A_2 that are on interdicted paths to 0, and, hence, every interdicted path's length is dropped from 4 to 3. So, if the answer to **SETCOVERd** is "no," the answer to the corresponding instance of **MAXMIN2d** must be "no."

On the other hand, suppose that the answer to **SETCOVERd** problem is "yes." Interdict arcs corresponding to the cover as above. Then, the interdicted but unexpedited length of each path is 4, and the $|N_2| - 1$ units of expediting resource only suffice to reduce those path lengths to $3+1/|N_2|$. So, the answer to the corresponding instance of **MAXMIN2d** is "yes." ∎

Note that Theorem 6 holds also in the special case of $c_k = d_k = 1$, $t_k \in \{0, 1\}$ for all $k \in A$. Since **MAXMAX2** is a (linear) ILP, it can be solved by a standard LP-based branch-and-bound algorithm. In addition, **MAXMAX2** motivates a solution approach for the *general* problem **MAXMIN0** as described in the following.

6. Alternative Technologies

The discussion at the end of Section 2 implies that adding alternative technologies into the mix, i.e., going from **MAXMIN2** to the completely general model **MAXMIN0**, may move us from the realm of NP-complete problems into NP-hard problems that may not be in NP. This will be the case if, for fixed $\mathbf{x} = \hat{\mathbf{x}}$, the solution of **MAXMIN0** requires the solution of an NP-complete ILP. That is, just checking whether the interdictor's objective $z(\hat{\mathbf{x}})$ exceeds a specified threshold for a candidate solution $\hat{\mathbf{x}}$ requires the solution of an NP-complete problem, rather than the application of some polynomial-time procedure.

However, if no expediting is allowed we would like to know the resulting complexity of evaluating $z(\mathbf{x})$. That is, faced with a fixed set of task lengths $\hat{t}_k = t_k + d_k \hat{x}_k$, the project manager would like to solve the **DCPM**, the "decision CPM problem," (Crowston and Thompson 1967), which selects a set of alternative technologies by choosing $\mathbf{w} \in \mathbf{W}$ to minimize project completion time. We state **DCPMd**, the decision version of **DCPM**, in terms of deleting technologies (and represent the remaining technologies after deleting $\mathbf{w}$ by $\mathbf{1} - \mathbf{w}$) to help show its NP-completeness:

DEFINITION 7 **DCPMd**. *Given: A project network $G = (N, A)$ with arc lengths $\hat{t}_k \in Z^+$; constraints $\mathbf{w} \in \mathbf{W}$ indicating feasible sets of alternative technologies; and threshold $\underline{z} \in Z^+$. Question: Does there exist a set of technologies represented by $\mathbf{w}'$, with $\mathbf{1} - \mathbf{w}' \in \mathbf{W}$, such that the longest path in $G' = G - N'$ is no longer than $\underline{z}$, given $N' \equiv \{i \in N | w'_i = 1\}$?* ∎

We will show that **DCPMd** is strongly NP-complete through a transformation of **VERTEXCOVERd** (Garey and Johnson 1979, pp. 79, 190). We note that De et al. (1997) prove the NP-completeness of the "discrete time-cost tradeoff problem for project networks" (i.e., optimal project crashing with discrete expediting quantities), and that proof can be applied to **DCPMd**. However, our proof is substantially shorter than that of De et al., and we believe its inclusion is warranted for that reason, as well as for the sake of completeness.

DEFINITION 8 **VERTEXCOVERd**. *Given: An undirected graph $G = (N, A)$ and threshold $\bar{n}$. Question: Does there exist a set of nodes, (a "vertex cover," or "node cover"), $N' \subset N$, with $|N'| \leq \bar{n}$, such that every edge $k \in A$ is incident to at least one node in N'?* ∎

Note that N' is a node cover if $G' = G - N'$ consists of a set of completely disconnected nodes.

THEOREM 9 **DCPMd** *is strongly NP-complete.*
Proof: We are given an instance of **VERTEXCOVERd** with $G = (N, A)$ and will show how to construct an instance of **DCPMd** with

project network $G'' = (N'', A'')$ such that N', with $\bar{n} \equiv |N'|$, is a node cover for G if and only if the longest path in $G'' - N'$ has length $\bar{n}$ (where N' has been translated into G'' appropriately). $G'' - N'$ is the solution to an instance of **DCPMd** where $\mathbf{w} \in \mathbf{W}$ simply requires $\sum_{i \in N''} w_i = |N'| - \bar{n}$, $w_i \in \{0, 1\}$ for all $i \in N''$ and $w_a = w_b = 1$.

1 Convert G into a directed acyclic graph $G' = (N, A)$ by orienting arcs appropriately, and place the nodes N in topological ordering $N = \{1, 2, \ldots, n\}$

2 Create N'' by adding to N a set of "parallel" nodes $\{1', 2', \ldots, n'\}$ plus an extra node denoted $(n+1)'$. Node $1'$ will be the project start node, and node $(n+1)'$ will be the project completion node.

3 Define $t_k = 1$ for all $k \in A$.

4 Create A'' by adding to A the following arcs, all with $t_k = 0$;
   (a) $(i', (i+1)')$ for $i = 1, \ldots, n$,
   (b) (i', i) for $i = 1, \ldots, n$, and
   (c) $(i, (i+1)')$ for $i = 1, \ldots, n$.

This construction creates a directed acyclic graph that may be interpreted as a project network. And, a small example should convince the reader that N' is a node cover for G if and only if $G'' - N'$ has a longest path length of 0: (i) If N' is a cover, then $G'' - N'$ contains none of the original edges from A and all paths must have length 0 (and such paths do exist), and (ii) if $G'' - N'$ has a path of length greater than 0, then at least one edge $k = (i, j)$ remains in $G - N'$ so that N' is not a cover.

The fact that we transform a strongly NP-complete problem into **DCPMd**, and do not substantially change the size of the data required to describe the problem, implies that **DCPMd** is strongly NP-complete. ∎

So, **MAXMIN0**, even without expediting, is NP-hard and may not be a member of NP. But, when the number of alternative technologies is limited to a few (e.g., Spears 2001), **MAXMIN0** can be solved by solving the ILP **MINMAX2** just a few times. Specifically, enumerate all possible combinations of technologies, i.e., for each feasible vector $\hat{\mathbf{w}} \in \mathbf{W}$, solve the resulting instances of **MAXMAX2**, and choose the best interdiction plan from among those solutions. The instances of **MAXMAX2** would be polynomially solvable, pseudo-polynomially solvable or would be ILPs with exponential worst-case complexity. However, the most difficult of these solution techniques, solving a few ILPs, is likely to be easier than devising an effective, and completely general algorithm for **MAXMIN0**.

7. Conclusions

This paper has investigated the computational complexity of variants of an interdiction model that uses limited resources to delay tasks of an adversary's project in order to delay the project's overall completion time. We show that the most general "project-interdiction problem," and certain variants, are NP-hard. However, we also show that potentially useful variants may be strongly NP-complete, weakly NP-complete, or even solvable in polynomial time.

Furthermore, in practice, the NP-hard problems may not be as difficult as they appear at first glance. Their complexity derives from binary variables that model alternative technologies; however, in the real world, the number of such options will often be quite small. For example, if the project's manager must use one of, say, three mutually exclusive technologies, then only three instances of a simpler project-interdiction problem need be solved. Each of these would be an integer-linear program, a dynamic program, or a simple network-optimization problem.

Acknowledgments

Kevin Wood thanks the Naval Postgraduate School and the University of Auckland for their research support. Gerald Brown and Kevin Wood thank the Office of Naval Research and the Air Force Office of Scientific Research for their research support. Johannes Royset expresses his thanks for financial support from the National Research Council's Associateship program.

References

Ahuja, R.K., Magnanti, T.L. and Orlin, J.B. (1993). *Network Flows: Theory, Algorithms, and Applications*: Upper Saddle River, NJ: Prentice-Hall.

Brown, G.G, Carlyle, W.M., Harney, R. C., Skroch, E. M. and Wood, R. K., (2004). "Interdicting a Nuclear Weapons Project," draft, May 9.

Carlyle, W.M. and Wood, R.K., (2003). "Lagrangian Relaxation and Enumeration for Solving Constrained Shortest Paths," *Proceedings of the 38th Annual ORSNZ Conference*, University of Waikato, Hamilton, New Zealand, 21-22 November, pp. 3–12.

Crowston, W. and Thompson, G.L. (1967). "Decision CPM: A Method for Simultaneous Planning, Scheduling, and Control of Projects," *Operations Research*, **15**, pp. 407–426.

De, P., Dunne, E.J., Ghosh, J.B., Wells, C.E. (1997). "Complexity of the Discrete Time-Cost Tradeoff Problem for Project Networks," *Operations Research*, **45**, pp. 302–306.

Israeli, E. and Wood, R.K. (2002). "Shortest-path Network Interdiction," *Networks*, **40**, pp. 97–111.

Elmaghraby, S.E. (1977). *Activity Networks*. New York: Wiley.

Garey, M.R. and Johnson, D.S. (1979). *Computers and Intractability. A Guide to the Theory of NP-Completeness.* New York: W. H. Freeman and Co.

Handler, G.Y. and Zang, I. (1980). A Dual Algorithm for the Constrained Shortest Path Problem, *Networks*, **10**, pp. 293–310.

Hindelang, T.J. and Muth, J.F. (1979) "A Dynamic Programming Algorithm for Decision CPM Networks" *Operations Research*, **27**, pp. 225–241.

Malcolm, D.G., Roseboom, J.H., Clark, C.E., and Fazar, W. (1959) "Application of a Technique for Research and Development Program Evaluation," *Operations Research*, **7**, pp. 646–669.

Moder, J.J., Phillips, C.R., and Davis, E.W. (1983). *Project Management with CPM, PERT and Precedence Diagramming*, 3rd ed. New York: Van Nostrand Reinhold Company Inc.

Moore, J.T. and Bard, J.F. (1990). "The Mixed Integer Linear Bilevel Programming Problem," *Operations Research*, **38**, pp. 911-921.

PERT. (1958). "Program Evaluation Research Task, Phase 1 Summary Report," Special Projects Office, Bureau of Ordinance, 7, Department of the Navy, Washington, D.C., pp. 646-669.

Reed, B. K. (1994). "Models for Proliferation Interdiction Response Analysis," Masters Thesis, Naval Postgraduate School, Monterey, California, September.

Skroch, E. (2004). "How to Optimally Interdict a Belligerent Project to Develop a Nuclear Weapon," Masters Thesis, Naval Postgraduate School, Monterey, California, September.

Spears, D. (ed.). (2001). "Technology R&D for Arms Control. Arms Control and Nonproliferation Technologies", US Department of Energy, National Nuclear Security Administration, Defense Nuclear Nonproliferation Programs, Washington, D.C.

von Stackelberg, H. (1952). *The Theory of the Market Economy*, (trans. from German). London: William Hodge & Co.

Wood, R.K. (1993). "Deterministic Network Interdiction," *Mathematical and Computer Modelling*, **17**, pp. 1–18.

A NOTE ON ESWARAN AND TARJAN'S ALGORITHM FOR THE STRONG CONNECTIVITY AUGMENTATION PROBLEM

S. Raghavan
The Robert H. Smith School of Business
University of Maryland, College Park

Abstract In a seminal paper Eswaran and Tarjan [1] introduced several augmentation problems and presented linear time algorithms for them. This paper points out an error in Eswaran and Tarjan's algorithm for the strong connectivity augmentation problem. Consequently, the application of their algorithm can result in a network that is not strongly connected. Luckily, the error can be fixed fairly easily, and this note points out the remedy yielding a "corrected" linear time algorithm for the strong connectivity augmentation problem.

1. Introduction

Approximately 30 years ago Eswaran and Tarjan introduced the strong connectivity augmentation problem that can be described as follows. Let $D = (N, A)$ be a directed graph with node set N and arc set A. The strong connectivity augmentation problem is the problem of finding a minimum cardinality set of arcs A^{AUG} such that $D = (N, A \cup A^{\text{AUG}})$ is strongly connected.

Eswaran and Tarjan also describe an elegant linear time algorithm for the problem. Their algorithm consists of three steps. It first condenses the directed graph by shrinking every strongly connected component of the directed graph to obtain an acyclic digraph. A node with no incoming arc in this acyclic graph is called a source, and a node with no outgoing arc in this acyclic graph is called a sink. The second step of their algorithm constructs a particular ordering of sources and sinks with a *desired* set of properties. Their third step then adds arcs to strongly connect this acyclic digraph. (They show that it suffices to solve the augmentation problem on the condensed graph.) The correctness of their procedure relies on the second step of their algorithm, where an ordering of sources and sinks with a set of desired properties is constructed.

In this note we point out an error in Eswaran and Tarjan's strong connectivity augmentation algorithm. Specifically, we show that the algorithm described in their paper *does not* provide an ordering of sources and sinks with the desired set of properties. Consequently, the application of their augmentation algorithm (as will be shown with a counterexample) can lead to a directed graph that is not strongly connected. We also provide a corrected procedure for the second step of their algorithm that runs in linear time.

2. The Algorithm and the Error

We now review Eswaran and Tarjan's algorithm and elaborate on the error within it. We note that our notation differs slightly from Eswaran and Tarjan.

Given a directed graph $D = (N, A)$ the first step of their procedure is to create its condensation $D^{\text{SCC}} = (N^{\text{SCC}}, A^{\text{SCC}})$, obtained by shrinking every strongly connected component of D. D^{SCC} contains one node for every strong component of D, and there is an arc (i, j) in D^{SCC} if there is an arc in D from any node in the strong component corresponding to node $i \in D^{\text{SCC}}$ to any node in the strong component corresponding to node $j \in D^{\text{SCC}}$. For notational convenience they define the two mappings α and β as follows. For every $v \in N$, let $\alpha(v)$ be the node in D^{SCC} corresponding to the strong component in D that contains node v. For every $v \in N^{\text{SCC}}$, $\beta(v)$ defines any node in the strongly connected component of D corresponding to node v. Eswaran and Tarjan show the following lemma, proving that it suffices to solve the augmentation problem on D^{SCC}.

LEMMA 1 *Let X be an augmenting set of arcs which strongly connects D. Then $\alpha(X) = \{(\alpha(v), \alpha(w)) | (v, w) \in X, \alpha(v) \neq \alpha(w)\}$ is a set of arcs which strongly connects D^{SCC}. Conversely, let Y be an augmenting set of arcs which strongly connects D^{SCC}. Then $\beta(Y) = \{(\beta(x), \beta(y)) | (x, y) \in Y\}$ is a set of arcs which strongly connects D.*

In the acyclic digraph D^{SCC}, a source is defined to be a node with outgoing but no incoming arcs, a sink is defined to be a node with incoming but no outgoing arcs, and an isolated node is defined to be a node with no incoming and no outgoing arcs. Let s, t and q denote the number of source nodes, sink nodes, and isolated nodes respectively in D^{SCC}, and assume without loss of generality $s \leq t$.

The second step of the algorithm finds an index p and an ordering $v(1), \ldots, v(s)$ of sources of D^{SCC}, $w(1), \ldots, w(t)$ of sinks of D^{SCC} with the following properties:

1 there is a path from $v(i)$ to $w(i)$ for $1 \leq i \leq p$;

2 for each source $v(i)$, $p + 1 \leq i \leq s$ there is a path from $v(i)$ to some $w(j)$, $1 \leq j \leq p$; and

3 for each sink $w(j)$, $p+1 \leq j \leq t$, there is a path from some $v(i)$, $1 \leq i \leq p$, to $w(j)$.

Let $x(1), \ldots, x(q)$ denote the set of isolated nodes of D^{SCC}. They show that a minimal augmentation of D^{SCC} is obtained from the arc set.[1]

$$
\begin{aligned}
A^{\text{ASC}} &= \{(w(i), v(i+1)) | 1 \leq i < p\} \cup \{(w(i), v(i)) | p+1 \leq i \leq s\} \\
&\cup \begin{cases}
(w(p), v(1)) & \text{if } q = 0 \text{ and } s = t; \\
(w(p), w(s+1)) \cup \{(w(i), w(i+1)) | s+1 \leq i < t\} & \\
\cup (w(t), v(1)) & \text{if } q = 0 \text{ and } s < t; \\
(w(p), w(s+1)) \cup \{(w(i), w(i+1)) | s+1 \leq i < t\} & \\
\cup (w(t), x(1)) \cup \{(x(i), x(i+1)) | 1 \leq i < q\} & \\
\cup (x(q), v(1)) & \text{otherwise.}
\end{cases}
\end{aligned}
$$

Eswaran and Tarjan show that $\max(s,t) + q$ is a lower bound on the number of arcs needed to augment D^{SCC} so that it is strongly connected.[2] Note that the augmenting set A^{ASC} contains $t + q$ arcs. To see that the addition of these arcs strongly connects D^{SCC} observe that by construction the nodes $v(1), \ldots, v(p), w(1), \ldots, w(p)$, $w(s+1), \ldots, w(t), x(1), \ldots, x(q)$ are on a directed cycle (denoted by $\mathcal{C}$) and thus strongly connected. For $v(p+1), \ldots,$ $v(s)$, $w(p+1), \ldots, w(s)$, the correctness of their procedure relies on Properties (2) and (3). Due to Property (2) there is a path from each $v(i)$, $p+1 \leq i \leq s$, to some node $w(j)$, $1 \leq j \leq p$, and thus by construction to every node in the cycle $\mathcal{C}$. From Property (3), and the addition of the arcs $(w(i), v(i))$, for $p+1 \leq i \leq s$, there is a path from every node in the cycle $\mathcal{C}$ to each $v(i)$, $p+1 \leq i \leq s$. A similar argument shows that there is a directed path from the nodes in the cycle to each $w(i)$, $p+1 \leq i \leq s$, and from each $w(i)$, $p+1 \leq i \leq s$ to the nodes in the cycle $\mathcal{C}$.

The Eswaran and Tarjan paper provides the algorithm ST shown in Figure 1. We note that for any ordering of sources and sinks satisfying Properties (1)–(3), arbitrarily permuting the ordering of sources $v(p+1), \ldots, v(s)$, sinks $w(p+1), \ldots, w(t)$, and isolated nodes $x(1), \ldots, x(q)$ results in an ordering that continues to satisfy Properties (1)–(3). Consequently the algorithm focuses on obtaining p and the ordering $v(1), \ldots, v(p)$ of sources and $w(1), \ldots, w(p)$ of sinks satisfying Property (1), while ensuring that the remaining sources and sinks satisfy the desired Properties (2) and (3). The authors state that it is obvious that the algorithm ST finds a sequence of sources

[1] There is a typographical error in the first line of the equation shown on page 657 of [1] that is corrected here.

[2] Since there are $s + q$ nodes with no incoming arcs, at least $s + q$ arcs are needed to augment D^{SCC} so that it is strongly connected. Similarly, as there are $t + q$ nodes with no outgoing arcs, at least $t + q$ arcs are needed to augment D^{SCC} so that it is strongly connected.

```
1.  algorithm ST: begin
2.      procedure SEARCH(x);
3.          if x is unmarked then
4.              begin
5.                  if x is a sink and (w = 0) then w := x;
6.                  mark x;
7.                  for each y such that (x, y) is an edge do SEARCH(y);
8.              end SEARCH
9.      initialize all nodes to be unmarked;
10.     i := 0;
11.     while some sink is unmarked do begin
12.         choose some unmarked source v;
13.         w := 0;
14.         SEARCH(v);
15.         if w ≠ 0 then begin
16.             i := i + 1;
17.             v(i) := v;
18.             w(i) := w;
19.     end end
20.     p := i;
21. end ST;
```

Figure 1. Eswaran and Tarjan's algorithm to find an ordering of sources and sinks that satisfies Properties (1)–(3).

and sinks satisfying Properties (1)–(3). We now show that this statement is not true, and the algorithm can produce an ordering of sources and sinks that does not satisfy Property (2). For convenience, we also display in Figure 2 Eswaran and Tarjan's algorithm for the strong connectivity augmentation problem.

Consider the example shown in Figure 3. The acyclic digraph shown in Figure 3(a) has two sources nodes a and c, and two sinks nodes b and d. There is a directed path from source node a to sink node b and the first arc on this path is (a, k). There is also a directed path from source node a to sink node d and the first arc on this path is (a, l). There is a directed path from source node c to sink node d, and the paths from a to d and c to d are identical from node m onwards. Suppose the search starts from a (i.e., a is the first unmarked source selected in line 12), and suppose that in line 7 of the algorithm arc (a, k) is considered before arc (a, l). Then the procedure will first find sink b and set

```
algorithm STRONGCONNECT: begin
SC1: Use depth-first search to form the condensation D^SCC of D,
     identifying the sources, sinks, and isolated nodes of D^SCC;
SC2: Apply algorithm ST to D^SCC to find a set of sources and sinks
     satisfying (1)–(3);
SC3: Construct the corresponding augmenting set of arcs A^ASC;
SC4: Convert A^ASC into an augmenting set of arcs A^AUG for D, using
     Lemma 1;
end STRONGCONNECT;
```

Figure 2. Eswaran and Tarjan's algorithm to solve the strong connectivity augmentation problem.

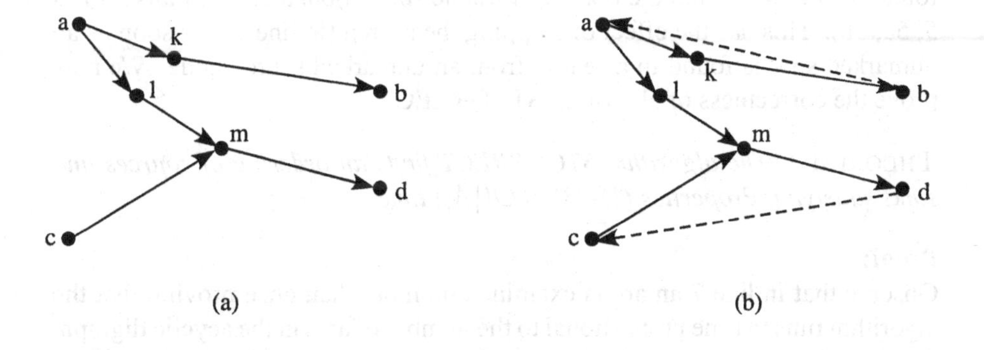

Figure 3. Counterexample that shows algorithm ST does not find an ordering of sources and sinks that satisfies Property (2). (a) Running algorithm ST gives $p = 1$, $v(1) = a$, $w(1) = b$, $v(2) = c$, $w(2) = d$. (b) Augmenting by adding arcs $(w(1), v(1)) = (b, a)$ and $(w(2), v(2)) = (d, c)$ results in a digraph that is not strongly connected.

$w := b$ (in line 5). It will then continue the search from a considering arc (a, l) in line 7 of the algorithm. This will result in traversing the path from a to d and marking nodes m and d. It will then set $v(1) := a$ and $w(1) := b$ in lines 17 and 18. Since all sinks are marked the procedure stops. Eswaran and Tarjan's procedure will also set $p := 1$, $v(2) := c$ and $w(2) := d$. Observe now that there is no path from c to node b, and thus this ordering does not satisfy Property (2). The augmentation procedure adds the arcs $(w(1), v(1)) = (b, a)$ and $(w(2), v(2)) = (d, c)$. The resulting digraph shown in Figure 3(b) is obviously not strongly connected.

3. The Correction

As the example indicates algorithm ST fails to find an ordering that satisfies Property (2). The problem within the algorithm is that the search continues from a source that has found an unmarked sink. This search may mark unmarked sinks (and thus these sinks would be ordered with an index of $p + 1$ or greater) that are the only sinks an unmarked source has a directed path to, leading to a violation of Property (2).

We now show how to rectify the problem by modifying algorithm ST so that it obtains an ordering of sources and sinks that satisfies Properties (1)–(3). The corrected algorithm is called STCORRECT and is displayed in Figure 4. It is identical to ST except for the following modest changes. A boolean variable *sinknotfound* is added that is true if the search from a source node has not yet encountered an unmarked sink. Further, in line 11 the search continues until all sources are marked (as opposed to ST where the search continues until all sinks are marked). At the start of a search from an unmarked source *sinknotfound* is set to true (line 13a). Within the procedure SEARCH, if an unmarked sink is found then $w := x$, and the boolean variable *sinknotfound* is set to false (lines 5, 5a, 5b). This has the effect of stopping the search (in line 7a) as soon as an unmarked sink is found in a search from an unmarked source node. We now prove the correctness of algorithm STCORRECT.

THEOREM 1 *The algorithm STCORRECT finds an ordering of sources and sinks satisfying Properties (1)–(3) in $\mathcal{O}(|A|)$ time.*

Proof:
Observe that in line 7 an arc is examined no more than once proving that the algorithm runs in time proportional to the number of arcs in the acyclic digraph. We now show a very useful property of algorithm STCORRECT that will be invaluable in our proof.

LEMMA 2 *All nodes marked by STCORRECT have a directed path to some sink node in $w(1), \ldots, w(p)$.*

Proof:
We show this by induction. Since the digraph is acyclic, the first time STCORRECT searches from an unmarked source it marks all nodes on a unique path to an unmarked sink node. At that point the search initiated from the unmarked source stops, and the unmarked source is set to $v(1)$ and the unmarked sink just found is set to $w(1)$. Observe that all marked nodes have a directed path to $w(1)$. Consider what happens at any later point in the algorithm when the search is initiated from an unmarked source. The search marks nodes along a path until it encounters a marked node x (in which case we do not search from x but the search from the unmarked source continues) or encounters an unmarked

```
1.   algorithm STCORRECT: begin
2.       procedure SEARCH(x);
3.           if x is unmarked then
4.               begin
5.                   if x is a sink then begin
5a.                      w := x;
5b.                      sinknotfound:=false;
5c.                  end;
6.                   mark x;
7.                   for each y such that (x, y) is an arc do
7a.                      if (sinknotfound) do SEARCH(y);
8.               end SEARCH
9.       initialize all nodes to be unmarked;
10.      i := 0;
11.      while some source is unmarked do begin
12.          choose some unmarked source v;
13.          w := 0;
13a.         sinknotfound:=true;
14.          SEARCH(v);
15.          if w ≠ 0 then begin
16.              i := i + 1;
17.              v(i) := v;
18.              w(i) := w;
19.      end end
20.      p := i;
21.  end STCORRECT;
```

Figure 4. A "corrected" linear time algorithm to find an ordering of sources and sinks that satisfies Properties (1)–(3).

sink (in which case the search stops). Assume the inductive argument is true at the conclusion of the search from the previous unmarked source in the algorithm. By induction the marked node x has a path to a sink in $w(1), \ldots, w(p)$, thus all nodes in the path to x have a path to a sink in $w(1), \ldots, w(p)$. In the case the procedure encounters an unmarked sink all nodes on the path to the unmarked sink are marked in the search from the unmarked source, and since the unmarked source and sink are now marked and added as $v(i)$ and $w(i)$, with $i \leq p$, the marked nodes on the path from $v(i)$ to $w(i)$ have a path to a sink in $w(1), \ldots, w(p)$. □

Consider the search from any unmarked source. The search successfully finds a directed path from the unmarked source to an unmarked sink (and these are added as $v(i)$ and $w(i)$ with $i \leq p$), or it fails to find a path to an unmarked sink (in which case the source has an index $i > p$). Failure occurs only if there is a marked node on every path from the unmarked source to every unmarked sink. By Lemma 2 these marked nodes have a path to some sink in $w(1), \ldots, w(p)$, proving that the ordering STCORRECT provides satisfies Property (2).

Suppose the ordering STCORRECT provides does not satisfy Property (3). Then there is an unmarked sink node with no directed path from any source in $v(1), \ldots, v(p)$. The unmarked sink node must then have a path from some source $v(j)$, for $p+1 \leq j \leq s$. Consider this path, and consider the last marked node y on this path to the unmarked sink. Node y could not have been marked by any source $v(i)$, with $p + 1 \leq i \leq s$. If it had been, then the search from $v(i)$ would have found the unmarked sink and both the source $v(i)$ and the unmarked sink would have been added to the ordering with an index less than or equal to p. But that means that y must have been marked in the search from one of the sources in $v(1), \ldots, v(p)$. Consequently, there is a directed path from a source in $v(1), \ldots, v(p)$ to the unmarked sink yielding a contradiction to our assumption. □

References

[1] K. P. Eswaran and R. E. Tarjan, *Augmentation problems*, SIAM Journal on Computing, 5 (1976), pp. 653–665.

II

INTEGER AND MIXED INTEGER PROGRAMMING

GENERATING SET PARTITIONING TEST PROBLEMS WITH KNOWN OPTIMAL INTEGER SOLUTIONS

Edward K. Baker[1], Anito Joseph[1] and Brenda Rayco[2]
[1]*Department of Management Science, University of Miami, Coral Gables, Florida 33124;*
[2]*Department of Mathematics, Southern Illinois University, Edwardsville, Illinois 62026*

Abstract: In this work, we investigate methods for generating set partitioning test problems with known integer solutions. The problems are generated with various cost structures so that their solution by well-known integer programming methods can be shown to be difficult. Computational results are obtained using the branch and bound methods of the CPLEX solver. Possible extensions are considered to the area of cardinality probing of the solutions

Key words: integer programming; test problems; branch and bound; cardinality probing

1. INTRODUCTION

The set partitioning (SP) problem considers a set of m objects that must be partitioned into mutually exclusive and collectively exhaustive subsets. Let $a_{ij} = 1$ if object i is contained in subset j, and equal to zero otherwise. Similarly, let $x_j = 1$ if subset j is used in the solution, and zero otherwise. Let c_j be the cost of subset j. The set partitioning problem may then be specified as follows:

Minimize $\Sigma\, c_j\, x_j$

Subject to: $\Sigma\, a_{ij}\, x_j = 1, i = 1, 2, \ldots, m$

$x_j = [0,1], j = 1, 2, \ldots, n$

The set partitioning model is quite flexible and has had many successful applications. Balinski and Quandt (1964), for example, used the set partitioning model to solve a version of the vehicle routing problem. Arabeyre et al. (1969), Marsten (1974), and Mingozzi et al. (1999) among others, have used the model to solve the airline crew scheduling problem. Garfinkel and Nemhauser (1970) and, more recently, Mehrotra, Johnson, and Nemhauser (1998) considered the model's use in the solution of political districting problems, while Yildiz (2004) used the model in the solution of large scale coloring problems.

Because of its general applicability, the solution of the set partitioning problem has attracted the attention of a number of researchers and numerous methods have been proposed for its solution. Garfinkel and Nemhauser (1969) proposed an enumeration algorithm for the set partitioning problem. Marsten (1974) considered a linear programming relaxation–based branch and bound algorithm for the problem, while Thiriez (1969) developed a solution method based on group theoretic techniques. Crainic and Rousseau (1987) proposed a column generation technique for the model and Hoffman and Padberg (1993) described a branch-and-cut solution procedure.

The set partitioning problem continues to be used on a regular basis in the solution of many important real world applications. Major airlines, for example, use the model daily to solve their crew scheduling problems. As this one application alone has enormous cost implications, researchers continue to seek methods that will find optimal solutions to ever larger problem instances. In the development of new algorithms for the problem, a need arises to be able to generate test problems of various sizes with known optimal integer solutions. This study represents an initial step in the development of a framework for the construction of such test problems.

1.1 Sources for mathematical programming test problems

The need for test problems in the demonstration of new algorithms is part of the experimental process. In many cases, the first sets of test data came from the applications that inspired the problem. In the set partitioning arena, the papers of Marsten and Shepherdson (1981) and Baker et al. (1979), for example, consist entirely of problems from industry. An alternate approach is to generate test problems by a random process. The papers of the Hillier (1969), Joseph (1995) and Joseph et al. (1998) provide such problems for various mixed integer programming scenarios. As interesting problems became known, noted researchers compiled libraries of these problems.

Libraries of test problems for various classes of mathematical programming models have been developed over the years from a number of sources. The Netlib of David Gay (Gay 1985) provided one of the first widely accepted libraries of test problems for linear programming algorithms. In 1990, John Beasley made the popular OR-Library (Beasley 1990) available for a variety of problem classes including assignment, scheduling, location, cutting stock, and multidimensional knapsack problems. MIPLIB (Bixby et al. 1996), developed soon thereafter, is a collection of real world and theoretical integer and mixed-integer programming problems contributed by number of researchers. The current version, MIPLIB 2003 is maintained by Thorsten Koch, Alexander Martin and Tobias Achterberg. MIPLIB 2003 may be found at http//miplib.zib.de. A comprehensive collection of test problems for constrained optimization algorithms was initially proposed by Floudas and Pardalos (1990) and has recently appeared in handbook form (Floudas et al. 1999).

Although these test problem libraries are extremely valuable in providing a consistent framework within which researchers may evaluate mathematical programming techniques (see Mulvey 1982), the instances available are limited. Test problems for the set partitioning problem, for example, are frequently obtained from real world crew scheduling applications. These problems possess a certain legitimacy, however, they also tend to possess similar structures. For example, the number of rows covered by a particular column in a crew scheduling problem is typically less than 10 and the density of the problem constraint matrix is typically less than five percent.

In this paper we consider the generation of test problems that have densities as great as 50% and that have cost structures that produce rather narrow duality gaps. It will be shown that these types of problems are not well suited for solution by traditional branch and bound methods and hence suggest the development and use of more flexible and adaptive algorithmic solutions.

2. GENERATING SET PARTITIONING TEST PROBLEMS

2.1 Optimal integer solutions

The generation of a set partitioning problem with a known optimal integer solution is, in one sense, straightforward. The unique structure of the set partitioning constraint matrix requires that each row in the matrix be covered exactly once. If one is free to assign costs to the columns of the problem, then any feasible solution, x_j, $j \in J^*$, to the problem can be made the optimal solution by assigning costs in the following manner:

1. Let t_j be the sum of the elements in column j. That is, $t_j = \Sigma_i a_{ij}$, $j = 1, 2, \ldots, n$.

2. Cost each of the columns j, $j \in J^*$, in the proposed optimal solution as $c_j = \alpha * t_j$.

3. Cost all other columns j, $j \notin J^*$, in the problem as $c_j = (\alpha * t_j) + \varepsilon$, $\varepsilon > 0$.

The optimal solution to the set partitioning problem, $x_j = 1, j \in J^*$, will have a value equal to $m*\alpha$.

The columns of a set partitioning constraint matrix contain m rows, the coefficients within which may be either 0 or 1. The number of possible columns one might generate in a set partitioning problem of m rows is then 2^m. For example, in a problem with m = 3 rows there would be $2^3 = 8$ possible columns. These would appear as shown in Table 1.

Table 1. Possible partitions of three rows

Col. 1	Col. 2	Col. 3	Col. 4	Col. 5	Col. 6	Col. 7	Col. 8
1	1	1	1	0	0	0	0
1	1	0	0	1	1	0	0
1	0	1	0	1	0	1	0

Since the last column of all zeroes is not a possibility in the partition, and would not be in any such enumeration, the number of possible columns in a set partitioning problem with m rows is then $2^m - 1$. In the initial part of this work, we will consider problems for which all possible columns have been generated for the set partitioning problem.

2.2 Test problems for simplex-based solution approaches

A number of methods proposed for the solution of the set partitioning problem are based upon finding an initial solution using the simplex algorithm for linear programming. The use of simplex-based procedures was perhaps encouraged by the seemingly favorable properties possessed by the set partitioning constraint matrix. Several interesting extreme point properties of partition solutions are discussed by Garfinkel and Nemhauser (1972) and by Padberg and Balas (1972).

2.3 Causing the simplex algorithm to find a non-integer optimal solution

If one wishes to divert a simplex-based solver away from an optimal integer solution, one must provide a feasible fractional solution of lower cost. This is easily accomplished. Consider, for example, the m columns containing m-1 values equal to one and exactly one zero. In an m = 5 row problem, these five columns would appear as in Table 2.

Table 2. Five columns allowing a fractional partition solution

Column 1	Column 2	Column 3	Column 4	Column 5
1	1	1	1	0
1	1	1	0	1
1	1	0	1	1
1	0	1	1	1
0	1	1	1	1

A feasible fractional solution to the set partitioning problem may be found by assigning each column a value of .25. To allow this solution to be the minimal cost solution to the relaxation of the set partitioning problem, cost each of the above columns $j \in J^f$, J^f the set of columns in the fractional solution, in the following manner:

1. Cost each of the columns $j, j \in J^f$, in the proposed optimal fractional solution $c_j = (\alpha * t_j) - \varepsilon_f$, where $0 < \varepsilon_f$,

2. Cost all other columns j in the problem $c_j = (\alpha * t_j)$.

The optimal fractional solution to the set partitioning problem will have a value equal to $m*\alpha - \varepsilon_f$.

3. SIMPLEX-BASED BRANCH AND BOUND PROCEDURES

Most commercial solver systems provide a simplex-based branch and bound procedure for the solution of integer programming problems. The branch and bound technique (Land and Doig, 1960) progressively divides the solution space into mutually exclusive subspaces by choosing a variable and fixing its value. In a binary linear program this is done by fixing a particular decision variable, say x_j, to one in the first subspace and to zero in the second. The bounds come about by solving the relaxation problem defined by each of the subspaces. By carefully controlling the enumeration process and by using the bounds in an intelligent way, the branch and bound method has been shown to be highly effective in solving many integer programming instances.

The success of any particular search, however, depends largely on the sequence of nodes selected for the search. Many approaches for the systematic selection of nodes have been investigated in the literature. We provide a brief overview of several of these methods in the paragraphs below.

Node selection strategies can be broadly classified into two groups: (1) methods based on simple selection rules, e.g. Best-First, Depth-First, and (2) methods that try to predict the effect on the objective function value for a given node, e.g., Best-Estimate.

The Best-First approach seeks to minimize the size of the search tree and hence, the overall solution time. In this approach, the subproblem with the best bound is chosen for processing, so that the search tree tends to be relatively smaller. One disadvantage of this approach is that it does not necessarily find feasible solutions quickly. If good bound information is not found early in the search, the candidate list of subproblems to be solved can become very large and lead to memory storage problems.

Where there are limitations on memory, Depth-First search is preferred. Depth-First search always selects the most recently created node for processing. Therefore, the candidate list of nodes for processing tends to be relatively small. This approach also tends to find feasible integer solutions quickly relative to other node selection strategies. Again, if good bound information is not obtained early in the search, it can result in very large search trees.

Estimation methods (Forrest et al. 1974, Mitra 1973, Benichou et al. 1971) choose nodes based on their potential for finding a better integer solution. To select the next node for processing, the Best-Projection criterion uses the objective function value of the relaxed subproblem combined with an estimate of the change in objective function that it would take to correct the integer feasibilities of the subproblem solution. The Best-Estimate criterion tries to account for the individual contribution of fractional variables to the objective function value. Each fractional integer variable of the subproblem solution is associated with two pseudo-costs. The pseudo-costs estimate the degradation in objective function value that would be observed by fixing the variable to one or zero. An estimate of the best solution possible at branching is then obtained and that determines the next node for processing.

Hybrid approaches combine different strategies to capitalize on the strengths of the chosen methods, e.g. Depth-First followed by Best-Estimate.

Linderoth and Savelsbergh (1999) conducted a comprehensive study of node selection strategies and found that while Depth-First search performed poorly in practice, "there is no node selection method which clearly dominates the others..." (p. 185). They recommended that many different node selection strategies should be available in sophisticated MIP solvers.

4. COMPUTATIONAL RESULTS

4.1 Generation of the test problems

In the computational testing that follows, each problem generated has $n = 2^m - 1$ variables, where m is the number of rows for the problem. In these preliminary tests, values of m = 10, 11, 12, 13, and 14 were used. The test problem characteristics are shown in Table 3.

Table 3. Test problem characteristics

Number of Rows, m	10	11	12	13	14
Number of Columns	1023	2047	4095	8191	16383
$\sum_i t_i$	5120	11264	24576	53248	114688

Three distinct cost structures were employed for each problem size. Each of the cost structures is described below.

4.1.1 Cost structure I

For constants α and β, the objective function coefficients, c_j, were set as follows.

$$c_j = \begin{cases} t_j + \alpha, & 1 \le j < \lfloor m/2 \rfloor \\ t_j, & \lfloor m/2 \rfloor \le j \le \lceil m/2 \rceil \\ t_j - \beta, & \lceil m/2 \rceil < j < m \\ t_j + \alpha, & j = m \end{cases}$$

In this implementation, the values $\alpha = 0.5$ and $\beta = 0.3$ were used. This cost structure creates a sequence of fractional solutions that leads the simplex solver "away" from the optimal integer solution with increasingly smaller fractional solution values.

4.1.2 Cost structure II

For constant θ, $0 \leq \theta \leq 1$, the objective function coefficients, c_j, were set as follows.

$$c_j = \begin{cases} t_j + \theta(\lfloor m/2 \rfloor - j), & 1 \leq j < \lfloor m/2 \rfloor \\ t_j, & \lfloor m/2 \rfloor \leq j \leq \lceil m/2 \rceil + 1 \\ t_j - \theta(j - (\lceil m/2 \rceil + 1)), & \lceil m/2 \rceil + 1 < j < m \\ t_j + \theta, & j = m \end{cases}$$

In this implementation, the value $\theta = 0.1$ was used. Cost structure II has the least cost fractional solution in the t_j = m-1 columns and leads the simplex solver "toward" the optimal integer solution with increasingly larger fractional solution values.

4.1.3 Cost structure III

The cost structure III coefficients are defined in the same manner as the cost structure II coefficients, except for the c_j's in the range $\lceil m/2 \rceil + 1 < j < m$. There are two cases; one for when the number of rows, m, is even and the other for when m is odd.

When m is even, we have:

$$c_j = \begin{cases} t_j - \theta(j - (\lceil m/2 \rceil + 1)), & \lceil m/2 \rceil + 1 < j < m, j \text{ is odd} \\ t_j, & \lceil m/2 \rceil + 1 < j < m, j \text{ is even} \end{cases}$$

And, when m is odd, we have:

$$c_j = \begin{cases} t_j - \theta(j - (\lceil m/2 \rceil + 1)), & \lceil m/2 \rceil + 1 < j < m, j \text{ is even} \\ t_j, & \lceil m/2 \rceil + 1 < j < m, j \text{ is odd} \end{cases}$$

In cost structure III, a "scalloped" cost effect is created in the sequence of fractional solutions that track toward the optimal integer solution.

Allowing c_j^k to denote the objective coefficient for column j in cost structure k, and again letting $t_j = \sum a_{ij}$, the objective function coefficients for the case when $m = 12$ are shown in Table 4.

Table 4. Objective function coefficients for test problems with $m = 12$ rows

t_j	c^1_j	c^2_j	c^3_j
1	1.5	1.5	1.5
2	2.5	2.4	2.4
3	3.5	3.3	3.3
4	4.5	4.2	4.2
5	5.5	5.1	5.1
6	6	6.0	6.0
7	6.7	7.0	7.0
8	7.7	7.9	8.0
9	8.7	8.8	8.8
10	9.7	9.7	10.0
11	10.7	10.6	10.6
12	12.5	12.1	12.1

4.2 Solving the test problems

The test problem instances were solved using the MIP Solver of CPLEX. All three of the solver node selection strategies were investigated. The possible strategies are as follows.

- The Depth-First (DF) Search Strategy chooses the most recently created node.
- The Best-Bound (BB) Strategy chooses the node with the best objective function for the associated LP relaxation.
- The Best-Estimate (BE) Strategy chooses the node with the best estimate of the integer value that would be obtained from a node once all integer feasibilities are removed.

Furthermore, CPLEX makes use of a *Backtrack* parameter that controls the frequency with which backtracking is performed during the branching process. At each node, the objective function value or estimated integer objective value is compared to the corresponding values at parent nodes. The Backtrack parameter value provides a measure of how much degradation is tolerated before backtracking. A small parameter value tends to increase the amount of backtracking, thereby making the search more of a Best-Bound strategy. Conversely, higher values make the search more of a pure Depth-First strategy.

In our experimentation, we found that all three node selection strategies produced very similar results across all of the test problem instances. To illustrate our findings, the results of the Best-Bound search strategy using a Backtrack parameter value of 0.01 are presented in Table 5. For the most part, the column labels are self-explanatory, the "Node, Best Integer" label of column four, however, is the first node in the search process at which the optimal integer solution was identified.

Table 5. Best-Bound results for the test problems

Number of Rows, m	Cost Structure	No. of Nodes	Node, Best Integer	Simplex Iterations	CPU (Seconds)
10	I	744	590	6220	2.7
10	II	320	257	1846	1.3
10	III	252	129	833	0.8
11	I	1207	584	13352	10.8
11	II	464	240	1877	2.9
11	III	365	305	2805	3.4
12	I	3419	2786	44511	60.1
12	II	1768	1489	17888	34.3
12	III	602	417	2299	6.8
13	I	4876	4171	89307	252.2
13	II	2278	2049	26503	111.1
13	III	2205	1867	24901	105.4
14	I	13286	12155	263757	1394.4
14	II	6816	6713	119246	1008.3
14	III	6881	6604	121666	833.7

The results of the initial tests indicated that Best-Bound and Best-Estimate searches were not able to outperform a basic Depth-First search, even when the cost structure might suggest that they should. This is caused partially because CPLEX uses an additional parameter to set the rule for selecting the branching variable at the node which has been selected for branching.

The variable selection options available in CPLEX are as follows:

- the Maximum Infeasibility rule which selects the variable with the largest fractional value;
- the Minimum Infeasibility rule which selects the variable with the smallest fractional value;
- a variable selection rule based on pseudo-costs which are derived from pseudo-shadow prices; and,

- a strong branching rule that causes variable selection based on partially solving a number of sub-problems with tentative branches to determine which branch is most promising.

When the default value is used, CPLEX automatically selects the rule based on the problem. In our experimentation, the default was used, and the Maximum Infeasibility rule was invariably selected.

In an attempt to allow cost Structure III to be exploited more fully by the algorithm, we solved the problems specifying the Strong Branching alternative using the Best-Bound search strategy and a Backtrack parameter value of 0.01. Although, the number of nodes examined and the number of simplex iterations required using the Strong Branching option was less, the computation times were significantly higher. In the case of the 14 row problems for example, cost structure I required 5.2 hours to solve.

4.3 Possible extensions

The apparent lack of success in solving the proposed relatively small test problems suggests several avenues into which this research may be extended. First, the proposed test problems are far denser than most test problems currently available for the set partitioning problem. An interesting exercise would be to test the efficiency of the solution approaches as a function of problem density. We are able to see some evidence of this relationship in the results of the airline crew scheduling problems solved in Table 7.

As a second possible extension, it is noted that the focus of this study was on branch and bound procedures and did not investigate the possibility of adding various cuts to the test problems. A number of branch and cut methods are available in CPLEX (see Bixby et al. 1999), although the density of our test problems tends to create many clique cuts for the problem that do not eliminate the fractional solutions. The success of Hoffman and Padberg (1993), for example, with branch and cut methods makes this a promising area for future research.

Finally, it is noted that the cardinality of the optimal integer solution to each of the proposed test problems was two. This suggests that, in addition to the local cardinality implications of cliques and odd-cycles, the cardinality of the final solution should be another dimension of the global search process. Joseph (2003) was able to show effective results by including a single optimal cardinality cut in a number of airline crew scheduling problems from the literature. In cases of small optimal cardinality, the probing techniques of Savelsbergh (1994) may be very effective as well.

To combine some of the effects of a branch and cut approach with the purpose of probing for a low order cardinality solution, the test problems were solved including a single cardinality cut at a value equal to the cardinality of the optimal integer solution. These results are presented in Table 6. The first six column labels are the same as in Table 5. The CPU Ratio in column seven represents the ratio of the CPU times reported in Table 6 divided by the corresponding CPU times reported in Table 5.

Table 6. Best-Bound branching with the optimal cardinality cut

Number of Rows, m	Cost Structure	Number of Nodes	Node, Best Int.	Simplex Iterations	CPU (Seconds)	CPU Ratio
10	I	499	425	3865	2.0	0.741
10	II	15	15	57	0.2	0.154
10	III	33	33	82	0.2	0.250
11	I	932	13	9275	5.7	0.528
11	II	358	358	780	1.9	0.655
11	III	357	357	784	1.9	0.559
12	I	1919	1332	16216	31.0	0.516
12	II	0	0	7	0.3	0.009
12	III	0	0	7	0.3	0.044
13	I	3853	1938	57737	147.1	0.583
13	II	1496	1496	3106	30.5	0.275
13	III	1862	1862	4361	39.7	0.377
14	I	7457	6773	120794	1020.2	0.732
14	II	1	1	41	1.3	0.001
14	III	1	1	41	1.3	0.002

In addition to the generated test problems, three of the airline crew scheduling problems, NW4, AA1 and AA4, from Hoffman and Padberg (1993) were solved. These real world problems had densities that varied from 20.2 to one percent. The computational results of these problems are presented in Table 7.

Table 7. Computational Results for Airline Crew Scheduling Problems

Problem	Rows, Columns	Number of Nodes	Node, Best Integer	Simplex Iterations	CPU (Seconds)
NW4	36, 87482	2080	1110	28465	348
NW4-Card	37, 87482	23	22	1177	44.8
AA1	823, 8904	191	175	54017	924.7
AA1-Card	824, 8904	169	118	50919	1167.7
AA4	426, 7195	257	236	37273	728.0
AA4-Card	427, 7195	177	100	29403	642.1

The results of these additional tests accomplish two purposes. First, the selected results presented show the extent to which the branch and bound procedure may be sharply curtailed with the addition of a single cardinality cut. In the instance where m=12 for cost structure II, the augmented linear programming relaxation solved to the optimal integer solution. Generally, the cardinality of the optimal integer solution is not known before the problem is solved; however, given the limited range of cardinalities in set partitioning integer solutions a "cardinality probe" may prove a fruitful endeavor.

The second point observed in these results is the effectiveness of the cardinality cut in the real world airline crew scheduling problems. Through the examination of the specially constructed test problems, a method of solution is suggested that may have important consequences in the solution of a number of real world problems.

5. CONCLUSIONS

A linear programming simplex-based solver is frequently an essential component in the solution of set partitioning problems. In this paper, we have shown that it is relatively easy to construct set partitioning test problems that a linear programming-based solver will find difficult. In examining the nature of the search process in these problems, it was observed that the addition of an optimal cardinality cut tends to focus the search and leads more quickly to the discovery of the optimal integer solution. Similar results were shown for three airline crew scheduling problems obtained from the literature.

Although the computational results of this paper are far from exhaustive, the preliminary conclusion one would draw from this work is that the selection of the most effective branch and bound procedure for a particular set partitioning problem instance requires extensive knowledge of the problem structure. Future research in this area would include, for example, the initial screening of the research problem to include data density as a measure of the problem's breadth. Additionally, preprocessing and probing the cardinality of the optimal integer solution was shown to offer a promising avenue of investigation. The incorporation of a robust cardinality probe within the solution process may prove to be very effective. Finally, it is noted that the test problem instances and cost structures considered here may be better suited for solution by parallel processing algorithms; however, this is an area for future research.

6. REFERENCES

Arabeyre, J., Fearnley, J., Steiger, F., and Teather, W., 1969, The airline crew scheduling problem: a survey, Transportation Science **3:** 140-163.

Baker, E.K., Bodin, L.D., Finnegan, W.F. and Ponder, R.J., 1979, Efficient heuristic solutions to an airline crew scheduling problem, AIIE Trans. **11**: 79-85.

Balinski, M. and Quandt, R., 1964, On an integer program for a delivery problem, Operations Research **12**: 300-304.

Beasley, J.E., 1990, OR-Library: Distributing test problems by electronic mail, Journal of the Operational Research Society **41**: 1069-1072.

Benichou, M., Gauthier, J.M., Girodet, P., Hentges, G., Ribiere, G. and Vincent, O., 1971, Experiments in mixed-integer linear programming, Math. Prog. **1**: 76-94.

Bixby, R., Boyd, A., Dadmehr, S., and Indovina, R., 1996, The MIPLIB mixed integer programming library, Mathematical Programming Society Committee on Algorithms, Bulletin **22**: 2-5.

Bixby, R.E., Fenelon, M., Zonghao, G., Rotheberg, E., and Wunderling, R., 1999, MIP: Theory and practice - closing the gap, *System Modeling and Optimization Methods, Theory and Applications*, M.J.D Powell and S. Scholtes, eds., Kluwer, The Netherlands.

Crainic, T.G. and Rousseau, J.M., 1987, The column generation principle and the airline crew scheduling problem, INFOR **25**:136-151.

Floudas, C.A. and Pardalos, P.M., 1990, A collection of test problems for constrained optimization algorithms, *Lecture Notes in Computer Science*, No. 455. Springer-Verlag, Heidelberg, Germany.

Floudas, C.A., Pardalos, P.M., Adjiman, C.S., Esposito, W.R., Gumus, S.T., Harding, Z.H., Klepeis, J.L., Meyer, C.A., and Schweiger C.A., 1999, *Handbook of Test Problems in Local and Global Optimization, Nonconvex Optimization and Its Applications.* Kluwer, Dordrecht, The Netherlands.

Forrest, J.J.H., Hirst J.P.H. and Tomlin, J.A., 1974, Practical solution of large scale mixed integer programming problems with UMPIRE, Management Science **20**: 736-773.

Garfinkel, R.S. and Nemhauser, G.L., 1969, The set–partitioning problem: set covering with equality constraints, Ops. Res. **17**: 848-856.

Garfinkel, R.S. and Nemhauser, G.L., 1970, Optimal political districting by implicit enumeration techniques, Management Science **14:** B495-B508.

Garfinkel, R.S. and Nemhauser, G.L., 1972, *Integer Programming.* J. Wiley and Sons. New York.

Gay, D.M. 1985. Electronic mail distribution of linear programming test problems. Mathematical Programming Society COAL Newsletter.

Hillier, F., 1969, Efficient heuristic procedures for integer linear programming with an interior, Operations Research **17:** 600-637.

Hoffman, K.L. and Padberg, M., 1993, Solving airline crew scheduling problems by branch-and-cut, Management Science **39:** 657-682.

Joseph, A., 1995, A parametric formulation of the general integer linear programming problem, Computers and Operations Research **22:** 883-892.

Joseph, A., Gass, S. I., and Bryson, N., 1998, An objective hyperplane search procedure for solving the general integer linear programming problem, European Journal of Operational Research **104:** 601-614.

Joseph, A., 2003, Cardinality corrections for set partitioning, Research Report, Mgt. Science Department. Univ. of Miami, Coral Gables, Florida.

Land, H. and Doig, A.G., 1960, An automatic method for solving discrete programming problems, Econometrica **28:** 497-520.

Linderoth, J.T. and Savelsbergh, M.W.P., 1999, A computational study of search strategies for mixed integer programming, INFORMS Journal on Computing **11**: 173-187

Marsten, R.E., 1974, An algorithm for large set partitioning problems, Management Science **20**: 774-787.

Marsten, R.E. and Shepherdson, F., 1981, Exact solution of crew scheduling problems using the set partitioning model: recent successful applications, Networks **11**: 165-177.

Mitra, G., 1973, Investigation of some branch and bound strategies for the solution of mixed integer linear programs, Math. Prog. **4:** 155-170.

Mingozzi, A., Boschetti, M.A., Ricciarde S., and. Bianco, L., 1999, A set partitioning approach to the crew scheduling problem, Operations Research **47:** 873-888.

Mehrotra, A., Nemhauser, G.L. and Johnson, E., 1998, An Optimization-Based Heuristic for Political Districting, Mgt. Sci. **44:** 1100 – 1114.

Mulvey, J. M. (ed.) 1982. *Evaluating Mathematical Programming Techniques*, Lecture Notes in Economics and Mathematical Systems 199, Springer-Verlag, Berlin.

Padberg, M. and Balas, E., 1972, On the set-covering problem, Operations Research **20:** 1152-1161.

Savelsbergh, M.W.P., 1994, Preprocessing and probing for mixed-integer models. Computational Optimization and Applications **3**: 317-331.

Thiriez, H., 1969, Airline crew scheduling: a group theoretic approach, MIT Department of Aeronautics and Astronautics Report R69-1. Cambridge, MA.

Yildiz, H., 2004, A large neighborhood search heuristic for graph coloring, Working Paper, Graduate School of Industrial Administration, Carnegie Mellon University, Pittsburgh, PA.

COMPUTATIONAL ASPECTS OF CONTROLLED TABULAR ADJUSTMENT:ALGORITHM AND ANALYSIS

Lawrence H.Cox, James P. Kelly, and Rahul J. Patil

National Center for Health Statistics, 3311 Toledo Road, Room 3211, Hyattsville, MD 20782; OptTek Systems,Inc, 1919 Seventh Street, Boulder, CO 80302 ; University of Colorado, Leads School of Business, Boulder, CO 80309-0419

Abstract: Statistical agencies have used complementary cell suppression to limit statistical disclosure in tabular data for four decades. Cell suppression results in significant information loss and reduces the usefulness of published data, significantly so for the unsophisticated user. Furthermore, computing optimal complementary cell suppressions is known to be an NP-hard problem. In this paper, we explore a recent method for limiting disclosure in tabular data, controlled tabular adjustment (CTA). Based on a mixed integer-programming model for CTA, we present a procedure that provides a lower bound on the objective, which is demonstrated to decrease the computational effort required to solve the model. We perform experiments to examine heuristics for CTA proposed elsewhere that can be used to convert the MIP problem into linear programming problems while preserving essential information.

Key words: Statistical Disclosure Limitation; Mixed Integer Programming

1. INTRODUCTION

Government agencies and commercial organizations that collect, store and publish data typically have a responsibility to protect the confidentiality of data pertaining to individual respondents. For tabular data, respondent risk is evaluated for each tabulation cell, based, e.g., on the number of respondents contributing to the cell (for *count data* such as totals by age, race or sex) or the distribution of the contributions (for *magnitude data* such as total retail sales or total farm acreage). Cells representing unacceptable risk are identified as *sensitive cells.* The Federal Committee on Statistical

Methodology (1994) provides an overview of the statistical disclosure and disclosure limitation methods.

Over the last four decades, different procedures have been proposed to protect sensitive cells in tabular data. *Complementary cell suppression* (Cox 1980) has been extensively practiced by statistical agencies to protect sensitive cells from disclosure through the manipulation of additive relationships in statistical tables. Though widely used, this methodology has many fundamental limitations, which undermine its efficiency. Cell suppression results in missing data, which complicates and can thwart thorough analysis. Though suppressed entries can be replaced by interval estimates of their hidden values, interval data present analytical challenges and destroy additivity to totals. Complementary cell suppression is an NP-hard problem (Kelly et al. 1992). Even worse, under cell suppression, a data intruder can estimate expected values of the suppressed entries, and often these estimates are close to the original values.

In this paper, we examine a recently proposed method for statistical disclosure limitation in tabular data that overcomes the drawbacks of the complementary cell suppression--*controlled tabular adjustment* (CTA) (Dandekar and Cox 2002; Cox and Dandekar 2003). A combination of mixed integer programming (MIP) and the CTA concept can be used to protect sensitive cells in a tabular data while releasing a full, unsuppressed table or set of tables of arbitrary dimensionality or complexity. We introduce a procedure that computes a useable lower bound for a typical data loss criteria, expressed as a linear objective function of the adjustment variables. We demonstrate that this bound can improve computational efficiency for the MIP solver by providing better information for the branch and bound algorithm. Still, not all MIPs can be solved optimally. So, we critically examine three heuristics proposed elsewhere for reducing the MIP to an LP. This is important because most statistical organizations are bound to resort to heuristics at least some of the time.

The literature on CTA is recent but growing. Cox (2000) published the first formal mathematical statement of the problem as an MIP, but provided little elaboration. Dandekar and Cox (2002, unpublished) is the first full paper devoted to the problem. That paper is based on heuristics for MIP to LP reduction, examined critically here. It also introduced a number of relevant measures of data distortion expressed as linear objective functions in the adjustment variables, including "total absolute cell adjustment," used here to illustrate our methods.

Section 2 presents the underlying concept of the controlled tabular adjustment and an associated mixed integer programming (MIP) formulation from Cox and Kelly (2004). We propose a method for computing a lower bound for the CTA objective, that we use both as a cutting plane and to improve the efficiency of the MIP branch-and-bound. Section 3 illustrates the use of general CTA formulation and a lower bounding procedure for a 2-dimensional table. Section 4 extends the approach to a 3-dimensional example. Section 5 examines three different heuristics for converting the MIP formulation to an LP formulation and compares their performance to the exact method. The 3-dimensional example and heuristics were introduced in Dandekar and Cox (2002). Section 6 provides some concluding remarks.

2. MATHEMATICAL MODEL FOR CONTROLLED TABULAR ADJUSTMENT AND LOWER BOUNDING ALGORITHM

2.1 Mathematical Model for CTA

Controlled tabular adjustment (CTA) aims to closely mimic original tabular data, subject to providing sufficient confidentiality protection to the sensitive cells. The original CTA formulation (Cox 2000; Dandekar and Cox 2002) accomplishes this by setting each sensitive cell to either its lower or its upper protection limit and adjusting the values of nonsensitive cells to restore tabular additivity. The original procedure controlled data by minimizing a global quality measure expressed in the form of a linear function in the adjustments such as total absolute cell adjustment, total absolute relative cell adjustment, etc. Adjustments to nonsensitive cells can be controlled in various ways. For example, by means of capacity constraints, selected nonsensitive cells such as zero cells can be exempted from change and other adjustments can be confined to within meaningful limits such as sampling variability (Cox and Dandekar 2003).

Mixed integer linear programming is used to model the CTA problem, as follows. Tabular systems with marginal totals are represented by a system of equations in matrix form: MX = 0, where column vector X represents the ordered set of values of the tabulation cells comprising the system and matrix M, the aggregation matrix, represents the tabular structure between the cells and contains only -1, 0, and +1 entries. Each row of M represents one tabular equation in which "+1" corresponds to a contributing internal cell and "-1" represents a contributing marginal cell. The following mixed integer programming formulation models the general mathematical structure of the CTA problem, analogous to that introduced in Cox (2000) and extended in Cox and Kelly (2004). Notation is as follows:

$i = 1,\ldots,p$: denote the p sensitive cells

$i = p+1,\ldots,n$: denote the $n - p$ nonsensitive cells

B_i = binary variable representing choice of downward (0) or upward (1) adjustment to sensitive cell $i = 1,\ldots,p$

$LPPROTECT_i$ = minimum downward adjustment necessary to protect sensitive cell $i = 1,\ldots,p$

$UPPROTECT_i$ = minimum upward adjustment necessary to protect sensitive cell $i = 1,\ldots,p$

y_i^+ = upward adjustment to cell value $i = 1,\ldots, n$

y_i^- = downward adjustment to cell value $i = 1,\ldots, n$

UB_i = upper capacity on upper adjustment y_i^+

LB_i = upper capacity on lower adjustment y_i^-

c_i = cost per unit adjustment to value of cell i

Unit costs c_i in CTA are nonnegative. Downward/upward adjustments generally have equal unit cost (as suggested by our notation), but not necessarily. Protection limits $LPPROTECT_i$ and $UPPROTECT_i$ are often equal, but not necessarily (see Table 1). Zero cells are often exempt from adjustment, but not necessarily. Cells of lower quality or importance can be targeted for adjustment (via $c_i = 0$) or permitted larger adjustment (via UB_i ,LB_i).

MILP for Controlled Tabular Adjustment

$$\text{Min} \sum_{i=1}^{n} c_i (y_i^+ + y_i^-) \tag{1}$$

Subject to:

$$M(y^+ - y^-) = 0 \tag{2}$$

For i = 1,…, n:

$$0 \le y_i^+ \le UB_i \quad (3)$$

$$0 \le y_i^- \le LB_i \quad (4)$$

For i = 1,…, p:

$$UB_i * B_i \ge y_i^+ \ge UPPROTECT_i * B_i \quad (5)$$

$$LB_i * (1-B_i) \ge y_i^- \ge LPPROTECT_i * (1-B_i) \quad (6)$$

Equation 1 describes the objective function, which minimizes the cost due to cell adjustments. Three linear cost functions commonly used in practice, each defined over adjustment variables $y_i^+ + y_i^-$, are as follows. The first has coefficients $c_i = 1$, corresponding to minimizing the distortion measure "total absolute cell adjustment" (see Table 1). The second has cost coefficient equal to the corresponding cell value, and the third has c_i = 1/(cell value) for nonzero cells, corresponding to minimizing total relative absolute adjustment. *Protection limit*s for a sensitive cell equal the minimum amounts, which must be added or subtracted from the true value to make the sensitive cell "safe". Cox (1980, 2001) and Willenborg and de Waal (2001) discuss protection limits theory in detail. CTA perturbs the sensitive cells until they are safe, i.e., sensitive cell values are adjusted sufficiently far from their original values. Equations 5 and 6 ensure that values for the sensitive cells are safe and that downward and upward adjustments to each sensitive cell are complementary. Adjustments to the sensitive cells create inconsistency in the tabular system, as sums may no longer be maintained; Equation 2 maintains tabular consistency. The process of restoring the tabular consistency can incur unacceptably large adjustments to the nonsensitive cells. For example, large negative adjustments can make the nonsensitive cell value negative. Equations 3 and 4 are used to constrain the nonsensitive cell adjustments to be feasible and acceptably small. Usually, these capacity constraints are computed using the estimated measurement errors for each nonsensitive cell. It can be noted that CTA provides additional confidentiality protection inherently, as sensitive cells are not highlighted or in any other way distinguished from nonsensitive cells. Moreover, in general, the intruder has no way to determine the direction of perturbation. After solving the model using MIP, the new tabular data $t = (t_i)$ is given by $t_i = x_i + y_i^+ - y_i^-$ where x_i denotes the original value of cell i.

2.2 Lower bounding algorithm

Solving the LP relaxation of CTA directly will result in values of 0.5 for all decision variables in the symmetric protection limit case, corresponding to a value of 0 for the objective. This weak LP relaxation fails to provide a tight lower bound for the branch and bound tree. The integer programming literature has widely recognized the importance of developing tighter lower bounds to speed up the solution process. A tight lower bound constraint introduced at the root node acts as a cutting plane for eliminating certain non-optimal solutions from the search. At each branching node, a lower bound based on the current states of the solution vector and constraint system can be computed. Whenever this bound is greater than (or equal to) the current best solution, progress along the branch can cease. This motivates us to develop a method that computes a useable lower bound.

The lower bound on the amount of perturbation required to protect a tabulation system is obtained as follows. To simplify notation, we consider first the case of a 2-dimensional table and later show how this can be generalized. The method is applicable to any data loss criterion expressed as a linear objective function in the adjustment variables. In Sections 3 and 4, we illustrate the method for the data loss criterion "total absolute cell adjustment" (all $c_i = 1$).

For any row or column equation, if there is only one sensitive cell (by abuse of notation, denoted i), a lower bound on perturbation required within this summation equals 2*min {$LPPROTECT_i$, $UPPROTECT_i$}. Otherwise, a lower bound is obtained by solving a small partition problem, as follows. Given two or more sensitive cells (indexed by i) in a row or column, protection limits $LPPROTECT_i$ and $UPPROTECT_i$ and directions for down/upward adjustment B_i, a lower bound on total perturbation required within the row or column subject to these choices equals the sum of the following quantities: the total of cost-weighted lower protections for cells selected for downward adjustment (denoted s_0), the total of cost-weighted upper protections for cells selected for upward adjustment (denoted s_1), and the absolute value of the difference $s_1 - s_0$ (representing net cost-weighted adjustment of nonsensitive cells required). A lower bound for the row or column equals the minimum over all choices of direction, namely, $z = \min \{s_0 + s_1 + | s_1 - s_0 |\}$. Let z_1 denote the sum of lower bounds over rows, and z_2 denote the sum of lower bounds over columns. Then, a lower bound on cost-weighted adjustment to the entire table is given by $\max \{z_1, z_2\}$. Consequently, for each row and column, we must solve the following small set partitioning problem.

$$Min\ \ s_0 + s_1 + z$$

$$z >= s_0 - s_1$$

$$z >= s_1 - s_0$$

$$s_1 = \sum_{ii} c_i * B_i * UPPROTECT_i$$

$$s_0 = \sum_{i} c_i * (1 - B_i) * LPPROTECT_i$$

For higher dimensional tables, the procedure is the same only the maximum is taken across each dimension. For complex tabulation structures, a similar approach can be applied based on linearly independent sets of summations treated like dimensions.

The set-partitioning model attempts to make the best mutual use of the sensitive cells to compute the minimum extra cost-weighted adjustment needed for a particular total. In practice, sensitive adjustments beyond minimum feasible safe values are undesirable and avoided for data quality purposes (by means of high per unit costs), but are sometimes necessary to ensure feasibility, necessitating minimal increases from protection limits. This situation does not affect the utility of our lower bound; however, as per unit adjustment costs are nonnegative. This is important; otherwise we must solve a nonlinear program ($LPPROTECT_i$ and $UPPROTECT_i$ replaced by the y_i^- and y_i^+) and not a MIP.

3. ILLUSTRATION OF THE MATHEMATICAL MODEL FOR A 2-DIMENSIONAL TABLE

Consider the following example, which illustrates how the mathematical programming formulation can be used to protect the sensitive cells in a 2-dimensional table as shown in Table 1. Cells (3, 1), (1, 2), and (3, 2) shown in bold have been identified as sensitive cells and the associated protection limits are shown in the brackets. The upper and lower capacities on adjustments to the nonsensitive cells are set at 30% of the original cell values. Note that the size and magnitude of the protection limits in relation to the cell values in this example are unrealistic. This was done deliberately for clarity of presentation. Table 2 shows the tabular data after solving the MIP. Cells with * indicate that they are perturbed.

Table 1. Original Table (Before CTA)

74	17[-17, +20]	85	176
71	51	30	152
1[-1, +20]	9[-9, +20]	36	46
146	77	151	374

Table 2. Table after CTA

64*	37*	75*	176
82*	40*	30	152
0*	0*	46*	46
146	77	151	374

The explicit mathematical programming formulation for "total absolute cell adjustment" is:

$$Minimize \quad : \sum_{i=1}^{4} \sum_{j=1}^{4} (y_{ij}^{+} + y_{ij}^{-})$$

$$subject \quad to :$$

$$\sum_{j=1}^{3} (y_{ij}^{+} - y_{ij}^{-}) - y_{i4}^{+} + y_{i4}^{-} = 0 \qquad for \quad i = 1..4$$

$$\sum_{i=1}^{3} (y_{ij}^{+} - y_{ij}^{-}) - y_{4j}^{+} + y_{4j}^{-} = 0 \qquad for \quad j = 1..4$$

$$y_{31}^{+} = 20 * B_{31} \qquad y_{31}^{-} = 1 * (1 - B_{31})$$

$$y_{12}^{+} = 20 * B_{12} \qquad y_{12}^{-} = 17 * (1 - B_{12})$$

$$y_{32}^{+} = 20 * B_{32} \qquad y_{32}^{-} = 9 * (1 - B_{32})$$

$$0 \le y_{11}^{+}, y_{11}^{-} \le 22 \qquad 0 \le y_{21}^{+}, y_{21}^{-} \le 21$$

$$0 \le y_{41}^{+}, y_{41}^{-} \le 43 \qquad 0 \le y_{22}^{+}, y_{22}^{-} \le 15$$

$$0 \le y_{42}^{+}, y_{42}^{-} \le 23 \qquad 0 \le y_{13}^{+}, y_{13}^{-} \le 25$$

$$0 \le y_{23}^{+}, y_{23}^{-} \le 9 \qquad 0 \le y_{33}^{+}, y_{33}^{-} \le 10$$

$$0 \le y_{43}^{+}, y_{43}^{-} \le 45 \qquad 0 \le y_{14}^{+}, y_{14}^{-} \le 52$$

$$0 \le y_{24}^{+}, y_{24}^{-} \le 45 \qquad 0 \le y_{34}^{+}, y_{34}^{-} \le 13$$

$$0 <= y_{44}^{+}, y_{44}^{-} <= 142 \qquad B_{ij} \in (0,1)$$

We now illustrate the lower bounding procedure for the 2-dimensional example. After solving the set partitioning problem (as necessary) for each row and column, lower bounds on the objective are 34, 0, 20, 0, respectively, for the rows, and 2, 40, 0, 0 for the columns. For example, for column 2 the following binary integer program yields an optimal objective value of 40 with $B_{12} = 1$, $B_{32} = 0$.

$$\begin{aligned} &\textit{Minimize} \quad \{s_0 + s_1 + Z\} \\ &Z \geq s_0 - s_1 \\ &Z \geq s_1 - s_0 \\ &s_1 = 20\,B_{12} + 20\,B_{32} \\ &s_0 = 17\,(1 - B_{12}) + 9(1 - B_{32}) \end{aligned}$$

In a similar manner, the optimal objective for row 3 is 20 with $B_{31} = B_{32} = 0$. The lower bound for the table is thus 54 (= 34 + 0 + 20 +0) compared to optimal total absolute cell adjustment of 82. The LP relaxation (without a lower bound constraint) gave an objective value of 0. Thus, the lower bound procedure provides a much tighter bound for the MIP.

4. ILLUSTRATION OF THE MATHEMATICAL MODEL FOR A 3-DIMENSIONAL TABLE

The preceding example involved a small problem that would not meaningfully benefit from speed-up (only 2^3 branches for complete enumeration) but did provide a simple, comprehensive illustration of the effectiveness of the lower bounding method. CTA can be applied to tables of arbitrary dimensionality or size. We next illustrate the method for a 3-dimensional table, containing 10 columns, 6 rows and 4 levels from Dandekar and Cox (2002). The table contains 191 non-zero cells, of which 24 cells are sensitive cells. As is customary, we exempt zero cells from adjustment, and assume symmetric protection, viz., $LPPROTECT_i = UPPROTECT_i$. The location of the sensitive cells, their cell values and required cell protection limits are illustrated in Table 3.

Table 3. Sensitive cells and protection limits for (10x6x4) 3-dimensional table

col	row	Lev	Val	prot	col	row	Lev	Val	Prot	col	row	Lev	Val	
2	1	1	714	39	2	1	2	539	59	2	4	3	644	35
4	1	2	70	7	4	1	3	614	34	4	2	2	786	87
4	2	3	928	51	4	4	2	382	42	4	6	2	1238	17
5	1	1	140	7	6	2	2	1074	59	6	3	2	544	30
7	1	3	549	61	7	3	2	631	70	7	5	2	726	40
7	5	3	134	7	8	1	3	92	10	8	4	2	1050	58
8	5	1	664	36	8	5	4	664	36	9	2	1	1042	57
9	3	3	820	91	9	5	2	1598	88	9	5	4	1598	88

Using traditional complementary cell suppression techniques, Dandekar and Cox (2002) report that, following Kelly et al. (1992), this example required 39 complementary suppressions to protect the 24 sensitive cells. This results in significant information loss, reducing the usefulness and usability of the table useless for many practical applications.

To generate a CTA table that mimics the original table while limiting disclosure as specified in Table 3, we use the procedure described in Section 2. Again, all costs equal 1. Table 3 summarizes the cell locations and magnitude of the controlled adjustments to true cell values. We have marked sensitive cells with a symbol *w*, so that readers can easily verify that adjustments to sensitive cells are at either of their respective protection limits. The original table appears in Table 4.

Table 4. Example (10x6x4) 3-dimensional table – "w" mark sensitive cells

6764		3356	4067	140w	--	3932	1478	--	**20451**
1994	--	5593	--	3022	3504	--	3220	1042w	**18375**
3744	--	3708	--	3678	2502	--	--	--	**13632**
2810		--	2445	--	--	2313	2978	7548	**28726**
3682	--	--	--	4667	1988	1748	664w	--	**12749**
18994	**11346**	**12657**	**6512**	**11507**	**7994**	**7993**	**8340**	**8590**	**93933**
--	539w	--	70w	--	7472	715	3832	--	**12628**
2253	--	4948	786w	472	1074w	1830	5030	--	**16393**
640	--	986	--	--	544w	631w	48	750	**3599**
1334	--	1016	382w	3175	3302	3803	1050w	--	**14062**
1648	2814	--	--	--	2102	726w	--	1598w	**8888**
5875	**3353**	**6950**	**1238w**	**3647**	**14494**	**7705**	**9960**	**2348**	**55570**
--	3552	3476	614w	1916	1131	549w	92w	1772	**13102**
--	--	3222	928w	--	--	308	429	87	**4974**
4145	--	--	3692	2115	4196	414	3804	820w	**19186**
5995	644w	--	--	2410	1677	--	1912	4134	**16772**
2016	--	--	2212	2826	1627	134w	--	--	**8815**
12156	**4196**	**6698**	**7446**	**9267**	**8631**	**1405**	**6237**	**6813**	**62849**
6764	4805	6832	4751	2056	8603	5196	5402	1772	**46181**
4247	--	13763	1714	3494	4578	2138	8679	1129	**39742**
8529	--	4694	3692	5793	7242	1045	3852	1570	**36417**
10139	11276	1016	2827	5585	4979	6116	5940	11682	**59560**
7346	2814	--	2212	7493	5717	2608	664w	1598w	**30452**
37025	**18895**	**26305**	**15196**	**24421**	**31119**	**17103**	**24537**	**17751**	**212352**

Table 5. Controlled tabular adjustments to (10x6x4) 3-dimensional table

--	-39w	--	5	7w	--	--	13	--	-14
--	--	--	--	-7	-46	--	--	57w	4
--	--	--	--	--	12	--	--	--	12
--	35	--	-5	--	--	--	23	-53	--
--	--	--	--	--	34	--	-36w	--	-2
--	**-4**		**--**	**--**	**--**	**--**	**--**	**4**	**--**
--	59w	--	7w	--	--	-49	-3	--	14
--	--	--	-87	--	59	3	25	--	--
--	--	--	--	--	-30w	70w	36	-88	-12
--	--	--	42w	--	--	16	-58w	--	--
--	-20	--	--	--	-29	-40w	--	88w	-1
--	**-39**	**--**	**-38w**	**--**	**--**	**--**	**--**	**--**	**1**
--	--	--	-34w	--	--	61w	-10w	-17	--
--	--	--	51w	--	--	2	--	-57	-4
--	--	--	--	--	4	-70	-25	-91w	--
--	-35w	--	--	--	--	--	35	--	--
--	--	--	--	--	-4	7w	--	--	3
--	**-35**	**--**	**17**	**--**	**--**	**--**	**--**	**17**	**-1**
--	20	--	-22	7	--	12	--	-17	--
--	--	--	-36	-7	13	5	25	--	--
--	--	--	--	--	-14	--	11	3	--
--	--	--	37	--	--	16	--	-53	--
--	-20	--	--	--	1	-33	-36w	88w	--
--	**--**	**--**	**-21**	**--**	**--**	**--**	**--**	**21**	**--**

Adjustments to Table 4 appear in Table 5. The full table after controlled tabular adjustment appears in Table 6. Zero cells appear as blanks for readability. In a real application only the CTA values are published (viz., no *w's*).

In the CTA adjusted Table 6, true values are published for 106 cells. For the remaining 85 cells, published cell values are adjusted sufficiently from their true values to protect the sensitive cell values from disclosure within their protection intervals. Most of the cell values of the marginal cells are unaffected in the CTA table, and the table is additive in all dimensions.

To assess the effectiveness of the lower bounding algorithm described in Section 2.2, two experiments are performed. The first experiment simply applies a single lower bound at the root of the branch and bound tree. A single lower bound constraint reduces the computation time by 65% compared to the model without a lower bound constraint. The number of nodes explored in the branch and bound tree is reduced by 64%. In the second experiment, a lower bound is calculated at each node of the branch and bound tree. Although more computation time is spent on lower bounding calculations, the search is more efficient due to the tighter dynamic bounds. The number of nodes explored in the branch and bound tree is reduced by 67% compared to the model without lower bounds.

Unfortunately, due to the computational cost of the lower bounding algorithm applied at each node, the overall time is only reduced by 25%. More efficient data structures may result in reduced computation times. It is anticipated that as the problem size increases, the effectiveness of the lower bounding algorithm will also increase extending this approach to larger tables.

Table 6. CTA on (10x6x4) 3-dimensional table

	675w	3356	4072	147w	--	3932	1491	--	**20437**
1994	--	5593	--	3015	3458	--	3220	1099w	**18379**
3744	--	3708	--	3678	2514	--	--	--	**13644**
2810	10667	--	2440	--	--	2313	3001	7495	**28726**
3682	--	--	--	4667	2022	1748	628w	--	**12747**
18994	**11342**	**12657**	**6512**	**11507**	**7994**	**7993**	**8340**	**8594**	**93933**
--	598w	--	77w	--	7472	666	3829	--	**12642**
2253	--	4948	699w	472	1133w	1833	5055	--	**16393**
640	--	986	--	--	514w	701w	84	662	**3587**
1334	--	1016	424w	3175	3302	3819	992w	--	**14062**
1648	2794	--	--	--	2073	686w	--	1686w	**8887**
5875	**3392**	**6950**	1200w	**3647**	**14494**	**7705**	**9960**	**2348**	**55571**
--	3352	3476	580w	1916	1131	610w	82w	1755	**13102**
--	--	3222	979w	--	--	310	429	30	**4970**
4145	--	--	3692	2115	4200	344	3779	911w	**19186**
5995	609w	--	--	2410	1677	--	1947	4134	**16772**
2016	--	--	2212	2826	1623	141w	--	--	**8818**
12156	**4161**	**6698**	**7463**	**9267**	**8631**	**1405**	**6237**	**6830**	**62848**
6764	4825	6832	4729	2063	8603	5208	5402	1755	**46181**
4247	--	13763	1678	3487	4591	2143	8704	1129	**39742**
8529	--	4694	3692	5793	7228	1045	3863	1573	**36417**
10139	11276	1016	2864	5585	4979	6132	5940	11629	**59560**
7346	2794	--	2212	7493	5718	2575	628w	1686w	**30452**
37025	**18895**	**26305**	**15175**	**24421**	**31119**	**17103**	**24537**	**17772**	**212352**

5. COMPARISON OF THREE HEURISTIC METHODS FOR CTA WITH EXACT METHOD

In general, the mixed integer linear programming exact method is suitable for solving small to medium-sized problems (less than 1,000 entries). The exact method takes a prohibitively long time to solve larger size instances to optimality, even with the lower bounding speed-up. Cox et al. (2004) argue that statistical agencies are interested in finding good solutions albeit not necessarily optimal solutions in reasonable amounts of time. Once the desiderata and essential elements of the problem (e.g., lower bounds) have been expressed as constraints, most if not all feasible solutions will provide an acceptable solution to the data confidentiality or confidentiality/quality problem. In this context, MIP techniques would appear to remain a viable solution mechanism.

Heuristic methods for selecting the individual directions for change of sensitive cells are still of interest, however. If a heuristic method is used to specify the binary directional variables then the remaining mathematical program becomes a linear programming problem. Linear programming software can be used to solve problems with hundreds of thousands of variables in reasonable amounts of time, and also to solve many problems related to the same publication. Heuristics were proposed in Dandekar and Cox (2002) to set the directions for the sensitive cells. We consider the following heuristics.

The "*Ordering Heuristic*" first sorts the sensitive cells in descending order and then assigns the directions for the sensitive cells in an alternating fashion. In our example from the last section, the sorting procedure gives the following sequence for the sensitive cells: (1,2), (3,2), (3,1). We set the alternating directions for these sensitive cells as 1, 0, 1 respectively. We can convert the MIP formulation into an LP formulation by setting the binary variables as $B_{12}=1, B_{32}=0, B_{31}=1$.

The "*Max*" heuristic sets the "UP" directions for all the sensitive cells. In our example, it can be accomplished by setting binary variables as $B_{12}=1, B_{32}=1, B_{31}=1$.

The "*Min*" heuristic sets the "DOWN" direction for all the sensitive cells. In our example, it can be done by assigning the binary variables values as $B_{12}=0, B_{32}=0, B_{31}=0$.

We examine the performance of these heuristics on our two examples. After applying each heuristic, we used the ILOG CPLEX solver to solve the resulting linear programs. Table 7 shows the performance of the heuristics compared to the exact MIP method. The "Ordering" and "Min" heuristics performed reasonably well, given the solution efficiency. Unfortunately, the "Max" heuristic forced changes to the sensitive cells that in some cases were too large to be offset by adjustments to nonsensitive cells within their respective capacities, resulting in an infeasible system of constraints.

Table 7. Performance of exact and three heuristics on the 2- dimensional table

Method	Solution Vector	Objective Value	%from Optimal
Exact	(0,1,0)	82	0
Order Heuristic	(1,1,0)	102	24.4
Min Heuristic	(0,0,0)	106	29.3
Max Heuristic	(1,1,1)	Infeasible	infeasible

A similar analysis was performed on the 3-dimensional example. The results are shown in Table 8.

Table 8. Performance of exact and three heuristics on the 3-dimensional example

Method	Objective Value	%from Optimal
Exact	2424	0%
Order Heuristic	3728	54%
Min Heuristic	3762	55%
Max Heuristic	3762	55%

Tables 7 and 8 indicate that better heuristics are needed for this problem. This is the focus of on-going research.

6. CONCLUDING COMMENTS

We have examined computational and algorithmic aspects of controlled tabular adjustment as a method for statistical disclosure limitation in tabular data. CTA efficiently generates useful, nonconfidential tabular data for publication. It can be extended to preserve the analytical and statistical usability of the original data (Cox and Kelly 2004; Cox et al. 2004). CTA offers a superior option for disseminating information contained in tabulations than the conventional method, complementary cell suppression.

A lower bounding procedure provided a tighter lower bound on a wide class of data loss criterion in the MIP formulation, illustrated here for "total absolute cell adjustment," and has been demonstrated to reduce the required computational effort need to solve the MIP. Heuristics proposed elsewhere for converting the MIP problem into an LP problem were compared to the exact method. It would appear that a combination of improved algorithmic and heuristic methods is necessary to solve the range of problems encountered in practice. This is under investigation.

REFERENCES

Cox, L.H., 1980, Suppression Methodology and Statistical Disclosure control, *Journal of the American Statistical Association* **75**: 377-385.

Cox, L.H., 1995, Network Models for Complementary Cell Suppression, *Journal of the American Statistical Association* **90**: 1453-1462.

Cox, L.H., 2000,Discussion (on Session 49: Statistical Disclosure Control for Establishment Data), ICES II: The Second International Conference on Establishment Surveys-Survey methods for businesses, farms and institutions, Invited Papers, Alexandria, VA: American Statistical Association, 904-907.

Cox, L.H., 2001, Chapter 8: Disclosure Risk for Tabular Economic Data, in: Confidentiality, Disclosure and Data Access: Theory and practical applications for statistical agencies,

P.Doyle, J.I. Lane, J.J.M. Theeuwes, and L.V. Zayatz, eds, Amsterdam: North-Holland, 167-184.

Cox, L.H., and Dandekar, R.A., 2003,A New Disclosure Limitation Method for Tabular Data That Preserves Accuracy and Ease of Use, Proceedings of the 2002 FCSM Statistical Policy Seminar, Washington, DC: U.S. Office of Management and Budget, in press.

Cox, L.H., and Kelly, J.P., 2004, Balancing Quality and Confidentiality for Tabular Data, *Monographs of Official Statistics*, Luxembourg: Euro stat, in press.

Cox, L.H., Kelly, J.P., and Patil, R.J., 2004, Balancing Quality and Confidentiality for Multi-Variate Tabular Data, *Lecture Notes in Computer Science* **3050**, New York: Springer Verlag, to appear.

Dandekar, R.A., and Cox, L.H., 2002, Synthetic Tabular Data-An Alternative to Complementary Cell Suppression, unpublished manuscript.

Federal Committee on Statistical Methodology, 1994, Statistical Policy Working Paper 22: Report on Statistical Disclosure ad Statistical Disclosure Limitation Methodology, Washington, D.C.: US Office of Management and Budget.

Kelly, J .P., Golden, B.L., and Assad A.A., 1992, Cell Suppression: Disclosure Protection for Sensitive Tabular Data, *Networks* **22**:397-417.

Willenborg, L., and Waal, T. de., 2001, Elements of Statistical Disclosure Control, *Lecture Notes in Statistics,* **155**, New York: Springer.

THE SYMPHONY CALLABLE LIBRARY FOR MIXED INTEGER PROGRAMMING

Ted K. Ralphs
Dept. of Industrial and Systems Engineering, Lehigh University, Bethlehem PA 18015
tkr2@lehigh.edu

Menal Güzelsoy
Dept. of Industrial and Systems Engineering, Lehigh University, Bethlehem PA 18015
megb@lehigh.edu

Abstract SYMPHONY is a customizable, open-source library for solving mixed-integer linear programs (MILP) by branch, cut, and price. With its large assortment of parameter settings, user callback functions, and compile-time options, SYMPHONY can be configured as a generic MILP solver or an engine for solving difficult MILPs by means of a fully customized algorithm. SYMPHONY can run on a variety of architectures, including single-processor, distributed-memory parallel, and shared-memory parallel architectures under MS Windows, Linux, and other Unix operating systems. The latest version is implemented as a callable library that can be accessed either through calls to the native C application program interface, or through a C++ interface class derived from the COIN-OR Open Solver Interface. Among its new features are the ability to solve bicriteria MILPs, the ability to stop and warm start MILP computations after modifying parameters or problem data, the ability to create persistent cut pools, and the ability to perform rudimentary sensitivity analysis on MILPs.

Keywords: Integer Programming, Software, Branch and Bound, Branch and Cut, Sensitivity Analysis

1. Introduction

As recently as a decade ago, the software available for solving generic mixed-integer linear programs (MILPs) was relatively limited. In the last 10 years, this has changed dramatically. There are now more than a dozen solvers available, many of which are open source. Among the academic and research codes available for solving generic MILPs are MINTO [Nemhauser et al., 1994], MIPO [Balas et al., 1996], bc-opt [Cordier et al., 1997], SBB [Forrest, 2004], GLPK [Makhorin, 2004] bonsaiG [Hafer, 1999], PARINO [Linderoth, 1998] and FATCOP [Chen and Ferris, 2001]. Commercial offerings include ILOG's CPLEX, IBM's OSL (soon to be discontinued), and Dash's XPRESS. In addition, there are a number of frameworks available, including BCP [Ladányi and Ralphs, 2001], ABACUS [Jünger and Thienel, 2001], ALPS [Ralphs et al., 2004a], and PICO [Eckstein et al., 2000].

The Computational Infrastructure for Operations Research (COIN-OR) Foundation is a recently formed non-profit foundation that evolved from an initiative launched by IBM in 2001 [Lougee-Heimer, 2003]. The primary goal of COIN-OR is to promote the development of open source software for operations research. The COIN-OR software repository currently hosts a dozen open source projects, all available for free download. SYMPHONY is an open-source callable library for solving MILPs that originated as a framework authored by Ralphs and Ladányi for solving difficult combinatorial problems. The original has since spawned two derivative frameworks, SYMPHONY and BCP [Ladányi and Ralphs, 2001]. BCP is a C++ framework that is more general than SYMPHONY, but has a steeper learning curve and cannot be used "out of the box." SYMPHONY has recently been integrated with the COIN-OR libraries and outfitted as a generic MILP solver. The source code is available for download from `www.BranchAndCut.org/SYMPHONY`.

The core solution methodology of SYMPHONY is a branch, cut, and price algorithm that incorporates most of the solution management features available in other codes. Features not yet included, but under development, include an integer presolver, a primal heuristic, and better support for column generation. The absence of the first two features hurt SYMPHONY's performance as a generic MILP solver, but it is otherwise full-featured and well-suited for implementing the customized algorithms required for solving very difficult classes of problems. It also performs well in parallel [Ralphs et al., 2003]. SYMPHONY depends on several other open source libraries for specific functionality, including the Cut Generation Library, the Open Solver Interface, and the MPS file parser maintained by COIN-OR, GLPK's GMPL file parser, and a third-party solver for linear-programming problems (LPs), such as the one maintained by COIN-OR (CLP).

2. The Application Program Interface

SYMPHONY 5.0 is the first version of SYMPHONY to be implemented as a callable library with a new interface derived from the COIN-

OR Open Solver Interface (OSI). This change markedly improves SYMPHONY's usability and flexibility. SYMPHONY and solvers built using SYMPHONY have been the subject of a number of papers, most recently [Ralphs et al., 2003], [Ralphs, 2003a], and [Ralphs et al., 2004b]. SYMPHONY's legacy features are well-detailed in the SYMPHONY User's Manual [Ralphs, 2003b], so we focus here on new features, such as the application program interface (API), the bicriteria solver, the ability to warm start MILP computations, and the ability to perform rudimentary sensitivity analysis. To our knowledge, these features are not yet available in other MILP codes and should be of interest to potential users. Below, we briefly describe the new C API, the C++ interface, and the use of the user callback functions. We assume the reader is familiar with the fundamentals of mixed-integer linear programming.

2.1 The Callable Library

SYMPHONY's callable library consists of a complete set of subroutines for loading and modifying problem data, setting parameters, and invoking solution algorithms. The user invokes these subroutines through the API specified in the header file `symphony_api.h`. Some of the basic commands are described below. For the sake of brevity, the arguments have been left out.

`sym_open_environment()`: Opens a new environment, and returns a pointer to it. This pointer then has to be passed as an argument to all other API subroutines (in the C++ interface, this pointer is maintained for the user).

`sym_parse_command_line()`: Invokes the built-in command-line parser for setting commonly used parameters.

`sym_load_problem()`: Reads the problem data and sets up the root subproblem (see Section 2.3).

`sym_solve()`: Solves the currently loaded problem from scratch. This method is described in more detail in Section 3.1.

`sym_warm_solve()`: Solves the currently loaded problem from a warm start. This method is described in more detail in Section 3.2.

`sym_mc_solve()`: Solves the currently loaded problem as a multicriteria problem. This method is described in more detail in Section 3.3.

`sym_close_environment()`: Frees all problem data and deletes the environment.

```
int main(int argc, char **argv)
{
   sym_environment *env = sym_open_environment();
   sym_parse_command_line(env, argc, argv);
   sym_load_problem(env);
   sym_solve(env);
   sym_close_environment(env);
}
```

Figure 1. A generic MILP solver with implemented with SYMPHONY in C.

By default, SYMPHONY reads an MPS or GMPL file specified by the user, although this behavior can be overridden by implementing a user callback that reads the data from a file in a customized format (see Section 2.3). SYMPHONY can also be used easily with FLOPC++ [Hultberg, 2004], an open-source modeling system that accesses solvers through the OSI. As an example of the use of the library functions, Figure 1 shows the code for implementing a generic MILP solver with default parameter settings. Note that the user does not have to invoke a command to read the MPS file. During the call to `sym_parse_command_line()`, SYMPHONY determines that the user wants to read in an MPS file. During the subsequent call to `sym_load_problem()`, the file is read and the problem data stored. To read an MPS file called `sample.mps` and solve it using this program, the following command would be issued:

```
symphony -F sample.mps
```

The code of Figure 1 is identical for both sequential and parallel computations. The choice between sequential and parallel execution modes is made at compile-time. In addition to the parts of the API just described, there are a number of standard subroutines for accessing and modifying problem data and parameters. These can be used between calls to the solver to change the behavior of the algorithm or to modify the instance being solved.

2.2 The OSI Interface

The OSI is a C++ interface class maintained by COIN-OR that provides a standard API for accessing a variety of solvers for mathematical programs. A code implemented using calls to the methods in the OSI base class can be linked with any solver for which there is an OSI implementation. This allows development of solver-independent codes and eliminates many portability issues. The current incarnation of OSI sup-

```
int main(int argc, char **argv)
{
   OsiSymSolverInterface si;
   si.parseCommandLine(argc, argv);
   si.loadProblem();
   si.initialSolve();
}
```

Figure 2. A generic MILP solver implemented with SYMPHONY using OSI.

ports only solvers for linear and mixed-integer linear programs. A new version supporting a wider variety of solvers is currently under development.

We have implemented an OSI interface for SYMPHONY 5.0 that allows any solver built with SYMPHONY to be accessed through the OSI. For each method in the OSI base class, there is a corresponding method in the C API. The OSI methods are implemented simply as wrapped calls to the C library. When an OSI object is constructed, `sym_open_environment()` is called and a pointer to the environment is stored. When the OSI object is destroyed, `sym_close_environment()` is called and the environment is deleted. To fully support SYMPHONY's capabilities, we have extended the OSI interface to include some methods not in the base class, such as a `parseCommandLine()` method. Figure 2 shows the program of Figure 1 implemented using the OSI interface. The code would be exactly the same for accessing any customized SYMPHONY solver, sequential or parallel.

The current version of the OSI is geared primarily toward support of LP solvers. One reason for this is that LP solvers based on the simplex algorithm support much richer functionality than do typical MILP solvers. In SYMPHONY 5.0, we have begun to extend some of this functionality to the realm of MILP solvers. For example, our OSI implementation supports warm starting and some basic sensitivity analysis. The implementation of this functionality is rudimentary at the moment, but will be improved in future versions.

2.3 User Callback Functions

The user's main avenues for customization are the tuning of parameters and the implementation of one or more of over 50 callback functions. The callback functions allow the user to override SYMPHONY's default behavior for many of the functions performed as part of its algorithm, including branching, cutting-plane generation, management of the cut pool, management of the LP relaxation, search and diving strategies, program output, etc. Callbacks in SYMPHONY are implemented slightly differently than in other popular libraries. Each callback function is called from a SYMPHONY *wrapper function* that interprets the user's return value and determines what action should be taken. If the user performs the required function, the wrapper function exits without further action. If the user requests that SYMPHONY perform a certain default action, then this is done. All callback functions have names that begin with the prefix "`user_`" Files containing default function stubs for the callbacks are provided along with the SYMPHONY source code. These can then be modified by the user as desired. Makefiles and Microsoft Visual C++ project files are provided for automatic compilation. Below is a sampling of commonly used callback functions.

`user_initialize_root_node()`: The user can specify a *core relaxation* consisting of cuts and variables that are to be present in every subproblem. These cuts and variables are never considered for removal and need not be included in the description of each search-tree node, so specifying a core can potentially save memory and increase efficiency.

`user_display_solution()`: The user can specify a custom output format for feasible solutions. This is useful for combinatorial problems where a simple list of variables and values is not interpretable by a human.

`user_create_subproblem()`: Rather than specifying the model directly using an MPS or GMPL file, the user can write a function that creates the initial LP relaxations at each node "on the fly."

`user_find_cuts()`: The user can generate custom classes of cutting planes by separating the current relaxed solution.

`user_is_feasible()`: The user can determine whether a given solution is feasible or not. This is needed in cases where integrality does not necessarily imply feasibility.

`user_select_candidates()`: The user can select candidates for strong branching.

`user_compare_candidates()`: After presolving, the user can choose a candidate to be used for branching.

`user_generate_column()`: The user can generate columns using this function.

`user_logical_fixing()`: The user can tighten bounds or fix variables based on implicit problem structure.

A full list of callbacks is contained in the SYMPHONY User's Manual [Ralphs, 2003b].

3. Solution Procedures

Because SYMPHONY is designed to allow parallel execution, both the internal library and the set of user callback functions are divided along functional lines into five separate modules. This modularization facilitates the parallel implementation and eases code maintenance. The five modules are the *master*, *tree manager*, *cut generator*, *cut pool*, and *node processor* modules. Only the master module is persistent and the environment pointer described earlier is a pointer to the master module. Other modules encapsulate the specific functionality needed to execute the algorithms and exist only while a solve call is active. Each module can function as an independent remote process for parallel execution. A more complete description of the modular design of SYMPHONY can be found in [Ralphs et al., 2003].

For LPs, the OSI has two function calls for solving the loaded model, `initialSolve()` and `resolve()`. The first call is used when solving a problem from scratch and the second is used when re-solving after having modified the problem in some way. SYMPHONY's OSI implementation extends this idea to MILPs. We have also implemented a third solve call for solving bicriteria MILPs. In the next few sections, we describe some of the details of how these methods are implemented.

3.1 Initial Solve

Calling `initialSolve()` solves a given MILP from scratch, as described above. The first action taken is to create an instance of the tree manager module that will control execution of the algorithm. If the algorithm is to be executed in parallel on a distributed architecture, the master module spawns a separate tree manager process that will autonomously control the solution process. The tree manager, in turn, creates the modules for processing the nodes of the search tree, generating cuts, and maintaining cut pools. These modules work in concert to execute the solution process, communicating either through shared memory or through a message-passing protocol, such as PVM [Geist et al., 1994].

The overall flow of the algorithm is similar to other branch-and-bound implementations and is described in detail in [Ralphs et al., 2003]. A priority queue of candidate subproblems available for processing is maintained at all times and the candidates are processed in an order determined by the search strategy. The algorithm terminates when the queue is empty or when another specified condition is satisfied. A new feature in SYMPHONY 5.0 is the ability to stop the computation based on exceeding a given time limit, exceeding a given limit on the number of processed nodes, achieving a target percentage gap between the upper and lower bounds, or finding the first feasible solution. After halting prematurely, the computation can be restarted after modifying parameters or problem data. This enables the implementation of a wide range of dynamic solution algorithms, as we describe next.

3.2 Solve from Warm Start

Among the utility classes maintained by COIN-OR is a base class for describing the data needed to warm start the solution process for a particular algorithm. To support this option in SYMPHONY, we have implemented such a warm start class for MILPs. The main content of the class is a compact description of the search tree at the time the computation was halted. This description contains complete information about the subproblem corresponding to each node in the search tree, including the branching that created the node, the list of active variables and constraints, and warm-start information for the subproblem itself (which is a linear program). All information is stored compactly using SYMPHONY's native data structures, which store only the differences between a child and its parent. In addition to the tree itself, other relevant information regarding the status of the computation is recorded, such as the current bounds and best feasible solution found so far. Using the warm start class, the user can save a warm start to disk, read one from disk, or restart the computation at any point after modifying parameters or the problem data itself. This allows the user to easily build in fault tolerance by periodically backing up warm-start information to disk, to design dynamic algorithms in which the parameters are modified after the gap reaches a certain threshold, or to modify problem data during the solution process if needed.

The ability to re-solve after modifying problem data has a wide range of applications in practice. One obvious application is to allow modification of problem data after the solution procedure has already been initiated. Another obvious application arises when the solution of a family of related MILPs is required, as occurs, for instance, in decomposition algorithms, in parametric and stochastic programming algorithms, in multicriteria optimization algorithms, and in algorithms for analyzing infeasible mathematical models.

```
int main(int argc, char **argv)
{
   OsiSymSolverInterface si;
   si.parseCommandLine(argc, argv);
   si.loadProblem();
   si.setSymParam(OsiSymFindFirstFeasible, true);
   si.setSymParam(OsiSymSearchStrategy, DEPTH_FIRST_SEARCH);
   si.setSymParam(OsiSymKeepWarmStart, true);
   si.initialSolve();
   si.setSymParam(OsiSymFindFirstFeasible, false);
   si.setSymParam(OsiSymSearchStrategy, BEST_FIRST_SEARCH);
   si.resolve();
}
```

Figure 3. Implementation of a dynamic MILP solver with SYMPHONY.

Modifying Parameters. The most straightforward use of the warm start class is to restart the solver after modifying problem parameters. The master module automatically records the warm-start information resulting from the last solve call and restarts from that point if a call to `resolve()` is made, unless external warm-start information is loaded manually. To start the computation from a given warm start when the problem data has not been modified, the tree manager simply traverses the tree and adds those nodes marked as candidates for processing to the node queue. Once the queue has been reformed, the algorithm is then able to pick up exactly where it left off. Figure 3 shows the code for implementing a solver that changes from depth first search to best first search after the first feasible solution is found. The situation is more challenging if the user modifies problem data in between calls to the solver. We address this situation next.

Modifying Problem Data. If the user modifies problem data in between calls to the solver, SYMPHONY must make corresponding modifications to the leaf nodes of the current search tree to allow execution of the algorithm to continue. Changes to the original data that do not invalidate the subproblem warm-start data, i.e., the basis information for the LP relaxation, are the easiest to accommodate. Our current procedures only handle modifications to the right-hand side and objective function vectors of the original MILP. Note that modifications may invalidate valid inequalities that have been previously generated. Currently, we discard such cuts. Methods for handling other modifications, such

```
int main(int argc, char **argv)
{
   OsiSymSolverInterface si;
   CoinWarmStart* ws;
   si.parseCommandLine(argc, argv);
   si.loadProblem();
   si.setSymParam(OsiSymNodeLimit, 100);
   si.setSymParam(OsiSymKeepWarmStart, true);
   si.initialSolve();
   ws = si.getWarmStart();
   si.setSymParam(OsiSymNodeLimit, -1);
   si.resolve();
   si.setObjCoeff(0, 100);
   si.setObjCoeff(200, 150);
   si.setWarmStart(ws);
   si.resolve();
}
```

Figure 4. Use of SYMPHONY's warm starting capability.

as the addition and deletion of columns and rows or the modification of the constraint matrix itself, will be added in the future. To initialize the algorithm, each leaf node, regardless of its status after termination of the previous solve call, must be inserted into the queue of candidate nodes and reprocessed with the modified input data. After this reprocessing, the computation can continue as usual.

Code illustrating the use of the warm start facility is shown in Figure 4. In this example, the solver is allowed to process 100 nodes and then save the warm-start information. Afterward, the original problem is solved to optimality, then is modified and re-solved from the saved warm start. As an illustration of the use of warm starting procedures in practice, Table 1 shows the results of solving a set of 2-stage stochastic integer programming instances modified from [Holmes, 2004; Felt, 2004; Ahmed, 2004] with the dual decomposition algorithm of [Caroe and Schultz, 1999]. We used a straightforward implementation of the subgradient algorithm to solve the Lagrangian duals and SYMPHONY to solve the subproblems, with and without warm starting from one iteration to the next. SUTIL [Linderoth, 2004] was used to read in the instances. The presence of a gap indicates that the problem was not solved to within the gap tolerance in the time limit. Although the running times are not competitive overall because of the slow convergence of

Problem	Tree Size Without WS	Tree Size With WS	% Gap Without WS	% Gap With WS	CPU min Without WS	CPU min With WS
storm8	1	1	-	-	14.75	8.71
storm27	5	5	-	-	69.48	48.99
storm125	3	3	-	-	322.58	176.88
LandS27	71	69	-	-	6.50	4.99
LandS125	37	29	-	-	15.72	12.72
LandS216	39	35	-	-	30.59	24.80
dcap233_200	39	61	-	-	256.19	120.86
dcap233_300	111	89	0.387	-	1672.48	498.14
dcap233_500	21	36	24.701	14.831	1003	1004
dcap243_200	37	53	0.622	0.485	1244.17	1202.75
dcap243_300	64	220	0.0691	0.0461	1140.12	1150.35
dcap243_500	29	113	0.357	0.186	1219.17	1200.57
sizes3	225	165	-	-	789.71	219.92
sizes5	345	241	-	-	964.60	691.98
sizes10	241	429	0.104	0.0436	1671.25	1666.75

Table 1. Results of using warm starting to solve stochastic integer programs.

our subgradient algorithm, one can clearly see the improvement arising from the use of warm starting.

Persistent Cut Pools. To complement the ability to save the search tree, the user can also save and reuse the global cut pool. When saving the search tree, only the cuts that are currently active in some leaf node and are needed to restart the search process are saved. At times, however, it may be advantageous to save the entire global cut pool, including cuts that were generated, but are not currently active. If this is desirable, the user can direct SYMPHONY to maintain one or more persistent cut pools. Such pools exist as part of the master module and are attached to the tree manager whenever a solve call is made.

3.3 Bicriteria Solve

A bicriteria MILP is a generalization of a standard MILP in which one considers a second objective function. One notion of solving a bicriteria MILP consists of generating all *Pareto outcomes.* An *outcome* is the pair of objective function values corresponding to a given feasible solution. The Pareto outcomes are those for which there is no other outcome for which both components are at least as small, and at least one is strictly smaller. In some cases, we are only be interested in the *supported outcomes*, which are those corresponding to solutions to a MILP with a single objective function formed by a convex combination of the two original objectives. For those readers not familiar with bicriteria integer programming, surveys of methodology are provided in [Climaco et al., 1997] and more recently in [Ehrgott and Gandibleux, 2000; Ehrgott and Gandibleux, 2002] and [Ehrgott and Wiecek, 2004].

In [Ralphs et al., 2004b], we describe an asymptotically optimal algorithm for solving bicriteria MILPs. SYMPHONY 5.0 contains a generic implementation of this algorithm, along with a number of methods for approximating the set of Pareto outcomes. To support these capabilities, we have extended the OSI interface so that it allows the user to define a second objective function and have also added a method for invoking the bicriteria solver called `multiCriteriaBranchAndBound()`. Implementing this algorithm requires the underlying solver to have the ability to generate, among all optimal solutions to a MILP with a primary objective, a solution minimizing a given secondary objective. We added this capability to SYMPHONY through the use of optimality cuts, as described in [Ralphs et al., 2004b].

The algorithm itself consists of the solution of a sequence of MILPs with identical feasible region, but differing objective functions. Thus, it is possible in principle to use warm starting to improve efficiency. Although the objective function is nonlinear in the case of generating Pareto outcomes, it can be linearized through a standard reformulation. This reformulation does require modification of the constraint matrix from iteration to iteration, but it is easy to show that these modifications do not invalidate the basis, allowing the warm start to be loaded very efficiently.

In [Ralphs et al., 2004b], we report on our experience using the bicriteria solver to analyze the tradeoff between fixed and variable costs for a class of network routing problems. Applying our rudimentary version of warm starting to this problem over the same test set, we have achieved promising results, improving solution times in almost all cases. A summary of results is shown in Figure 5, with the dark bars representing running times with warm starting and the light bars representing running times without warm starting over two data sets described in [Ralphs et al., 2004b]. The effect is evident, although it is also clear that further refinements to our procedures are still needed.

4. Sensitivity Analysis

Besides yielding the ability to closely examine the tradeoffs between competing objectives, the bicriteria solver can be used to solve *parametric MILPs*, which are families of MILPs parameterized by a single scalar. Typically, parametric MILPs are obtained by parameterizing either the objective function or the right-hand side, replacing the usual single vector with a combination of two vectors. The goal is to determine the complete set of optimal values that occur as the parameter is varied over a given interval. This set characterizes how the optimal value varies

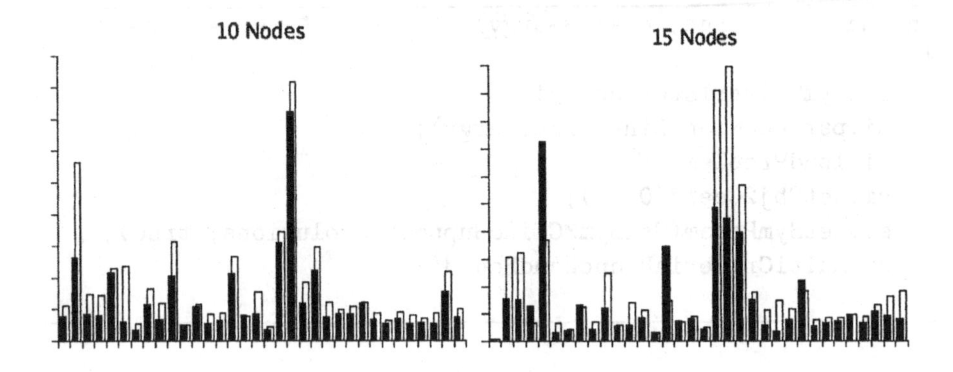

Figure 5. Results of using warm starting to solve multicriteria optimization problems.

as a function of change in either the objective function or the right-hand side in one dimension and is an elementary form of global sensitivity analysis (see [Bertsimas and Tsitsiklis, 1997] for a discussion of this in the case of linear-programming models).

As an example, consider the following simple parametric MILP.

$$\begin{aligned}
\max\ & 8x_1 + \theta x_2, \\
\text{s.t.}\quad & 7x_1 + x_2 \leq 56, \\
& 28x_1 + 9x_2 \leq 252, \\
& 3x_1 + 7x_2 \leq 105, \text{and} \\
& x_1, x_2 \geq 0, \text{integral.}
\end{aligned} \tag{1}$$

Taking the first objective function to be $(8, 1)$ and the second objective function $(0, 1)$, we can determine how the optimal value of the MILP varies as a function $p(\theta)$ of the second objective coefficient simply by invoking the bicriteria solution algorithm to enumerate all supported solutions. Figure 6 shows the code for performing this analysis. Applying the bicriteria solver of Figure 6 results in the function $p(\theta)$ shown in Table 2.

In addition to the sensitivity analysis that can be undertaken by using SYMPHONY's bicriteria solver, we have also implemented the method suggested in [Schrage and Wolsey, 1985] for performing approximate sensitivity analysis on the right-hand side vector and some related procedures. The method in [Schrage and Wolsey, 1985] is based on constructing an approximate dual price function from the dual solutions obtained while solving the LP relaxations in each search-tree node. The

```
int main(int argc, char **argv)
{
   OsiSymSolverInterface si;
   si.parseCommandLine(argc, argv);
   si.loadProblem();
   si.setObj2Coeff(0, 1);
   si.setSymParam(OsiSymMCFindSupportedSolutions, true);
   si.multiCriteriaBranchAndBound();
}
```

Figure 6. Performing sensitivity analysis with SYMPHONY's bicriteria solver.

θ range	$p(\theta)$	x_1^*	x_2^*
$(-\infty, 1.333)$	64	8	0
$(1.333, 2.667)$	$56 + 6\theta$	7	6
$(2.667, 8.000)$	$40 + 12\theta$	5	12
$(8.000, 16.000)$	$32 + 13\theta$	4	13
$(16.000, \infty)$	15θ	0	15

Table 2. Price function for example MILP (1).

price function does not have a simple closed form, and must be computed for each change in the right-hand side. This price function can be used to obtain approximate sensitivity information quickly when there is not enough time for a complete re-solve. Figure 7 shows a program that uses this sensitivity analysis function. This code will produce a lower bound for a modified problem with new right-hand side values of 7000 and 6000 in the 4th and 7th rows. Similar functions are provided for obtaining quick upper and lower bounds after changing either the right-hand side or objective function vectors.

5. Conclusions

We have described the main features of the SYMPHONY 5.0 callable library. SYMPHONY includes implementations of a number of techniques useful for performing sensitivity analysis, re-solving MILPs from a warm start, and analyzing bicriteria MILPs. To our knowledge, these techniques are not available in other solvers. The computational results presented here are very preliminary, but show promise. These capabilities are still being refined and new techniques developed, and we hope to improve them in future versions of the library. This is an area of active research that we believe has a great deal of potential and has

```
int main(int argc, char **argv)
{
   OsiSymSolverInterface si;
   si.parseCommandLine(argc, argv);
   si.loadProblem();
   si.setSymParam(OsiSymSensitivityAnalysis, true);
   si.initialSolve();
   int ind[2];
   double val[2];
   ind[0] = 4;    val[0] = 7000;
   ind[1] = 7;    val[1] = 6000;
   lb = si.getLbForNewRhs(2, ind, val);
}
```

Figure 7. Performing sensitivity analysis with SYMPHONY

received relatively little attention in the literature. However, it remains to be seen how well these methods will work in practice. In future work, we plan to extend and generalize the methods presented here to allow greater flexibility on the type of problem modifications and sensitivity analyses that can be performed and to further improve the power of the bicriteria solver.

Acknowledgments. This research was partially supported through NSF grant ACI-0102687 and the IBM Faculty Partnership Program.

References

Ahmed, S. (2004). SIPLIB. Available from http://www.isye.gatech.edu/ sahmed/siplib.

Balas, E., Ceria, S., and Cornuéjols, G. (1996). Mixed 0-1 programming by lift-and-project in a branch-and-cut framework. *Management Science*, 42:1229–1246.

Bertsimas, D. and Tsitsiklis, J. (1997). *Introduction to Linear Optimization*. Athena Scientific, Belmont, MA, USA.

Caroe, C. and Schultz, R. (1999). Dual decomposition in stochastic integer programming. *Operations Research Letters*, 24:37–45.

Chen, Q. and Ferris, M. C. (2001). FATCOP: A fault tolerant Condor-PVM mixed integer program solver. *SIAM Journal on Optimization*, 11:1019–1036.

Climaco, J., Ferreira, C., and Captivo, M. E. (1997). Multicriteria integer programming: an overview of different algorithmic approaches. In Climaco, J., editor, *Multicriteria Analysis*, pages 248–258. Springer, Berlin.

Cordier, C., Marchand, H., Laundy, R., and Wolsey, L. (1997). bc-opt: A branch-and-cut code for mixed integer programs. *Mathematical Programming*, 86:335.

Eckstein, J., Phillips, C., and Hart, W. (2000). PICO: An object-oriented framework for parallel branch and bound. Technical Report RRR 40-2000, Rutgers University.

Ehrgott, M. and Gandibleux, X. (2000). A survey and annotated bibliography of multiobjective combinatorial optimization. *OR Spektrum*, 22:425–460.

Ehrgott, M. and Gandibleux, X. (2002). Multiobjective combinatorial optimization—theory, methodology and applications. In Ehrgott, M. and Gandibleux, X., editors, *Multiple Criteria Optimization—State of the Art Annotated Bibliographic Surveys*, pages 369–444. Kluwer Academic Publishers, Boston, MA.

Ehrgott, M. and Wiecek, M. M. (2004). Multiobjective programming. In Ehrgott, M., Figueira, J., and Greco, S., editors, *State of the Art of Multiple Criteria Decision Analysis*, Boston, MA. Kluwer Academic Publishers.

Felt, A. (2004). Stochastic linear programming data sets. Available from http://www.uwsp.edu/math/afelt/slptestset.html.

Forrest, J. (2004). Simple branch and bound. Available from http://www.coin-or.org.

Geist, A., Beguelin, A., Dongarra, J., Jiang, W., Manchek, R., and Sunderam, V. (1994). *PVM: Parallel Virtual Machine.* The MIT Press, Cambridge, MA.

Hafer, L. (1999). bonsaiG: Algorithms and design. Technical Report SFU-CMPTTR 1999-06, Simon Frazer University Department of Computer Science.

Holmes, D. (2004). Stochastic linear programming data sets. Available from http ://users.iems.nwu.edu/ jrbirge/html/dholmes/post.html.

Hultberg, T. (2004). FlopC++. Available from http://www.mat.ua.pt/thh/flopc/.

Jünger, M. and Thienel, S. (2001). The ABACUS system for branch and cut and price algorithms in integer programming and combinatorial optimization. *Software Practice and Experience*, 30:1325–1352.

Ladányi, L. and Ralphs, T. (2001). *COIN/BCP User's Manual.* Available from http://www.coin-or.org.

Linderoth, J. (1998). *Topics in Parallel Integer Optimization.* PhD thesis, School of Industrial and Systems Engineering, Georgia Institute of Technology, Atlanta, GA.

Linderoth, J. (2004). SUTIL.

Lougee-Heimer, R. (2003). The common optimization interface for operations research. *IBM Journal of Research and Development*, 47:57–66.

Makhorin, A. (2004). Introduction to GLPK. Available from http://www.gnu.org/software/glpk/glpk.html.

Nemhauser, G. L., Savelsbergh, M., and Sigismondi, G. (1994). MINTO, a Mixed INTeger Optimizer. *Operations Research Letters*, 15:47–58.

Ralphs, T. (2003a). Parallel branch and cut for capacitated vehicle routing. *Parallel Computing*, 29:607–629.

Ralphs, T. (2003b). SYMPHONY Version 4.0 User's Manual. Technical Report 03T-006, Lehigh University Industrial and Systems Engineering.

Ralphs, T., Ladányi, L., and Saltzman, M. (2003). Parallel branch, cut, and price for large-scale discrete optimization. *Mathematical Programming*, 98:253–280.

Ralphs, T., Ladányi, L., and Saltzman, M. (2004a). A library hierarchy for implementing scalable parallel search algorithms. *Journal of Supercomputing*, 28:215–234.

Ralphs, T., Saltzman, M., and Wiecek, M. (2004b). An improved algorithm for biobjective integer programming and its application to network routing problems. To appear in Annals of Operations Research.

Schrage, L. and Wolsey, L. A. (1985). Sensitivity analysis for branch and bound linear programming. *Operations Research*, 33:1008–1023.

III

HEURISTIC SEARCH

HYBRID GRAPH HEURISTICS WITHIN A HYPER-HEURISTIC APPROACH TO EXAM TIMETABLING PROBLEMS

Edmund Burke[1], Moshe Dror[2], Sanja Petrovic[1] and Rong Qu[1]
[1]Automated Scheduling Optimization & Planning Group, School of CSiT, University of Nottingham, Nottingham, NG8 1BB, UK; [2]Department of Management Information Systems Eller College of Business and Public Administration, University of Arizona, Arizona 85721

Abstract: This paper is concerned with the hybridization of two graph coloring heuristics (Saturation Degree and Largest Degree), and their application within a hyper-heuristic for exam timetabling problems. Hyper-heuristics can be seen as algorithms which intelligently select appropriate algorithms/heuristics for solving a problem. We developed a Tabu Search based hyper-heuristic to search for heuristic lists (of graph heuristics) for solving problems and investigated the heuristic lists found by employing knowledge discovery techniques. Two hybrid approaches (involving Saturation Degree and Largest Degree) including one which employs Case Based Reasoning are presented and discussed. Both the Tabu Search based hyper-heuristic and the hybrid approaches are tested on random and real-world exam timetabling problems. Experimental results are comparable with the best state-of-the-art approaches (as measured against established benchmark problems). The results also demonstrate an increased level of generality in our approach.

Key words: case based reasoning, exam timetabling problems, graph heuristics, hyper-heuristics, knowledge discovery, tabu search.

1. INTRODUCTION

1.1 Timetabling Problems

Timetabling problems have been attracting the attention of the scientific research community across Artificial Intelligence and Operational Research for more than 40 years[1-10].

A general timetabling problem includes assigning a set of events (exams, courses, sports matches, meetings, etc) into a limited number of timeslots (time periods), while satisfying a set of constraints. These constraints are usually grouped into two types, which are described below:

- Hard constraints which cannot be violated under any circumstances. For example, a person cannot be assigned to two different events at the same time. Solutions which do not violate any of the hard constraints are called feasible solutions.
- Soft constraints are desirable but not essential. In most real world situations no solutions can be found which satisfy all of stipulated soft constraints.

In the early days of timetabling research, graph heuristics[11, 12, 13] were widely studied. They represent simple techniques but tend to be impractical for complex problems (at least when implemented on their own). Integer linear programming[14] is an exact solution method which tends to be computationally very expensive when solving large timetabling problems. Over the years, constraint programming methods have also been investigated at some length[15, 16, 17]. Meta-heuristics[18] have been shown to be very successful on a variety of timetabling problems. Examples include Tabu Search[19, 20], Simulated Annealing[21, 22] and Evolutionary Algorithms[23, 24]. Other new methods studied for timetabling problems include Case Based Reasoning[25] (for educational timetabling[26, 27, 28] and nurse scheduling[29]).

The work presented in this paper investigates the benefits of hybridizing two well-studied graph heuristics[11, 12, 13] by using a hyper-heuristic on timetabling. The term hyper-heuristic can be taken as a 'heuristic that searches for heuristics'[30]. A hyper-heuristic searches a space of heuristics rather than problem solutions. Our hyper-heuristic approach searches from a set of lower level heuristics according to different problem solving situations that might occur, and then applies those heuristics to the particular problem in hand. Different higher level heuristics/techniques employed within a hyper-heuristic framework include Case Based Reasoning[25], choice functions[31] and meta-heuristic methods[32, 33].

Many of the current state of the art approaches in exam timetabling employ specially tailored heuristic/meta-heuristics methods[20, 22, 34-39]. This kind of approach is also typical for other scheduling problems. The purpose of this paper is to describe our initial attempts at developing an approach which is fundamentally more general than the above methods. The goal is not necessarily to 'beat' those methods but to obtain comparable results by only employing general methods that can 'pick' appropriate heuristics and which would be applicable to a broader range of problems.

1.2 Case Based Reasoning (CBR)

CBR[25] is a knowledge-based technology that solves the problems in hand (target cases) by using knowledge obtained by solving previous similar problems. In a CBR system, a case base stores a set of previously solved problems with their good solutions or problem solving strategies (called source cases). A similarity measure, usually defined as a formula, is used to assess the similarities between the target case and source cases. The good solutions or problem solving strategies of the most similar source case are reused to tackle the target case.

The basic idea of CBR is to avoid solving new problems from scratch when the knowledge of solving similar problems is available. Our previous work using CBR on course and exam timetabling has presented successful results, either by reusing good partial solutions of problems whose constraints are structurally similar with current problems[26, 27], or by reusing good heuristics in similar problem solving situations[28]. This has provided the foundation for the research presented in this paper.

The next section presents our Tabu Search based hyper-heuristic (TSHH) on two graph heuristics. The results obtained by TSHH are utilized to propose two hybrid graph heuristics (including one which employs CBR). This is followed by experiments on both random and real-world problems. We conclude by briefly discussing the impact of the work and potential future research directions.

2. A TABU SEARCH BASED HYPER-HEURISTIC

In our previous work CBR was studied as the higher level searching technique to suggest different constructive heuristics during the exam timetabling problem solving process[28]. At each step of the solution construction, we select the heuristic (stored in the case base) that made the least penalty schedule in a previous *similar* situation. Employing this knowledge can help in finding good heuristics in new similar situations and in generating better quality solutions compared with those generated using single heuristics.

2.1 Tabu Search in a Hyper-heuristic Framework

In this paper we will employ Tabu Search as the higher level searching algorithm within a hyper-heuristic methodology that searches for the best combinations of heuristics (heuristic lists) for constructing the solutions for exam timetabling problems. This means that the heuristic list found not only represents good heuristics at each particular step (as before[28] where the least penalty schedules are made), but also it represents problem solving context (heuristics used and costs occurred before the current step, etc). A Tabu Search methodology within a hyper-heuristic framework has already been demonstrated as a successful, general methodology across very different problems (for course timetabling and nurse rostering[32]).

The search space of the TSHH consists of all of the possible permutations of the Saturation Degree (SD) and Largest Degree (LD), as shown below. Starting from an initial list of heuristics, a move within the TSHH is to change one of the heuristics in the heuristic list. The heuristics in the list are then employed, one by one, to construct the solution for the problem. The objective of TSHH is to find the heuristic list that generates the best quality solutions. TSHH stops after a certain number of iterations (5 times the number of the exams in the problem being considered).

SD and LD are two widely studied graph heuristics for applications to timetabling problems[11, 12, 13]. They are sequential methods that order the exams to be scheduled according to the difficulty of scheduling them. They then assign them one by one into feasible timeslots without violating any hard constraint and with the lowest penalty (i.e. the lowest total number of violations of soft constraints). They can be described as follows:

- Largest Degree (LD): Exams are ordered decreasingly by the number of conflicts they have with other exams. This heuristic aims to schedule the most conflicting exams first.
- Saturation Degree (SD): Exams that are not yet scheduled are ordered increasingly by the number of feasible timeslots available at that time. The priorities of the exams thus change dynamically according to the situations encountered at each step of the solution construction.

The heuristic lists selected to construct the solutions may not generate feasible solutions once they are performed, because the moves in the TSHH concern the changes of heuristics in the heuristic list, not the actual assignment of each exam. The search space of the TSHH is thus very large, containing a large number of non-valid heuristic lists. We add three mechanisms into the TSHH to reduce the size of the search space. They are described below:

- The parts of heuristic lists that generate infeasible assignment are stored in the searching process of the TSHH. At each move, the heuristic list selected will be checked before it is applied to see if it contains any stored infeasible heuristic lists. For example, if a heuristic list with the part 'SD LD ..' is stored because LD make an infeasible schedule at that step, all the heuristic lists selected later such as 'SD LD SD ..' or 'SD LD LD ..' will be ignored in the searching process. This mechanism significantly cuts the size of the search space by ignoring non-valid sections.

- At each step of the solution construction, we schedule a number of exams at once (we choose 3 here) by the given heuristic in the heuristic list. This is motivated by the observation[28] that the heuristics in the best heuristic lists tend to switch to others after a number of events have been scheduled. This mechanism also significantly reduces the size of the search space of TSHH.
- The initial heuristic list of TSHH is set as a list of SD only. We observe that in most cases SD is superior to LD, thus it is expected that the appearance of SD will be higher than that of LD in the best heuristic lists.

2.2 Experiments on Random Data Sets

The data sets we use are generated by using the same process as that of Carter et al[38] which simulates real-world exam timetabling problems. Each time a student is created and r exams are assigned (following a discrete uniform distribution in [2, 6]). This process is repeated until the defined density of conflict matrix is met, which is calculated as the number of conflicts among exams to the total number of exams. This generates 6 types of problems of 200 to 400 exams with density of 0.05, 0.15 and 0.25, namely '200-5', '200-15', '200-25', '400-5', '400-15' and '400-25'.

The hard constraints consider the 'conflict' between exams with students in common. The soft constraints under consideration concern spreading out the students' exams evenly. The cost function that evaluates the solutions is the same as that of Carter et al[38]. The objective is to minimize the cost per student. For each problem type, 20 distinct problems are tested on using the SD and LD heuristics alone, and the TSHH. The average costs and time spent by these approaches are presented in Table 1.

Table 1. SD, LD alone, two hybrid approaches and TSHH on random problems

prob-lem	SD		LD		SD+23%LD		SD+CBR		TSHH		den-sity
	cost	time	cost	time	cost	time	cost	time	cost	time	
200-5	10.1	1	10.2	0.02	9.9	1	10	2	**9.6**	195	0.24
200-15	9.5	5	9.4(18)	0.05	9.5	3	9.4	7	**9.2**	604	0.23
200-25	5.9	13	6.1(14)	0.08	5.9	10	5.9	28	**5.8**	3881	0.23
400-5	9.3	12	9.3(16)	0.06	9.3	10	9.3	23	**9.1**	5011	0.26
400-15	4.7	73	4.7(16)	0.31	4.7	57	4.7	73	**4.5**	20074	0.24
400-25	3.7	186	n/f(20)	0.59	3.7	158	3.7	204	**3.7**	73462	0.26

The best results in the table are presented in bold. For all the problems, the TSHH works much better than using SD and LD alone. The values in '()' represents the number of times infeasible solutions are obtained by the corresponding approach. Compared with SD or LD alone, TSHH takes a much longer time, especially for larger problems with higher densities. Note that in real world situations, exam timetables are usually generated weeks (or months) in advance and thus the time does not usually have much impact on the usefulness of the methodology.

The column under the title of 'density' in Table 1 presents the average densities of the LD in the best heuristic lists obtained by TSHH. We can see that for all of the problem types, the densities of LD in the best heuristic lists are almost the same, ranging from 0.23-0.26. This motivated us to build one of the hybrid approaches, whose details are presented in the next section.

3. HYBRID GRAPH BASED APPROACHES

By investigating the TSHH, we propose two hybrid approaches. They are SD injected with LD by 23%, and by a CBR system built using knowledge discovery techniques that extract the knowledge of TSHH. They are presented in the following two sub-sections.

3.1 SD Injected with 23% LD

In the first hybrid approach, LD is randomly injected into the heuristic list (of SD) to form 23% of it. This hybridization is proposed by using the density of LD that appears in the best heuristic lists. Our aim is to investigate whether such a heuristic list is good for all of the problem instances.

The results of this approach on all types of problems are presented in the column entitled SD+23%LD in Table 1. They are compared with the results obtained by using SD and LD heuristics alone. Please note that it presents the cost per student thus small differences from different approaches may indicate large differences in costs.

Compared with the results obtained by using SD and LD alone, the hybrid approaches perform on all of the problems, except for '400-25', where the same results (3.70) are obtained by SD+23%LD and SD+CBR. For problem type '400-25', LD failed to obtain feasible solutions for any pf the problem instances (indicated by 'n/f' in the table). The values in Table 1 present the average penalties for only the feasible solutions obtained.

Among the approaches, LD takes the least amount of time as the ordering of exams will not change during the problem solving process. The time spent by SD+23%LD is less than that of SD, which orders the exams that are not yet scheduled at each step of the problem solving process. This hybrid approach is superior to SD on both the solution quality and the problem solving time.

3.2 SD Injected with LD by CBR

We present another hybrid approach using a CBR system developed by investigating the heuristic lists obtained from the TSHH described above. The idea is to inject the LD by the suggestions from the CBR system which stores the appropriate heuristics in different problem solving situations. At each step of the scheduling process, the current problem solving situation is input into the CBR system as the target case, and the most similar source case is retrieved. The heuristic of this retrieved source case is employed to select and schedule the exams in the next step of solution construction.

In the CBR system, we use a list of feature-value pairs to represent the cases. The similarity measure employs a nearest neighborhood approach to sum up the differences of values for each pair of features in the target and source cases. The most similar source cases will be retrieved and the corresponding heuristics will be suggested for use in the next step of scheduling.

Knowledge discovery is carried out by using the best heuristic lists obtained from the TSHH. The objective is to discover the most relevant features to be used in the list of features to represent cases so that the correct heuristic can be selected by CBR. We collected a set of initial training features that describe the problem solving situations. They can be grouped into two types, which are presented below.

1. Simple features: this can also be grouped into two types:
   - Features that describe the problems. These include: the no. of exams, students, timeslots, the total no. of conflicts among all the exams, the density of the conflict matrix, the no. of the conflicts for the most conflicting exams and the no. of the most conflicting exams.
   - Features with values that are changeable during problem solving. These include: the no. of exams that have been scheduled in a particular timeslot, the heuristic employed before the current step, the increased penalty occurred by the last step schedule, and the cost of the partial solution concerning the violations of only the soft constraints.
2. A Combination of the simple features: the ratios between each pair of the simple features.

At each particular scheduling step during the TSHH, the problem solving situations (values of all of the training features presented above) are recorded for each problem solved. These situations along with the best heuristics at that step in the process form the *cases* to be used in the knowledge discovery process. All of the cases obtained are randomly divided into two groups: one group is stored as source cases in the case base and the other will be the training cases used just in the knowledge discovery process for training purposes.

A Tabu Search (not the same one used in TSHH but just for knowledge discovery) is used to discover the best feature list for the case representation. All of the possible feature lists form the search space of Tabu Search. An initial feature list is randomly selected and a move is a change of a feature and its weight. All of the training cases (whose heuristic is already obtained beforehand by TSHH) are input one by one into the system. If the heuristic of the most similar source case retrieved (by similarity measure upon the set of features for the training) is the same as that of the training case obtained beforehand, it will be seen as a successful retrieval. The total number of the successful retrievals indicates the system performance. The objective of the Tabu Search is to find the feature list upon which the highest system performance is obtained for all of the training cases. The most relevant features found by the first stage of knowledge discovery are: 'the increase in penalty in the previous step of scheduling' with weight 100, and 'the number of exams already scheduled at that step' with weight 1.

The second stage of knowledge discovery aims to refine the case base. The best feature list is obtained from the first stage each time a source case is removed from the case base. If the system performance is decreased, the removed source case will be added back into the case base as it contains useful information for the heuristic selection in that particular problem solving situation; otherwise the source case will be removed permanently as it contains either redundant or wrong information that is harmful for the heuristic selection.

The column entitled SD+CBR in Table 1 presents the results obtained by the hybrid SD with CBR on the same problems tested using other approaches. We can see that it outperforms SD and LD alone on all problems except on problem '400-25', where the same result is obtained by SD alone. Compared with SD+23%LD, it obtained slightly better solutions but with longer time occurred on searching the case base. On all of the problem types except '400-25', both of the two hybrid approaches outperform the SD and LD alone. Among all of the approaches tested TSHH works the best. Note, though, that the two hybrid approaches are much quicker than the TSHH although (as we already mentioned) time is not usually a critical issue in exam timetabling.

These results show that by embedding knowledge of employing different heuristics during problem solving, the hybrid approaches work better than those of the single heuristics. The hybrid approaches have the ability to choose appropriate heuristics in different situations thus have significant potential for being more generally applicable than the current state of the art.

4. EXPERIMENTS ON REAL-WORLD PROBLEMS

We carried out another set of experiments on 4 real-world benchmark exam timetabling problems presented by Carter et al[38]. These problems cover a range of characteristics (i.e. on number of exams and conflict matrix density). Table 2 presents the best results for these 4 problems by all of the approaches presented above, except LD which failed to generate feasible solutions for all of the problems.

Table 2. SD, two hybrid approaches, TSHH and state-of-the-art on real-world problems

problem	ute92	uta93	sta83	ear83
SD	37.63	n/f	191.93	n/f
SD+23% LD	38.37	5.09	n/f	n/f
SD+CBR	37.53	5.06	173.82	47.42
TSHH	35.40 (0.37)	4.52 (0.12)	158.2 (0.07)	45.60 (0.48)
Asmuni et al[34]	27.78	3.57	160.42	37.02
Burke et al[35]	25.7	3.4	159.1	35.4
Burke & Newall[36]	25.83	**3.20**	168.73	37.05
Caramia et al[37]	**24.4**	3.5	158.2	**29.3**
Carter et al[38]	25.8	3.5	161.5	36.4
Casey & Thompson[22]	25.4	n/f	**134.9**	34.8
Di Gaspero & Schaerf[20]	31.3	4.5	166.8	46.7
Merlot et al[39]	25.1	3.5	157.3	35.1

We can observe that for real-world problems, the results obtained from the hybrid approaches show different characteristics compared with those on random data sets. The reason may be that the knowledge discovered from the random data sets may not cover enough problem solving situations of real world problems with different characteristics.

SD+CBR shows promising results and is reasonably reliable over both random and real-world problems, as the injection of LD is made by using knowledge concerning different problem solving situations and thus can help to solve more types of problems. However, to be able to solve more types of problems, the CBR system needs to be trained to store more knowledge of problem solving over a wider range of problems (including both the random and real-world problems) with a variety of problem features.

One observation is that the densities of LD (presented in '( )' in Table 2) in the best heuristic lists found are different for different problems, and none of them has a value that is within 0.23-0.26. Thus SD+23%LD will not be the appropriate approach for solving the real-world problems presented here.

The TSHH shown in Table 2 outperforms all of the other approaches described in this paper on real problems. Table 2 also presents the current best published results on these benchmarks by 8 other approaches reported in the literature[20, 22, 34-39]. The best results are presented in bold. Note that TSHH gets into the same region as these sophisticated approaches which are 'tailor made' for exam timetabling.

Also note that except Carter et al[38] and Asmuni et al[34] all the other approaches are improving approaches that are based on initial solutions obtained beforehand. Our approaches are simple constructive methods that are independent of the initialization process and obtained comparable results from the reported approaches on the benchmark problems, for which most simple constructive methods failed to obtain feasible solutions. Moreover, this approach is far simpler and more generic than those approaches. It selects appropriate simple heuristics during the search process. These simple heuristics can be employed in many other timetabling and scheduling problems.

5. CONCLUSIONS AND FUTURE WORK

The overall goal of this paper is to investigate the development of approaches/systems which can operate at a higher level of generality than current approaches/systems. The TSHH uses only very simple heuristics (SD and LD) and clearly outperforms the heuristics on their own and the other two hybrid approaches we have described in this paper.

The heuristic selection methods described here represent a framework which can easily be applied to other timetabling and scheduling problems. They take simple heuristics and we demonstrated that those heuristics can be better employed by intelligent selection at appropriate points in the solution construction process. These methods are comparable to the bespoke methods even though the overall goal of this approach is to be more generally applicable rather than to produce the 'best' results on benchmark problems. Note also that the methods employed here use only the generally applicable graph coloring heuristics that can be easily employed for many timetabling and scheduling problems. The work can be extended in two ways: 1) extra graph heuristics can be added to the framework to give more choices; and 2) the same framework can be extended to other scheduling problems as little domain specific knowledge is employed. The searching time of TSHH needs further improvement upon larger problems with a higher number of constraints. This may be investigated and compared on other meta-heuristics such as Simulated Annealing and evolutionary approaches, etc.

For the randomly generated data sets, the two hybrid approaches produce better results than those obtained by using the graph heuristics alone, meaning that the knowledge extracted from the TSHH on random data helps in solving problems, avoiding the time and effort required for development of problem specific algorithms for timetabling problems. SD+CBR is able to provide appropriate heuristics within particular problem solving situations using the knowledge discovered beforehand, enabling it to underpin a more general approach for a wider range of problem types. More dedicated knowledge discovery techniques and machine learning methods can be investigated to discover more accurate knowledge within the critical area of learning in hyper-heuristic methodology for solving general timetabling problems.

REFERENCES

1. V. Bardadym. Computer-Aided School and University Timetabling: The New Wave. In: [2], pp. 22-45. (1995).
2. E. Burke and P. Ross eds. Selected Papers from the 1st International Conference on the Practice and Theory of Automated Timetabling, LNCS 1153, Springer-Verlag, 1996).
3. E. Burke and M. Carter eds. Selected Papers from the 2nd International Conference on the Practice and Theory of Automated Timetabling, LNCS 1408. (Springer-Verlag, 1998).
4. E. Burke and W. Erben, W. eds. Selected Papers from the 3rd International Conference on the Practice and Theory of Automated Timetabling, LNCS 2079, (Springer-Verlag, 2001).
5. E. Burke, K. Jackson, J Kingston and R. Weare. Automated Timetabling: The State of the Art. The Computer Journal, **40**(9): 565-571, (1997).
6. E. Burke and P. Causmaecker, eds. Selected Papers from the 4th International Conference on the Practice and Theory of Automated Timetabling, LNCS 2740. (Springer-Verlag, 2003).
7. M. Carter and G. Laporte. Recent Developments in Practical Examination Timetabling. In: [2], pp. 3-21. (1995).
8. M. Carter and G. Laporte. Recent Developments in Practical Course Timetabling. In: [3], pp. 3-19.
9. A. Schaerf. A Survey of Automated Timetabling. Artificial Intelligence Review. **13**(2): 87-127. (1999).
10. E. Burke and S. Petrovic, Recent Research Directions in Automated Timetabling. EJOR, **140**(2): 266-280. (2002).
11. D. Brelaz, New Methods to Color the Vertices of a Graph. Communications of the ACM, **22**(4): 251-256. (1979).
12. de Werra, Graphs, Hypergraphs and Timetabling. Methods of Operations Research. 49: 201-213. (1985).
13. E. Burke, J. Kingston and D. de Werra, Applications to Timetabling, Handbook of Graph Theory, (J. Gross and J. Yellen eds.), pp. 445-474, (Chapman Hall/CRC Press, 2003).
14. M. Carter, A Lagrangian Relaxation Approach to the Classroom Assignment Problem. IFOR **27**(2): 230-246. (1986).
15. B. Deris, S. Omatu, H. Ohta and D. Samat. University Timetabling by Constraint-based Reasoning: A Case Study. JORS. 48(12): 1178-1190. (1997).
16. K. Nonobe T. and Ibaraki. A Tabu Search Approach to the Constraint Satisfaction Problem as a General Problem Solver. EJOR. 106: 599-623. (1998).

17. D. Banks, P. Beel and A. Meisles. A Heuristic Incremental Modelling Approach to Course Timetabling. Proceedings of the Canadian Conference on Artificial Intelligence, pp. 16–29. (1998).
18. F. Glover, and G. Kochenberger, Handbook of Metaheuristics, Kluwer. 2003.
19. D. Costa. A Tabu Search for Computing an Operational Timetable. EJOR. 76: 98-110. (1994).
20. L. Di Gaspero and A. Schaerf, Tabu Search Techniques for Examination Timetabling, In: [4], pp. 104-117. (2000).
21. K. Dowsland, Off the Peg or Made to Measure", In: [3], 37-52. (1998).
22. S. Casey, J. Thompson, A Hybrid Algorithm for the Examination Timetabling Problem. In: [6], pp. 205-230. (2002).
23. E. Burke, J. Newall and R. Weare, R. Initialization Strategies and Diversity in Evolutionary Timetabling. Evolutionary Computation, **6**(1): 81-103. (1998).
24. E. Burke and J. Newall. A Multi-Stage Evolutionary Algorithm for the Timetabling Problem. The IEEE Transactions on Evolutionary Computation. **3**(1): 63-74. (1999).
25. D. Leake ed. Case-based Reasoning: Experiences, Lessons and Future Directions. (AAAI Press, Menlo Park, CA. 1996).
26. E. Burke, B., MacCarthy, S. Petrovic and R. Qu, Structured Cases in Case-Based Reasoning - Re-using and Adapting Cases for Time-tabling Problems. Knowledge-Based Systems, **13**(2-3): 159-165. (2000).
27. E. Burke, B. MacCarthy, S. Petrovic and R. Qu, Multiple-Retrieval Case-Based Reasoning for Course Timetabling Problems. Technical Report NOTTCS-TR-2004-3, School of CSiT, University of Nottingham, U.K. (accepted by JORS, 2004).
28. E. Burke, S. Petrovic and R. Qu, Case Based Heuristic Selection for Examination Timetabling. Technical Report NOTTCS-TR-2004-2, School of CSiT, University of Nottingham, U.K. (To appear in Journal of Scheduling, 2005).
29. S. Petrovic, G. Beddoe and G. Vandem Berghe, Storing and Adapting Repair Experiences in Employee Rostering. In: [6], pp. 148-165. (2003).
30. E. Burke, E. Hart, G. Kendall, J. Newall, P. Ross and S. Schulenburg, Hyper-heuristics: an Emerging Direction in Modern Search Technology. In: F. Glover and G. Kochenberger eds., Handbook of Meta-Heuristics, (Kluwer, 2003), pp. 457-474.
31. G. Kendall, P. Cowling and E. Soubeiga, Choice Function and Random HyperHeuristics. Proceedings of SEAL'02, pp. 667-671. (2002).
32. E. Burke, G. Kendall. G and E. Soubeiga, A Tabu Search Hyperheuristic for Timetabling and Rostering. Journal of Heuristics. **9**(6). (2003).
33. L. Han and G. Kendall. Investigation of a Tabu Assisted Hyper-Heuristic Genetic Algorithm. Congress on Evolutionary Computation, Canberra, Australia, 2230-2237. (2003).
34. H. Asmuni, E. Burke, and J. Garibaldi. Fuzzy Multiple Ordering Criteria For Examination Timetabling. To appear in the 5th International Conference on the Practice and Theory of Automated Timetabling. Pittsburgh, USA. Aug 2004.
35. E. Burke, Y. Bykov, J. Newall and S. Petrovic. A Time-Predefined Local Search Approach to Exam Timetabling Problems. IIE Transactions on Operations Engineering, 36(6), 509-528, (2004).
36. E. Burke and J. Newall, Enhancing Timetable Solutions with Local Search Methods. In: [6], pp. 195-206. (2002).

37. M. Caramia, P. Dell'Olmo and G. Italiano, New Algorithms for Examination Timetabling. In: S. Naher and D. Wagner eds. LNCS 1982, pp. 230-241. (2001).
38. M. Carter, G. Laporte and S. Lee, Examination Timetabling: Algorithmic Strategies and Applications, JORS, **47**: 373-383. (1996).
39. L. Merlot, N. Boland, B. Hughes and P. Stuckey. A Hybrid Algorithm for the Examination Timetabling Problem. In: E. Burke and P. De Causmaecker (eds.) Proceedings of the 4th International Conference on Practice and Theory of Timetabling, pp. 348-371. (2002).

METAHEURISTICS COMPARISON FOR THE MINIMUM LABELLING SPANNING TREE PROBLEM

Raffaele Cerulli,[1] Andreas Fink,[2] Monica Gentili[1] and Stefan Voß[2]

[1] *Computing Science Department, University of Salerno*
Via Ponte don Melillo - 84084, Fisciano (Salerno), Italy
raffaele@unisa.it, mgentili@unisa.it

[2] *Institute of Information Systems, University of Hamburg*
Von-Melle-Park 5, 20146 Hamburg, Germany
andreas.fink@uni-hamburg.de, stefan.voss@uni-hamburg.de

Abstract We study the *Minimum Labelling Spanning Tree Problem*: Given a graph G with a color (label) assigned to each edge (not necessarily properly) we look for a spanning tree of G with the minimum number of different colors. The problem has several applications in telecommunication networks, electric networks, multimodal transportation networks, among others, where one aims to ensure connectivity by means of homogeneous connections. For this NP-hard problem very few heuristics are presented in the literature giving good quality solutions. In this paper we apply several metaheuristic approaches to solve the problem. These approaches are able to improve over existing heuristics presented in the literature. Furthermore, a comparison with the results provided by an exact approach existing in the literature shows that we may quite easily obtain optimal or close to optimal solutions.

Keywords: Minimum Labelling Spanning Tree Problem, Simulated Annealing, Reactive Tabu Search, Pilot Method, Variable Neighborhood Search

1. Introduction

Many real-world problems can be modelled by means of graphs where a label or a weight is assigned to each edge and the aim is to optimize a certain function of these weights. In particular, we can think of problems where the objective is to find homogeneous subgraphs (respecting certain connectivity constraints) of the original graph. This is the case, e.g., for telecommunication networks (and, more generally, any type of communication networks) that are managed by different and competing companies. The aim of each company is to ensure the service to each terminal node of the network by minimizing the cost (i.e., by minimizing the use of connections managed by other companies). This kind of problem can be modelled as follows. The telecommunication network is represented by a graph $G = (V, E)$ where with each edge $e \in E$ is assigned a set of colors L_e and each color denotes a different company that manages the edge. The aim of each company is to define a spanning tree of G that uses the minimum number of colors. When the graph represents a transportation network and the colors, assigned to each edge, represent different modes of transportation, then looking for a path that uses the minimum number of colors from a given source s to a given destination t means to look for a path connecting s and t using the minimum number of different modes of transportation.

Many other problems may be modelled by looking for multicolored cycles, i.e., cycles with edges of different color (see, e.g., [3]), or monochromatic cycles (i.e., cycles whose edges have the same color). In this paper we focus on the *Minimum Labelling Spanning Tree Problem* (MLST): Given a graph G with a label (color) assigned to each edge (not necessarily properly) we look for a spanning tree of G with the minimum number of different colors. This problem was initially addressed by Broersma and Li [2]. They proved, on the one hand, that the MLST is NP-hard by reduction from the Minimum Dominating Set Problem, and, on the other hand, that the "opposite" problem of looking for a spanning tree with the maximum number of colors is polynomially solvable. Independently, Chang and Leu [5] provided a different NP-hardness proof of the problem by reduction from the Set Covering Problem. They also developed two heuristics to find feasible solutions of the problem and tested the performance of these heuristics by comparison with the results of an exact approach based on an A^* algorithm.

Krumke and Wirth [11] formulated an approximation algorithm with logarithmic performance guarantee and showed also that the problem cannot be approximated within a constant factor (in the sequel this algorithm will be referred to as the MVCA heuristic; see Section 2). Wan et al. [14] provided a better analysis of the greedy algorithm given in [11] by showing that its worst case performance ratio is at most $ln(n-1)+1$. Recently, Xiong et al. [16] obtained the better bound $1 + lnb$ where each color appears at most b times. Moreover, Xiong et al. [15] proposed a genetic algorithm to solve the MLST and provided some experimental results. A variant of the problem has been studied by Brüggemann et al. [4], where the MLST with Bounded Color Classes has been addressed. In this variant, each color of the graph is assumed to appear at most r

times. This special case of the MLST is polynomially solvable for $r = 2$, and NP-hard and APX-complete for $r \geq 3$. Local search algorithms for this variant, that are allowed to switch up to k of the colors used in a feasible solution have been studied, too. For $k = 2$, they showed that any local optimum yields an $\frac{(r+1)}{2}$-approximation of the global optimum, and this bound is best possible. For every $k \geq 3$, there exist instances for which some local optimum is a factor of $\frac{r}{2}$ away from the global optimum.

In this paper we present several metaheuristic approaches to solve the problem (namely, Simulated Annealing, Reactive Tabu Search, the Pilot Method and Variable Neighborhood Search) and compare them with the results provided by the MVCA heuristic presented in [11, 15]. The sequel of the paper is organized as follows. Section 2 describes the metaheuristics we implemented. Section 3 reports some computational results. Conclusions and further research are considered in Section 4.

2. Different Metaheuristic Approaches

The purpose of our study is to show that immediate adaptations of existing metaheuristics may lead to high quality results for the MLST. For achieving this goal, in this section we briefly describe the implemented metaheuristics: Simulated Annealing (SA), Reactive Tabu Search (RTS), the Pilot Method and Variable Neighborhood Search (VNS). All these metaheuristics should be implemented without time-consuming parameter tuning to get some insight into the different approaches and their robustness. To achieve this goal we follow an ad-hoc implementation for the VNS as well as an approach of applying an existing metaheuristics framework without time-consuming calibration (HOTFRAME; see [7]) for the remaining ones.[1]

Before going into detail we introduce some notation and the description of the neighborhood structure used in all our proposed strategies. Given an undirected graph $G = (V, E)$ with V being the set of nodes and E denoting the set of edges, let c_e be the color (label) associated with edge $e \in E$ and $L = \{c_1, c_2, \ldots, c_l\}$ be the set of all the colors. We denote by $C(S) = \bigcup_{e \in S} c_e$ the set of colors assigned with edges in $S \subseteq E$. Any spanning tree T of G can be represented by the set of its colors $C(T)$. Given a set of colors C, it is *feasible* for the MLST if and only if the corresponding set of edges defines a connected subgraph G_C that spans all the nodes of G.

[1]We assume that the reader is familiar with some basic concepts of metaheuristics. For a recent survey the reader is referred to [13].

We use the following family of neighborhoods, already defined in [4]. Let k be a fixed integer, and let $C = \{c_{i_1}, c_{i_2}, \dots, c_{i_q}\}$ be a feasible solution for some instance of the MLST, we can define the following different feasible neighbors of C:

***k* - Switch Neighborhood** $N_k(C)$**:**
A set $C' \in N_k(C)$ if and only if we can get the color set C' from the color set C by removing up to k colors from C and adding up to k new colors. That is, $N_k(C) = \{C' \subseteq L : |C' \backslash C| \leq k \text{ and } |C \backslash C'| \leq k\}$.

If we assume $k = 1$ and allow one color either to be added or to be removed we obtain the elementary "bit-flip" neighborhood which is based on representing the selection of colors as a binary vector. That is, if $C = \{c_{i_1}, c_{i_2}, \dots, c_{i_q}\}$ is a solution for some instance of the MLST, this selection of colors may be represented by a binary vector $x = (x_1, ..., x_l)$ of length l with $x_i = 1$ if $c_i \in C$, and $x_i = 0$, otherwise. With this representation we can define the following set of neighbors of C:

Bit-Flip Neighborhood $N_1^0(C)$**:**
$N_1^0(C) = \{C' : \exists i \in \{1, ..., l\} \text{ with } x'_i = 1 - x_i \text{ and } x'_j = x_j \ \forall j \neq i\}$.

A transition from one solution to another is called a move. A move in accordance with the 1 - Switch Neighborhood corresponds to two moves in accordance with the Bit-Flip Neighborhood. For example, if $l = 4$ and $C = \{1, 2, 3\}$ represents the current solution, $C' = \{1, 2, 4\}$ is within the 1 - Switch Neighborhood of C, which corresponds to two bit flips (i.e., color 3 is removed and color 4 is added).

In the following we use the k - Switch Neighborhood for the VNS and the Bit-Flip Neighborhood for SA, RTS and the Pilot method. In connection with the Bit-Flip Neighborhood we also consider infeasible solutions as neighbors. Infeasibilities are taken into account by adding appropriate penalty values to the objective function. In particular, if the subgraph induced by the activated current colors is separated into $c > 1$ connected components, the penalty value is calculated as $(1+\epsilon) \times (c-1)$ for a small value $\epsilon > 0$.

2.1 MVCA

Assuming that we have a binary representation for the problem in the sense that every color is either chosen or not we may apply simple steepest descent approaches using the bit-flip neighborhood. The first one (*add* approach) starts with an infeasible solution $(0, 0, ..., 0)$ and adds

Let $C = \emptyset$ be the set of used colors.
Repeat
Let H be the subgraph of G restricted to edges with colors from C.
for all $c_i \in L \setminus C$ **do**
Determine the number of connected components when inserting all edges with color c_i in H.
end for
Choose color c_i with the smallest resulting number of components and do: $C = C \cup \{c_i\}$.
until H is connected.

Figure 1. The MVCA heuristic.

colors as long as necessary to reduce the number of disconnected components with respect to the corresponding vector and its resulting graph. The second one (*drop* approach) could start with all colors included, i.e. $(1, 1, ..., 1)$, and consecutively drops colors as long as connectivity of the remaining graph is ensured.

The MVCA (Maximum Vertex Covering Algorithm) or some of its modifications, respectively, corresponds to the add approach (see, e.g., [5, 11, 15]). A more detailed description of the heuristic is given in Figure 1. That is, this strategy follows the idea of a simple steepest descent oriented construction procedure. Note that at the end it could be beneficial to try the drop approach while retaining feasibility.

2.2 Simulated Annealing

Simulated Annealing extends basic local search by allowing moves to worse solutions [10]. The basic concept of SA may be described as follows: Starting from an initial solution (we started in our implementation from an empty set of activated colors), successively, a candidate move is randomly selected; this move is accepted if it leads to a solution with a better objective function value than the current solution, otherwise the move is accepted with a probability that depends on the deterioration Δ of the objective function value. The acceptance probability is computed according to the Boltzmann function as $e^{-\Delta/T}$, using a temperature T as control parameter.

Various authors describe robust realizations of this general SA concept. Following [9], the value of T is initially high, which allows many worse moves to be accepted, and is gradually reduced through multiplication by a parameter *coolingFactor* according to a geometric cooling schedule. Given a parameter *sizeFactor*, *sizeFactor* $\times\ l$ candidate moves are tested (note that l denotes the neighborhood size) before the temperature is reduced. The starting temperature is determined as follows: Given a parameter *initialAcceptanceFraction* and based on an abbreviated trial run, the starting temperature is set so that the fraction of accepted moves is approximately *initialAcceptanceFraction*. A further parameter, *frozenAcceptanceFraction* is used to decide whether the annealing process is *frozen* and should be terminated. Every time a temperature is completed with less than *frozenAcceptanceFraction* of the candidate moves accepted, a counter is increased by one, while this counter is re-set to 0 each time a new best solution has been obtained. The whole procedure is terminated when this counter reaches a parameter *frozenLimit*. For our implementation we follow the parameter setting of [9], which was reported to be robust for different problems. Namely, we use $\alpha = 0.95$, *initialAcceptanceFraction* $= 0.4$, *frozenAcceptanceFraction* $= 0.02$, *sizeFactor* $= 16$ and *frozenLimit* $= 5$.

2.3 Reactive Tabu Search

The basic paradigm of tabu search is to use information about the search history to guide local search approaches to overcome local optimality (see [8] for a survey on tabu search). In general, this is done by a dynamic transformation of the local neighborhood. Based on some sort of memory certain moves may be forbidden, we say they are set tabu (and appropriate move attributes such as a certain index indicating a specific color put into a list, the so-called tabu list). As for SA, the search may imply acceptance of deteriorating moves when no improving moves exist or all improving moves of the current neighborhood are set tabu. At each iteration a best admissible neighbor may be selected. A neighbor, respectively a corresponding move, is called admissible, if it is not tabu.

Reactive TS aims at the automatic adaptation of the tabu list length [1]. The idea is to increase the tabu list length when the tabu memory indicates that the search is revisiting formerly traversed solutions. A possible specification can be described as follows: Starting with a tabu list length s of 1, it is increased to $\min\{\max\{s+2, s \times 1.2\}, b_u\}$ every time a solution has been repeated, taking into account an appropriate upper bound b_u (to guarantee at least one admissible move). If there has been no repetition for some iterations, we decrease it to $\max\{\min\{s - 2, s/1.2\}, 1\}$. To accomplish the detection of a repetition of a solution, one may apply a trajectory based memory using hash codes.

For RTS [1], it is appropriate to include means for diversifying moves whenever the tabu memory indicates that we are trapped in a certain region of the search space. As a trigger mechanism one may use, e.g., the combination of at least three solutions each having been traversed three times. A very simple escape strategy is to perform randomly a number of moves (depending on the average of the number of iterations between solution repetitions). For our implementation of RTS we consider as initial solution (as for the SA) an empty set of activated colors. As termination criterion we consider a given time limit.

2.4 Pilot Method

Using a basic algorithm such as a greedy construction heuristic (e.g., MVCA) as a building block or application process, the *pilot method* [6] is a meta-heuristic with the primary idea of performing repetition exploiting the application process as a *look ahead* mechanism. In each iteration (of the pilot method) one tentatively computes for every possible local choice (i.e., move to a neighbor of the current so-called *master solution*) a so-called *pilot solution*, recording the best results in order to extend at the end of the iteration the master solution with the corresponding move. One may apply this strategy by successively performing, e.g., a construction heuristic for all possible local choices (i.e., starting a new solution from each incomplete solution that can result from the inclusion of any not yet included element into the current incomplete solution).

We apply the pilot method in connection with a greedy local search strategy which operates on a solution space that includes incomplete (infeasible) solutions and a neighborhood that considers the addition of colors (alike the MVCA). We take into account infeasibilities by adding appropriate penalty values computed as mentioned above. The pilot strategy successively chooses the best local move (regarding the additional activation of one color) by evaluating such neighbors with a steepest descent until a local optimum, and with that a feasible solution, is obtained. (Note that as for the MVCA, at the end it may be beneficial to greedily drop colors while retaining feasibility.) As there are up to l iterations of the outer-level search loop and for each master solution there up to l local choices that are to be evaluated, the overall time complexity results as $O(l^2)$ times the time complexity of the MVCA.

2.5 Variable Neighborhood Search

Variable Neighborhood Search goes back to Mladenovic and Hansen [12]. The underlying idea of VNS is to generalize the classical local search based approaches by considering a multi-neighborhood structure, i.e., a set of pre-selected neighborhood structures $\mathcal{N} = \{N_1, N_2, \ldots, N_q\}$ such that $N_j(C)$, $j = 1, 2, \ldots, q$ is the set of solutions in the j-th neighborhood

Step 1. Consider an initial feasible solution $C \subseteq L$ and set $k \leftarrow 1$;
Step 2. Generate at random a solution $C' \in N_k(C)$;
Step 3. Apply a local search algorithm, starting from the initial solution C', to obtain a local optimum C'';
Step 4. If $\lvert C''\rvert < \lvert C\rvert$ then: $C \leftarrow C''$ and set $k \leftarrow 1$ otherwise $k \leftarrow k+1$;
Step 5. If $k \leq k_{max}$ then goto Step 1, else Stop.

Figure 2. The basic VNS algorithm.

of C. The basic VNS algorithm, applied to solve the MLST, is described in Figure 2.

In particular, we consider the k-switch neighborhood as described above, where $k = 1, 2, \ldots, k_{max}$. The procedure starts from an initial feasible solution C provided by the MVCA heuristic. At each generic iteration the VNS: (i) selects at random a feasible solution C' in the neighborhood $N_k(C)$, (ii) applies a local exchange strategy that, for a maximum number h_{max} of iterations, tries to decrease the size of C' to obtain a possible better solution C'', and, (iii) defines the new neighborhood to be explored in the next iteration. In our implementation of VNS, we let parameter k_{max} vary during the execution, that is $k_{max} = \min\{|C|, \frac{l}{4}\}$, where $|C|$ is the size of the current feasible solution whose neighborhood is being explored. The overall time complexity results as $O(l^2 * h_{max})$. Indeed, the VNS strategy explores sequentially k_{max} neighborhoods. The first neighborhood $N_1(C)$ is explored so many times as the number of improvements of the initial solution C; the total number of improvements is at most l, that, in the worst case, leads us to explore at most $l * k_{max}$ neighborhoods. Moreover, since for each neighborhood we carry out at most h_{max} iterations and $k_{max} \leq \frac{l}{4}$, we have the above mentioned time complexity.

As our idea here was to test the overall option of using robust implementations of different metaheuristics we keep content with the above implementation of the VNS for this paper.

3. Experimental Results

Computational results for the MLST for different algorithms have been reported in only a very limited number of references. In our paper we follow the idea of Xiong et al. [15] to compare the results of any heuristic (a genetic algorithm in their case) with the results of the MVCA. Unfortunately, data from other researchers were not provided to us so that we re-implemented the data generation mechanism to have at least a certain comparability with respect to the results of [15], who report that their genetic algorithm is able to produce competitive results and typically outperforms MVCA in many cases by a small percentage.

Below we report some of our computational results. We considered different groups of instances (according to the ones considered in [15]) in order to evaluate how the performance of the algorithms is influenced by both (i) the structures of the networks and (ii) the distribution of the labels on the edges. In particular, we defined two different groups of scenarios based on different parameter settings: n: number of nodes of the graph, l: total number of colors assigned to the graph, m: total number of edges of the graph computed by $m = \frac{d(n-1)n}{2}$ where d is a measure of density of the graph. Parameter settings for scenarios in Group 1 are: $l = n = 20, 30, 40, 50$ and $d = 0.2, 0.5, 0.8$, for a total of 12 different scenarios. Instances in Group 2 are characterized by $n = 50, 100$, $l = 0.25 * n, 0.5 * n, n, 1.25 * n$ and $d = 0.2, 0.5, 0.8$, for a total of 24 different scenarios. For each scenario we generated ten different instances. All the generated data are available upon request from the authors.

Results are reported in Tables 1 – 6. In each table the first three columns show the parameters characterizing the different scenarios (n, l and d). The remaining columns give the results of the MVCA heuristic and of our metaheuristics: Variable Neighborhood Search, Simulated Annealing, Reactive Tabu Search and the Pilot Method, respectively. More specifically, Table 1 gives the objective function values found by the algorithms for the Group 1 scenarios; all the values are average values over ten different instances. Tables 2 and 3 give the objective function values for the Group 2 scenarios with $n = 50$ and $n = 100$, respectively (again all the values are average values over 10 different instances). Computational times of the algorithms are reported in Table 4 for the instances of Group 1 and in Tables 5 and 6 for the instances of Group 2.

Looking at Tables 4 – 6 we can see that the computational times of all the algorithms (but RTS) increase with larger graphs and decrease with more dense graphs, as it was expected.[2] From Tables 1 – 3 we can see that all the algorithms either get the same average solution value of MVCA or return a better average value (but the SA approach that for the instances of Group 2 with $n = 100$ and $l = 100, 125$ performs a little worse). In particular, by considering the detailed results obtained for the whole set of instances (a total of 120 instances for Group 1 scenarios and of 240 instances for Group 2 scenarios, that are not reported

[2]The MVCA and the VNS run on Xeon 2.8Ghz Linux SUSE9, while the SA, RTS and Pilot run on a Pentium 4 / 2.4 GHz PC.

here for space reason) we can observe that: both the VNS strategy and the Pilot method never return a worse solution value than the one computed by the MVCA; the SA approach is worse than MVCA on 14 % instances of Group 2 with $n = 100$ and on 2% instances of Group 2 with $n = 50$; finally, the RTS strategy performs worse than MVCA on 1% instances of Group 1 and 6 % instances of Group 2 with $n = 100$; for all the other cases, the four algorithms either return the same solution value or a better solution value than the one provided by MVCA. Note that for dense graphs (i.e., $d = 0.8$) in most cases all the algorithms find the same solution. This happens also for some instances with $d = 0.5$, see, in particular, Table 1 where $n = 30$, Table 2 where $l = 12, 25$ and Table 3 where $l = 25$. The pilot method seems to obtain the best results in most of the scenarios. In particular, it attains the best results (not considering those instances where all the approaches perform the same) on 17 instances, SA on 10 instances, VNS on 8 instances and RTS on 7. More in detail, SA seems to be more effective on low density graphs (see Table 2 with $d = 0.2$) while the pilot strategy and the VNS work better on larger instances (see Table 3). In particular, the VNS method starts from the feasible solution provided by MVCA and explores different neighborhoods to improve it. The best improvement is obtained for the instances of Group 2 with $n = 50$, $d = 0.8$ and $l = 62$ (up to 12%). On the other hand, the solution cannot be improved for the most dense instances (those with $d = 0.8$) where all the algorithms perform the same and likely MVCA already finds a good solution. Moreover, we can see that, generally, improvements are attained for instances where the number k_{max} of different neighborhoods ($k_{max} = \min\{|C|, \frac{l}{4}\}$) explored is large enough (e.g., those instances with a feasible solution value ranging between 10 and 12).

For the smaller problem instances of Group 1 our re-implementation of the exact approach of [5] reveals that at least one of the metaheuristics finds the optimal solution. If the optimal solution is not found, it usually means that the found solution just differs by one color. CPU times for this approach are within the range of a few seconds with one exception for $n = 30, d = 0.2$. The improvements of our algorithms over MVCA are in the range of those of the genetic algorithm reported in [15] or better, especially for the pilot method. Our approaches follow the same idea of having a robust (and somewhat auto-adaptive) approach and it seems that we succeeded in reaching this goal. Of course the pilot method operates with a considerable number of computational repetitions and our robust ad-hoc application of the method using the HotFrame optimization framework leaves many options for improving those CPU-times.

Table 1. Computational results for Group 1

n	l	d	MVCA	VNS	SA	RTS	Pilot
20	20	0.8	2.6	2.4	2.4	2.4	2.4
20	20	0.5	3.5	3.2	3.1	3.1	3.2
20	20	0.2	7.1	6.9	6.7	6.7	6.7
30	30	0.8	2.8	2.8	2.8	2.8	2.8
30	30	0.5	3.7	3.7	3.7	3.7	3.7
30	30	0.2	8.0	7.8	7.4	7.4	7.5
40	40	0.8	2.9	2.9	2.9	2.9	2.9
40	40	0.5	3.9	3.9	3.9	4.0	3.7
40	40	0.2	8.6	8.3	7.4	7.9	7.7
50	50	0.8	3.0	3.0	3.0	3.0	3.0
50	50	0.5	4.4	4.1	4.2	4.2	4.0
50	50	0.2	9.2	9.1	8.7	8.8	8.6

Table 2. Computational results for Group 2 with n=50

n	l	d	MVCA	VNS	SA	RTS	Pilot
50	12	0.8	1.1	1.1	1.1	1.1	1.1
50	12	0.5	2.0	2.0	2.0	2.0	2.0
50	12	0.2	4.1	4.0	3.8	3.8	3.8
50	25	0.8	2.0	2.0	2.0	2.0	2.0
50	25	0.5	3.0	3.0	3.0	3.0	3.0
50	25	0.2	6.8	6.4	5.9	6.1	5.9
50	50	0.8	3.0	3.0	3.0	3.0	3.0
50	50	0.5	4.4	4.1	4.2	4.2	4.0
50	50	0.2	9.2	9.1	8.7	8.8	8.6
50	62	0.8	3.4	3.0	3.2	3.3	3.0
50	62	0.5	4.9	4.8	4.9	4.8	4.8
50	62	0.2	10.5	9.9	9.3	9.9	9.7

4. Conclusions and Further Research

In this paper we have considered the Minimum Labelling Spanning Tree Problem. This problem has been recently studied and has many important applications in the field of communication networks. Previous research provides algorithms with approximation guarantee ratio and some local search schemes based on the simple k-switch neighborhoods. Recently, a genetic algorithm has been presented in [15] and some computational results are given. In this paper we show that one may successfully adapt existing metaheuristics and provide computational experience based on the comparison of various metaheuristics tested on different scenarios. We compared our approaches with the results pro-

Table 3. *Computational results for Group 2 with n=100*

n	l	d	MVCA	VNS	SA	RTS	Pilot
100	25	0.8	1.8	1.8	1.8	1.8	1.8
100	25	0.5	2.1	2.0	2.0	2.0	2.0
100	25	0.2	4.8	4.5	4.5	4.6	4.5
100	50	0.8	2.1	2.0	2.0	2.1	2.0
100	50	0.5	3.2	3.0	3.1	3.2	3.1
100	50	0.2	7.5	6.8	6.9	7.1	6.9
100	100	0.8	3.3	3.1	4.0	3.4	3.0
100	100	0.5	5.1	5.0	5.2	5.1	4.7
100	100	0.2	11	10.7	10.7	10.9	10.3
100	125	0.8	4.0	4.0	4.1	4.0	4.0
100	125	0.5	5.9	5.8	6.0	5.8	5.4
100	125	0.2	12.2	11.9	11.9	12.3	11.3

Table 4. *Computational times (sec.) for Group 1*

n	l	d	MVCA	VNS	SA	RTS	Pilot
20	20	0.8	0.01	1.67	0.69	10.00	0.11
20	20	0.5	0.01	1.92	0.79	10.00	0.13
20	20	0.2	0.02	1.69	0.91	10.00	0.26
30	30	0.8	0.03	5.58	2.44	10.00	0.46
30	30	0.5	0.04	7.22	1.67	10.00	0.59
30	30	0.2	0.07	8.55	2.15	10.00	1.24
40	40	0.8	0.07	14.04	3.37	10.00	1.32
40	40	0.5	0.09	19.14	3.14	10.00	1.60
40	40	0.2	0.19	26.76	2.59	10.00	3.56
50	50	0.8	0.14	28.52	6.63	10.00	2.95
50	50	0.5	0.20	41.81	5.04	10.00	3.80
50	50	0.2	0.39	64.52	5.25	10.00	9.11

vided by the approximation guaranteed algorithm given in [11]. We considered different groups of instances in order to evaluate how the performance of the algorithms is influenced by both (i) the structures of the networks and (ii) the distribution of the labels on the edges.

As our implementations were addressing the option of having a robust instead of a specialized implementation there is considerable room for improving the computation times. Moreover, our further research is focused on (i) extending our computational experience in order to evaluate the algorithms on larger networks and on different graph classes reproducing real case networks; (ii) implementing different neighborhood structures that could lead to an improved performance when embedded,

Table 5. Computational times (sec.) for Group 2 with n=50

n	l	d	MVCA	VNS	SA	RTS	Pilot
50	12	0.8	0.03	2.77	1.74	50.00	0.07
50	12	0.5	0.04	3.26	1.13	50.00	0.08
50	12	0.2	0.06	4.80	1.21	50.00	0.12
50	25	0.8	0.06	8.39	3.22	50.00	0.43
50	25	0.5	0.08	11.14	2.41	50.00	0.58
50	25	0.2	0.16	16.96	2.58	50.00	1.13
50	50	0.8	0.14	28.52	6.63	50.00	2.95
50	50	0.5	0.20	41.81	5.04	50.00	3.80
50	50	0.2	0.39	64.52	5.25	50.00	9.11
50	62	0.8	0.19	41.76	8.01	50.00	4.83
50	62	0.5	0.28	62.98	6.76	50.00	7.33
50	62	0.2	0.55	95.56	5.67	50.00	16.69

Table 6. Computational times (sec.) for Group 2 with n=100

n	l	d	MVCA	VNS	SA	RTS	Pilot
100	25	0.8	0.28	22.73	9.53	50.00	1.04
100	25	0.5	0.31	29.09	6.63	50.00	0.97
100	25	0.2	0.52	54.04	4.56	50.00	1.86
100	50	0.8	0.48	73.46	19.00	50.00	5.56
100	50	0.5	0.64	104.99	14.35	50.00	6.95
100	50	0.2	1.31	209.57	10.39	50.00	15.15
100	100	0.8	1.25	258.93	53.29	50.00	37.19
100	100	0.5	1.76	381.55	27.66	50.00	50.86
100	100	0.2	3.24	710.55	19.36	50.00	122.42
100	125	0.8	1.75	390.67	45.24	50.00	79.68
100	125	0.5	2.55	578.48	35.05	50.00	99.21
100	125	0.2	4.99	1201.83	23.46	50.00	227.38

e.g., into a VNS (ideas may include new classes of neighborhoods in the sense that we exchange colors by taking into account the number of the connected components of the induced subgraphs of included colors as well as the size of the spanned node set); (iii) studying some variants of the problem inspired by real world applications. In particular, we will focus our attention on the Multilabel Spanning Tree Problem: looking for the minimum labelling spanning tree when multiple labels are assigned with each edge. When the colors assigned with edges have different weights, then the weighted version of the MLST and of the multilabel spanning tree problem are interesting objects of research. If edges are assigned also with costs (weights), one could think of the bi-objective version of these problems when both the total number of different colors and the total weight of the spanning tree have to be minimized.

References

[1] R. Battiti. Reactive search: Toward self-tuning heuristics. In V.J. Rayward-Smith, I.H. Osman, C.R. Reeves, and G.D. Smith, editors, *Modern Heuristic Search Methods*, pages 61–83. Wiley, Chichester, 1996.

[2] H. Broersma and X. Li. Spanning trees with many or few colors in edge-colored graphs. *Discussiones Mathematicae Graph Theory*, 17:259–269, 1997.

[3] H. Broersma, X. Li, and S. Zhang. Paths and cycles in colored graphs. *Electronic Notes in Discrete Mathematics*, 8, 2001.

[4] T. Brüggemann, J. Monnot, and G.J. Woeginger. Local search for the minimum label spanning tree problem with bounded color classes. *Operations Research Letters*, 31:195–201, 2003.

[5] R.-S. Chang and S.-J. Leu. The minimum labeling spanning trees. *Information Processing Letters*, 63:277–282, 1997.

[6] C.W. Duin and S. Voß. The pilot method: A strategy for heuristic repetition with application to the Steiner problem in graphs. *Networks*, 34:181–191, 1999.

[7] A. Fink and S. Voß. HotFrame: A heuristic optimization framework. In S. Voß and D. Woodruff, editors, *Optimization Software Class Libraries*, pages 81–154. Kluwer, Boston, 2002.

[8] F. Glover and M. Laguna. *Tabu Search*. Kluwer, Boston, 1997.

[9] D. S. Johnson, C. R. Aragon, L. A. McGeoch, and C. Schevon. Optimization by simulated annealing: An experimental evaluation; part I, graph partitioning. *Operations Research*, 37:865–892, 1989.

[10] S. Kirkpatrick, C.D. Gelatt Jr., and M.P. Vecchi. Optimization by simulated annealing. *Science*, 220:671–680, 1983.

[11] S.O. Krumke and H.-C.Wirth. On the minimum label spanning tree problem. *Information Processing Letters*, 66:81–85, 1998.

[12] N. Mladenović and P. Hansen. Variable neighbourhood search. *Computers & Operations Research*, 24:1097–1100, 1997.

[13] S. Voß. Meta-heuristics: The state of the art. In A. Nareyek, editor, *Local Search for Planning and Scheduling*, volume 2148 of *Lecture Notes in Artificial Intelligence*, pages 1–23. Springer, Berlin, 2001.

[14] Y. Wan, G. Chen, and Y. Xu. A note on the minimum label spanning tree. *Information Processing Letters*, 84:99–101, 2002.

[15] Y. Xiong, B. Golden, and E. Wasil. A one-parameter genetic algorithm for the minimum labeling spanning tree problem. Technical report, University of Maryland, 2003.

[16] Y. Xiong, B. Golden, and E. Wasil. Worst-case behavior of the MVCA heuristic for the minimum labeling spanning tree problem. Technical report, University of Maryland, 2003.

A NEW TABU SEARCH HEURISTIC FOR THE SITE-DEPENDENT VEHICLE ROUTING PROBLEM

I-Ming Chao[1] and Tian-Shy Liou[2]

[1]*Department of Industrial Engineering and Management, I-Shou University, 1 Sec. 1, Shiuecheng Rd., Dashu Township, Kaohsiung County, 840, Taiwaun, R.O.C.,* [2]*Department of Business Administration, Cheng Shiu University, 840 Chengcing Rd., Niaosong Township, Kaohsiung County, 833, Taiwaun, R.O.C.*

Abstract: The site-dependent vehicle routing problem takes into account some real-life applications of the basic vehicle routing problem when there are compatible dependencies between customers (sites) and vehicle types. Every customer is associated with a set of allowable vehicle types and has to select only one of them. A series of basic vehicle routing problems are solved over the customers that select the same vehicle type. The objective is to minimize the total distance traveled (or the total travel cost incurred) by the fleet and all constraints for the basic vehicle routing problem as well as the site-dependency constraints must be satisfied. In this paper, we present a new heuristic method based on tabu search combined with the deviation of the deterministic annealing method to carry out the intensification and diversification search by varying the values of the deviations within two different ranges respectively. We test the method on a set of 23 benchmark problems taken from literature, and the computational results show that the new method can solve the problem quite effectively.

Key words: Vehicle Routing, Site-dependent Vehicle Routing, Heuristic, Tabu Search, Deterministic Annealing

1. INTRODUCTION

In the basic vehicle routing problem (VRP), a set of routes is designed for dispatching a homogeneous or heterogeneous fleet of vehicles to service a set of customers from a single distribution depot or terminal. Each vehicle has a fixed capacity and each customer has a known demand that must be fully satisfied. Each customer must be serviced by exactly one visit of a single vehicle and each vehicle must depart from and return to the depot. Each route has a route length constraint that limits the distance traveled by each vehicle. The objective is to provide each vehicle with a sequence of visits so that all customers are serviced and the total distance traveled by the fleet (or the total travel cost incurred by the fleet) is minimized.

The site-dependent vehicle routing problem (SDVRP) extends the basic VRP to take a real-life application situation into account. In the SDVRP, the fleet of heterogeneous vehicles is dispatched to service a set of customers, but there exist compatible dependencies between customers (sites) and vehicle types. The fleet consists of several vehicle types in which each type has limited number of vehicles (for example, five small capacity, seven medium capacity, and three large capacity vehicles). In the VRP, any type of vehicle can visit each customer, but in the SDVRP not every type of vehicle can service every customer. Some customers with extremely high demands may require large vehicles; some customers located in congested areas may require small or medium vehicles, while the remaining customers can be serviced with any type of vehicle without site-dependent restrictions. For each customer *i*, there is a set of allowable vehicle types associated with it, denoted by A_i and called associated types of customers *i*. For example, there are A_1={Type 1, Type 3}, A_2 = {Type 1, Type 2, Type 3}, and A_3 = {Type 3}, etc. In the SDVRP, we select an allowable vehicle type for each customer *i* from its associated A_i, and then solve a basic VRP for each vehicle type over the customers selecting the vehicle type such that the total distance traveled by all types of vehicles is minimized.

As a VRP-variant application, the SDVRP can be summarized as follows. For N customers surrounding a single distribution depot and each requiring a service by a vehicle of its associated types, we seek to design a set of routes for each type of vehicle in the fleet that minimizes the total distance traveled by the fleet and satisfies the following constraints:

(1) each customer *i* selects exactly one allowable vehicle type from its associated A_i;
(2) a particular type of vehicle is able to visit a customer *i* only if the allowable vehicle type from its A_i is selected by i;
(3) no vehicle can travel between two customers unless both customers select the same vehicle type;
(4) if a vehicle visits a customer, then it leaves that customer;
(5) a vehicle can be used at most once and it must start and finish at the depot;
(6) the total load of a route cannot exceed the capacity of the assigned type of vehicle;

(7) the number of routes designed for a vehicle type is not more than the available number of the type of vehicles;
(8) If the model takes the route length constraint into account, the total distance of a route cannot exceed the bound.

The SDVRP can be shown to be NP-hard by starting with the VRP (see [7]). As the associated allowable vehicle types A_i for each customer i contain all vehicle types, namely, every type of vehicle can service every customer, the SDVRP becomes the VRP; therefore, the SDVRP is at least as hard as the VRP. To solve a practical size SDVRP, an exact method would require a large amount of computing time. Accordingly, we should tackle a moderate-size SDVRP problem using heuristics, and note that all existing SDVRP approaches are heuristics.

The SDVRP can be used to model many real-life application problems. Rochat and Semet [14] encountered the SDVRP-related problem in delivering pet food and flour that occurs in a major company producing pet food and flour in the French part of Switzerland. The great variety of location of all customers (center of a city, village, isolated farm, and so on) makes the access of all vehicles to all customers unlikely. They modeled the problem as an SDVRP-variant and solved it heuristically. Semet and Taillard [15] solved an SDVRP-variant, real-world problem that involved 45 grocery stores located in the cantons of Vaud and Lalais in Switzerland. Some stores received deliveries by a road train (a truck pulling a trailer) or by a truck only, while others received deliveries by a truck only and no road trains. This type of the practical problem can be modeled as a SDVRP or SDVRP-variant whenever there exist dependencies amongst customer sites and vehicle types.

The basic VRP and its many variants have attracted the attention of many researchers over the last two decades; however; to our knowledge only a few papers have presented solution approaches for solving the SDVRP. Nag [12] and Nag et al. [13] were the first to take dependencies amongst customers (sites) and vehicle types into account. They presented four heuristics for solving the SDVRP and a set of twelve SDVRP test problems. A description of the heuristics and the solutions to test problems are reviewed in detail in Chao et al. [2].

Chao, Golden, and Wasil [1,2] presented an SDVRP solution approach that is a two-phase heuristic. In the first phase, they generated an initial solution quickly by balancing the workload amongst the different vehicle types and accomplished this by solving a relaxation of an integer program. In the second phase, they improved the initial solution by deterministic annealing [9]. They used the record-to-record travel procedure of Dueck [5] and Dueck and Scheurer [6]. The improvement phase considers a series of uphill and downhill one-point movements that move one customer at a time.

Cordeau and Laporte [4] first viewed the SDVRP problem as a special case of the period Vehicle Routing Problem (PVRP), and adopted the Tabu Search Method for solving the PVRP to solve the SDVRP. Their method outperformed all existing methods in the literature.

In this paper, we develop a method that combines tabu search with deterministic annealing to solve the SDVRP. Besides the frequency-based tabu restriction, we define the deviation as a new type of tabu restriction. The values of deviation are self-adjusted within two ranges, one with the smaller deviation values to carry out the intensification search, and the other one with the larger deviation values to carry out the diversification search. In addition; customers are moved and exchanged strategically within or amongst vehicle types. We hope that by using the self-adjusted deviation values and strategically moving or exchanging customers, the new heuristic will consistently outperform all existing heuristics for the SDVRP.

The rest of this paper is organized as follows. In the next section, we describe the new heuristic method in detail. In the third section, we test the new method on 23 benchmark test problems taken from the literature. Our conclusions are given in section 4.

2. A NEW TABU SEARCH HEURISTIC FOR THE SDVRP

In this section, a new heuristic method based on the tabu search combined with a new type of tabu restriction is developed for solving the SDVRP. This new method has two phases: the *construction* phase for generating an initial solution easily and quickly, and the *improvement* phase for improving the initial solution with a variant of tabu search. Next, we describe each step in detail, and then integrate each of them into a new heuristic for solving the SDVRP.

2.1 The Construction Phase

In the construction phase, we generate an initial solution which is the starting point for the search procedure. First, we select an allowable vehicle type for each customer, and then solve a basic VRP for each vehicle type. In Chao, Golden, and Wasil [1,2], the vehicle type selection considers balancing the workload amongst the different vehicle types and solving a sequence of 0-1 integer programs to get a feasible solution. In this paper, the balance of workload amongst the vehicle types is no longer a consideration. Instead, the percentage of total capacity of each vehicle type is adjusted by solving a series of relaxed generalized assignment problems. The generalized assignment program is giving by

$$\min d_{ij} x_{ij} \tag{1}$$

$$\text{subject to} \quad \sum_{j \in A_i} x_{ij} = 1 \qquad i = 1,\ldots,N \tag{2}$$

$$\sum_{i=1}^{N} a_{ij} q_i x_{ij} \le k_j C_j \qquad j = 1,\ldots,M \tag{3}$$

$$x_{ij} \in \{0,1\} \qquad i = 1,\ldots,N;\ j \in A_i \tag{4}$$

where d_{ij} = the selection cost of customer *i* selecting vehicle type *j*, where $j \in A_i$,
$x_{ij} = 1$, if customer *i* chooses vehicle type *j*; and 0, otherwise,
$a_{ij} = 1$, if $j \in A_i$; and 0, otherwise,
C_j = the total capacity of type *j* vehicle,
M = number of vehicle types,
k_j = the used percentage of total capacity for vehicle type j.

For computing d_{ij}, a seed point is generated for each vehicle type *j* as the center of gravity for customers whose allowable vehicle types contain *j*. The coordinates of the seed point of the vehicle type *j* can be computed as $\overline{X}_j = \sum_{\forall i, j \in A_i} X_i \Big/ \sum_{\forall i, j \in A_i} 1$, and $\overline{Y}_j = \sum_{\forall i, j \in A_i} Y_i \Big/ \sum_{\forall i, j \in A_i} 1$, where $(X_i\ ,\ Y_i)$ is the Euclidean coordinate of customer *i*, and we compute the selection cost d_{ij} as

$$d_{ij} = c_{0i} + c_{ij} - c_{0j} \tag{5}$$

where c_{ij} is the distance between customer *i* and seed point *j*, and 0 denotes depot. The objective function (1) minimizes the total selection cost. Constraints (2) and (4) ensure that exactly one allowable vehicle type is selected for customer *i* from A_i. Constraint (3) ensures the maximum load of each vehicle type j to be less than or equal to k_j percent of its total capacity. In our LP relaxation, we replace the integrality restrictions (4) with

$$x_{ij} \geq 0. \tag{6}$$

As proved in Chao, Golden, and Wasil [1], solving the LP (1),(2),(3), and (6) produces a solution that contains at most 2M of the x_{ij}'s with fractional values whenever the LP is feasible and bounded. That is, at most M customers will not have integer x_{ij} valued solutions (either 0 or 1) and are not assigned a vehicle type in an optimal solution to the relaxed linear program. In most SDVRP problems, the number of vehicle types M is much smaller than the number of customers N. After solving the relaxed assignment problem, we round the largest x_{ij} for customer i to 1 in order to assign exactly one vehicle type to customer i.

After the relaxed assignment problem assigns one vehicle type for each customer, we use the Clarke and Wright savings algorithm [3] to solve the basic VRP for each vehicle type over customers that select it. Due to the rounding of noninteger variables or the tightness of capacity, the savings algorithm may generate an infeasible solution for some vehicle type; that is, a solution generated by the savings method may require more vehicles of a certain type than are available. If this occurs for a certain vehicle type j, we reduce the capacity rate k_j (initially, it's set equal to 1) by one percent and uses the adjusted k_j's to solve the LP (1), (2), (3), and (6). The savings algorithm is then used to generate a solution to the problem, and the of construction phase continues until a feasible solution is generated.

2.2 Improvement Phase

In this section, we describe the steps used by tabu search (TS) in the improvement phase. The basic form of TS for globally optimal solutions with the assistance of adaptive memory procedures usually explores the search space by moving from a solution to its best neighbor, even if this results in a deterioration of the objective value, in order to increase the likelihood of moving out of a poor local optimum. Solutions that were recently examined with a certain attribution are forbidden or declared tabu for a specified number of iterations unless it is overridden by *aspiration criteria*. This type of tabu restriction is usually called *frequency-based tabu restriction* and the number of iterations is called the *tabu duration*. An intensification strategy is carried out by moving from the current solution to its near neighbors. A diversification strategy usually moves solution the current to a further neighbor by performing some less-frequently executed candidates. Intensification and diversification are usually applied to accentuate and broaden the search in the solution space. Tabu search has enjoyed successes in a variety of problem settings such as scheduling, transportation, and layout and circuit design. Glover and Laguna [8] provided a detail survey and description of TS.

In this paper, we adopt the idea of the deviation in deterministic annealing [5,6,9] to define a new kind of tabu restriction called *objective-based tabu restriction.,* Our idea is to carry out intensification and diversification by using two different ranges of the values of the deviation. In the next several sections, we design two types of movements, one-point movement and two-point exchange, to generate neighbor candidates of a solution, and both tabu restrictions and aspiration criterion are used in both movements. We now describe each of them.

2.2.1 Tabu restriction and aspiration criterion

Two types of tabu restrictions are used in the new TS heuristic for the SDVRP. The first is the frequency-based tabu restriction (FTB) that forbids a customer being moved back to a certain route where it is just removed out for a specified of tabu duration. The second is the objective-based tabu restriction (OTB) that is the deviation or threshold in deterministic annealing. A candidate with the objective function value greater than the best objective function value plus the deviation will be forbidden as an OTB. In deterministic annealing, it is not easy to choose the value of deviation since a prescribed deviation does not work consistently well. We allow the deviation is designed to move between a range of values, smaller values for intensification and larger values for diversification. For intensification, we set the initial deviation ratio (D) equal to 0.01 and the deviation equal to D times the objective function value of the best solution obtained so far. If no candidate move can be implemented after trying all possible non-tabu candidates with current D, then D is increased by increment d=0.01, and the intensification terminates when D increases up to 0.2. For diversification, initially we set D equal to 0.2, and d equal to 0.05. The diversification terminates when at least two iterations have been tried and a new solution is generated, that is, the search procedure has reached a new solution region, and we restart to perform the intensification and descent steps. We explain this in more detail in the section on stopping rules.

In the tabu search algorithm, the aspiration criterion is introduced to determine when tabu restrictions can be overridden. We use the criterion aspiration by objective that overrides the tabu restriction whenever a movement produces a solution with an objective function value less than the current best value. To generate the candidate neighborhood of a current solution, we use two types of movements: (1) a one-point movement to move a customer from one route to another route feasibly, and (2) a two-point exchange to move two customers between two different routes feasibly. Both movements involve two types of tabu restrictions and the aspiration criterion and they take the site-dependencies and the other restrictions into account to preserve feasibility. The details of two types of movements are presented in the next subsections.

2.2.2 Neighborhood Generators

The One-Point Movement (OPM) tries to move one customer at a time from one route to another rout. In the intra-type OPM (*Intra OPM*), we try move a customer to another route in the same vehicle type while checking both tabu restrictions and the aspiration criterion. We outline the steps in Table 1.

Table 1. Intra-Type One Point Movement (Intra OPM)

For T = the first to the last vehicle type	(A loop)
For R = the first to the last route in T	(B loop)
For I = the first to the last customer in R	(C loop)
For R_a = the first to the last route in T, but $R_a \neq$ R	(D loop)
Remove I from R and insert it onto R_a in the cheapest way, if the movement is not in a tabu restriction, or it meets the aspiration criterion, then perform the movement; otherwise, the current I remains on the route R,and the next I is considered.	

The inter-type one point movement (Inter OPM) is performed similarly to the intra OPM. However, the inter OPM tries to move a customer to a route with a different allowable vehicle type. The detailed steps are listed in Table 2.

Table 2. Inter-Type One Point Movement (Inter OPM)

For T = the first to the last vehicle type	(A loop)
For R = the first to the last route in T	(B loop)
For I = the first to the last customer in R	(C loop)
For T_a = the first to the last vehicle type in A_I, but $T_a \neq$ T	(D loop)
For R_a = the first to the last route in T_a	(E loop)
Remove I from R and insert it onto R_a in the cheapest way, if the movement is not in a tabu restriction, or it meets the aspiration criterion, then perform the movement; otherwise, the current I remains on the route R, and the next I is considered.	

The difference between intra OPM and inter OPM is that the site-dependent constraints need to be checked in the inter OPM. We check two types of tabu restrictions and the aspiration criterion to carry out the cheapest insertion. The movement with the smallest objective function value but belonging to an FTB cannot be treated as a candidate unless it meets the aspiration criterion. The OTB is used as the second tabu restriction to make a decision of whether nor not to make the candidate move. The amount of the OTB deviation adjusts automatically in the small and large ranges of values to carry out intensification and the diversification search, respectively.

Our two-point exchange (TPE) also has intra-type and inter-type steps. The intra and inter TPE (intra TPE and inter TPE) exchange two customers in two different routes in the same vehicle type and in different vehicle types, respectively. The intra TPE does not need to take into account site-dependent constraints, but the inter TPE needs to check the site-dependent constraints in both routes to preserve the feasibility. The tabu restrictions and the aspiration criterion are used in the same way as those in the OPM. The intra TPE and the inter TPE are listed in detail in Table 3 and Table 4, respectively.

Table 3. Intra-Type Two Point Exchange (Intra TPE)

For T = the first to the last vehicle type	(A loop)
For R = the first to the last route in T	(B loop)
For I = the first to the last customer in R	(C loop)
For R_a = the first to the last route in T, but $R_a \neq R$	(D loop)
For I_a = the first to the last customer in R_a	(E loop)
Remove I from R and insert it onto R_a and remove I_a from R_a and insert it onto R in the cheapest way simultaneously, and if neither movement is in a TBR1 (or TBR2) or the exchange meets the aspiration criterion, then performing the exchange.	

Table 4. Inter-Type Two Point Exchange (Inter TPE)

For T = the first to the last vehicle type	(A loop)
For R = the first to the last route in T	(B loop)
For I = the first to the last customer in R	(C loop)
For T_a = the first to the last vehicle type in A_I, but $T_a \neq T$	(D loop)
For R_a = the first to the last route in T_a	(E loop)
For I_a = the first to the last customer in R_a	(F loop)
Remove I from R and insert it onto R_a and remove I_a from R_a and insert it onto R in the cheapest way simultaneously, and if neither one movement is in a TBR1 (or TBR2), or the exchange meets the aspiration criterion, then performing the exchange.	

2.2.3 Local Clean-up

To improve a solution locally, the 2-opt procedure (see Lin [10]) is applied to every route. The 2-opt will not accept any uphill move and the solution is improved without alternating the assignment of allowable vehicle type of any customer. Furthermore, the tabu restrictions and aspiration criterion are not applied in the local 2-opt procedure.

2.2.4 Stopping Rules

The improvement phase has three stages: the intensification stage, the descent stage, and the diversification stage (See Table 5). The intensification stage searches the best neighbor by using smaller values of the deviation; the descent stage allows no uphill moves. The diversification stage leads the searching procedure to explore a new solution region by using larger values of the deviation given the high quality solution generated in the intensification and the descent stages. Three stages are performed consecutively and each terminates by checking three different local stopping rules. The intensification stopping rule (ISR) has two parts, either the loop is running full, (namely, $k > K$ in Table 5) or the deviation ratio increases up to a certain prescribed threshold (namely, $D>0.2$ in Table 5). The diversification stopping rule (DSR) terminates the loop when at least two iterations have been performed and at least one candidate move has been made. The descent stopping rule (DSR) terminates the loop when no better solution has been generated.

Table 5. Improvement Steps in Tabu Search Heuristic for Solving the SDVRP

Set D= initial deviation ratio, and deviation = D times the best objective function value
- For k = 1,…,K
  - Perform intra OPM, inter OPM, intra TPE, then TPE
  - If no candidate move is made, increase D by d, and update the deviation
  - Check the local stopping rule (intensification, diversification, or descent)
- End K loop

Table 6. New Tabu Search Heuristic for Solving the SDVRP

Step 1. Construction phase
- Perform the construction steps to generate an initial solution

Step 2. Improvement phase
- Step 2.1 Perform the intensification stage and check LSR
- Step 2.2 Perform the descent stage and check DSR
- Step 2.3 Apply and check GSR
- Step 2.3 Replace solution by the best solution, perform the diversification stage, and check DSR.

The global stopping rule (GSR) is used to terminate the entire search procedure. We perform the three stages at least 30 times and if no better solution appears in 10 consecutive iterations, then we terminate the entire search procedure. At first, intensification is applied to improve initial solution produced in the construction phase. The descent stage is applied to locally improve the best solution obtained in the intensification stage. Finally, the diversification stage is applied to generate a new search region to generate a new initial solution for the next iteration. The iterations are performed consecutively in attempt to fine a global optimal solution to the problem. After trying at least 30 times and no better solution generated in 10 consecutive iterations, we terminate the procedure. Our new heuristic to solve the SDVRP is shown in Table 6.

3. COMPUTATIONAL TESTING

In this section, we apply the new tabu search method for the SDVRP (TSSDVRP) to the set of 23 test problems taken from literature. Problems 1 to 6 are due to Nag et al. [13], problems 7 to 18 are from Chao et al. [2], and problems 19 to 23 are converted from five well-known basic VRPs in Chao et al. [2]. The dimensions and summaries of all test problems are available in Chao et al. [2], where the number of customers ranges from 25 to 325, the site-dependent constraints are involved in all problems, the vehicle types are between two and three, and allowable vehicle type combinations ranges from 4 to 6 (see Tables 2, 6, and 7 in [2]).

The TSSDVRP was coded in Fortran, compiled with Digital Visual Fortran 5.0, and run on a 1.7 Ghz Pentium IV-based PC. The XMP code of Marsten [11] was used to solve the relaxation of the generalized assignment in the initialization phase.

We compare the results by the TSSDVRP with those produced by Cordeau and Laporte (CL)[4] whose approach outperforms each of previous methods due Nag et al. [13] and Chao et al. [2]. We used one set of parameter values in TSSDVRP to solve 23 benchmark problems. We compared the solutions obtained by TSSDVRP to the best solutions founded by CL [4] in Table 7.

We compare the performance of our new heuristic with one set of parameter values (denoted by TSSDVRP (A) in Table 7) to the performance of the heuristic of Cordeau and Laporte [4] with one set of parameters values (denoted by CL-one (B) in Table 7) on the 23 problems. TSSDVRP produces the best objective function value on 13 problems, CL-one produces the best objective function value on six problems, and the two heuristic are tied on four problems. On average, the objective function values of the solutions produced by TSSDVRP are 0.42% lower than those generated by CL-one. TSSDVRP is very fast-one average it takes 3.43 minutes to solve a problem. Overall, TSSDVRP finds five new best-known solutions (problems 13, 14, 15, 21, 23).

Table 7. Comparison of Results on Test Problems

No.	TSSDVRP(A)	CL-one(B)	CL-best[C]	(A-B)/B	(A-C)/C	(B-C)/C	Time in minutes
1	642.66	643.80	642.66	*-0.18%*	0.00%	0.18%	0.22
2	598.10	598.10	598.10	0.00%	0.00%	0.00%	0.38
3	971.85	977.40	959.36	*-0.57%*	1.30%	1.88%	0.27
4	879.56	861.93	854.43	2.05%	2.94%	0.88%	0.30
5	1020.98	1025.84	1020.22	*-0.47%*	0.07%	0.55%	0.93
6	1037.13	1053.03	1036.02	*-1.51%*	0.11%	1.64%	0.50
7	391.30	391.30	391.30	0.00%	0.00%	0.00%	0.01
8	664.46	664.46	664.46	0.00%	0.00%	0.00%	0.08
9	948.23	948.23	948.23	0.00%	0.00%	0.00%	0.51
10	1228.40	1223.88	1223.88	0.37%	0.37%	0.00%	0.93
11	1501.59	1497.90	1464.98	0.25%	2.50%	2.25%	2.03
12	1718.77	1735.55	1695.67	*-0.97%*	1.36%	2.35%	2.41
13	1196.60	1197.21	1196.73	*-0.05%*	*-0.01%*	0.04%	0.15
14	1960.88	2017.96	1962.66	*-2.83%*	*-0.09%*	2.82%	1.69
15	2727.67	2751.45	2751.45	*-0.86%*	*-0.86%*	0.00%	1.97
16	3542.78	3545.04	3491.18	*-0.06%*	1.48%	1.54%	6.98
17	4255.95	4230.96	4230.96	0.59%	0.59%	0.00%	13.73
18	5042.91	5012.68	4929.71	0.60%	2.30%	1.68%	33.82
19	854.35	852.05	850.39	0.27%	0.47%	0.20%	1.38
20	1053.89	1055.35	1046.14	*-0.14%*	0.74%	0.88%	3.33
21	1317.04	1338.65	1337.83	*-1.61%*	*-1.55%*	0.06%	5.21
22	1029.25	1060.45	1012.87	*-2.94%*	1.62%	4.70%	1.36
23	807.64	819.92	818.75	*-1.50%*	*-1.36%*	0.14%	0.71
			Average	*-0.42%*	0.52%	0.95%	3.43

TSSDVRP(A)-one set of parameter values
CL-one(B)-one set parameters for the procedure of Cordeau and Laporte [4]
CL-best(C)-best solutions found by Cordeau and Laporte [4] over all parameters

4. CONCLUSIONS

In this paper, we developed a new tabu search algorithm for solving the SDVRP and tested it on 23 test problems taken from literature. Besides the frequency-based tabu restriction, we also applied the objective-based tabu restriction to solve the SDVRP, and varied the amount of the objective-based tabu restriction to carry out the tabu intensification and diversification search. The new tabu search method is simple, easy to code, and works effectively and efficiency. Compared with the deterministic annealing method, our new method involves two types of tabu restrictions that avoid cycling and help to escape setting trapped in a poor local optimal solution. The values of deviation are more easily set to those in the threshold acceptance method. Clearly, the new TS method solves the SDVRP quite effectively. In the further work, we hope to apply our objective-based tabu search method to other optimization problems.

Acknowledgments

This research is supported by the National Science Council of the Republic of China in Taiwan under grants number NSC88-2416-h-145-001. This support is gratefully acknowledged.

References

1. I. M. Chao, B. L. Golden, and E. A. Wasil, A new algorithm for the site-dependent vehicle routing problem, in: *Advances in Computational and Stochastic Optimization, Logic Programming, and Heuristics Search*, edited by D. L. Woodruff (Kluwer Academic Publishers, The Netherlands, 1998), pp. 301-312.
2. I. M. Chao, B. L. Golden, and E. A.Wasil, A computational study of a new heuristic for the site-dependent vehicle routing problem, *INFOR: Information Systems and Operational Research* 37(3), 319-336 (1999).
3. G. Clarke, and J. W. Wright, Scheduling of vehicles from a central depot to a number of delivery points, *Operations Research* 12, 568-581(1964).
4. J. F. Cordeau, and G. Laporte, A Tabu Search Algorithm for The Site Dependent Vehicle Routing Problem with Time Windows, *INFOR: Information Systems and Operational Research* 39(3), 292-298 (2001).
5. G. Dueck, New optimization heuristics: the great deluge algorithm and the record-to-record travel, *Journal of Computational Physics* 104, 86-92 (1993).
6. G. Dueck, and T. Scheurer, Threshold accepting: a general purpose optimization algorithm, *Journal of Computational Physics* 90, 161-175 (1990).
7. M. R. Garey, and D. S.Johnson, *Computers and Intractability: A Guide to The Theory of NP-Completeness* (Freeman, San Francisco, 1979).
8. F. Glover, and M. Laguna, Tabu search, in: *Modern Heuristic Techniques for Combinatorial Problems,* edited by C. R. Reeves (John Wiley & Sons, New York, 1993) pp. 70-150.
9. B. L. Golden, E. A. Wasil, J. Kelly, and I. M. Chao, The Impact of Metaheuristics on Solving the Vehicle Routing Problem: Algorithms, Problem Sets, and Computational Results, in: *Fleet Management and Logistics*, edited by T. Crainic and G. Laporte (Kluwer Academic Publishers, Dordrecht, The Netherlands 1998), pp.33-56.
10. S. Lin, Computer Solutions of the Traveling Salesman Problem, *Bell System Technical Journal* 44, 2245-2269 (1965).
11. R. E. Marsten, The Design of the XMP Linear Programming Library, *ACM Transactions on Mathematical Software* 7(4), 481-497 (1981).
12. B. Nag, *Vehicle Routing in the Presence of Site/Vehicle Dependency Constraints* (Ph.D. Dissertation, College of Business and Management, University of Maryland at College Park, 1986).
13. B. Nag, B. L. Golden, and A. Assad, Vehicle Routing with Site Dependencies, in: *Vehicle Routing; Methods and Studies*, edited by B. L. Golden and A. Assad (North Holland, Amsterdam, 1988), pp.149-159.
14. Y. Rochat, and F. Semet, A Tabu Search Approach for Delivering Pet Food and Flour in Switzerland, *Journal of Operational Research Society* 45(11), 1233-1246 (1994).
15. F. Semet, and E. Tailard, Solving Real-life Vehicle Routing Problems Efficiently Using Tabu Search, *Annals of Operations Research* 41, 469-488 (1993).

A HEURISTIC METHOD TO SOLVE THE SIZE ASSORTMENT PROBLEM

Kenneth W. Flowers[1], Beth A. Novick[2], Douglas R. Shier[2]

[1]*Department of Mathematics and Computer Science, Georgia College and State University, Milledgeville, GA 31061*

[2]*Department of Mathematical Sciences, Clemson University, Clemson, SC 29634-0975*

Abstract This paper considers the *size assortment problem*, in which a large number of size distributions (e.g., for retail stores) need to be aggregated into a relatively small number of groups in an optimal fashion. All stores within a group are then allocated merchandise according to their common size distribution. A neighborhood search heuristic is developed to produce near-optimal solutions. We investigate the use of both random and "intelligent" starting solutions to initiate the heuristic. The intelligent starting solutions are based on efficiently solving one-dimensional versions of the original problem and then combining these into consensus solutions. Computational results are reported for some small specially structured test problems, as well as some large test problems obtained from an industrial client.

Keywords: Clustering, heuristic, matching, minimum clique, neighborhood search

1. Introduction

Merchandise optimization systems are increasingly being incorporated into large retail organizations to guide decisions using a variety of optimization models [15]. The current work is motivated by a particular merchandise optimization problem brought to our attention by an industrial client. Specifically, a nationwide company produces a particular item, sold in a variety of sizes by its many retail stores. Each store has a known demand for sales of the item by size. The parent company cannot supply the item to meet exactly the demand distribution at every individual store. Rather the n stores are to be grouped into a fixed number m ($m \ll n$) of *bins* with the stores comprising each bin being treated similarly. Namely, each store in a given bin is allocated merchandise according to the same size distribution. This grouping into a fixed number of bins should be done in an optimal or near-optimal fashion relative to an appropriate objective measure. (The allocation into a fixed number of bins reflects the constraints imposed by the industrial client.) As will be seen, this *size assortment problem* is a type of clustering problem that combines both the l_1 and l_2 distance metrics in a novel way.

To begin, we formulate the problem as follows. Suppose that there are n given stores, selling an item available in s sizes. The demand at store $i \in \{1, \ldots, n\}$ for items of size $k \in \{1, \ldots, s\}$ is denoted v_{ik}, so the total (sales) volume generated by store i is then $V_i = \sum_k v_{ik}$. The distribution of the sizes (or *profile*) for store i is given by the normalized values $p_i(k) = \frac{v_{ik}}{V_i}$, for $k = 1, \ldots, s$.

We wish to aggregate the stores into exactly m bins $\mathcal{B} = \{B_1, \ldots, B_m\}$, where the *center* c_j of bin B_j is given by the average of all store profiles in that bin, weighted by the store volumes V_i. That is, for $k = 1, \ldots, s$

$$c_j(k) = \frac{\sum_{i \in B_j} V_i p_i(k)}{\sum_{i \in B_j} V_i}. \tag{1}$$

Notice that the center c_j is itself a valid probability distribution (profile). To measure the degree of similarity between the profile for store i and the center c_j of bin B_j, define the *distance* d_{ij} as the weighted sum of absolute differences between these two distributions:

$$d_{ij} = V_i \, ||p_i - c_j||_1 = V_i \sum_k |p_i(k) - c_j(k)|. \tag{2}$$

Thus d_{ij} measures the number of items of store i that are misclassified by using the (common) distribution c_j instead of p_i. Then the total misallocation cost associated with the set $\mathcal{B} = \{B_1, \ldots, B_m\}$ of bins is given by

$$f(\mathcal{B}) = \sum_j \sum_{i \in B_j} d_{ij}.$$

The *size assortment problem* requires finding a collection of m bins $\mathcal{B}$ that minimizes the total misallocation cost $f(\mathcal{B})$.

We remark that requiring the center of bin B_j to itself be a distribution leads to the definition (1) of c_j as a centroid, which in turn minimizes the (weighted) sum of squared deviations between c_j and each p_i in bin B_j. On the other hand, using the sum of absolute values in (2) is appropriate since we want to measure the total number of misallocated items. Thus the current problem represents an interesting (and natural) blend of both l_2 and l_1 distances. This problem is different from other multidimensional clustering [8, 12, 14, 17] and location problems [1, 9] previously studied in the literature.

To illustrate these concepts, consider the example having $n = 8$ stores and $s = 3$ sizes, whose demands v_{ik} are displayed in Table 1a together with the store volumes V_i. The normalized store profiles are given in Table 1b. Consider the candidate set of bins $B_1 = \{1,3\}$, $B_2 = \{2,5,7\}$, $B_3 = \{4,6,8\}$. From (1) the center c_2 of bin B_2 is the weighted average of the profiles for stores 2, 5, 7: namely $c_2 = (0.2891, 0.5181, 0.1928)$. Within this bin, the distances from each store to the center of the bin are given by (2): $d_{22} = 4.095$, $d_{52} = 2.430$, $d_{72} = 3.808$, so the misallocation cost within bin B_2 is 10.333. The corresponding misallocation cost for bin B_1 is 2.286 and for bin B_3 is 2.265, giving a total misallocation cost $f(\mathcal{B}) = 14.884$, which turns out to be the smallest cost possible. Hence this given set of candidate bins is in fact optimal.

Table 1a. Demands v_{ik} for an 8 store example

store i	*size 1*	*size 2*	*size 3*	V_i
1	3	8	10	21
2	6	15	4	25
3	5	11	12	28
4	7	5	4	16
5	8	15	7	30
6	9	8	6	23
7	10	13	5	28
8	12	10	7	29

Table 1b. Profiles $p_i(k)$ for an 8 store example

store i	*size 1*	*size 2*	*size 3*
1	0.1429	0.3809	0.4762
2	0.2400	0.6000	0.1600
3	0.1786	0.3928	0.4286
4	0.4375	0.3125	0.2500
5	0.2667	0.5000	0.2333
6	0.3913	0.3478	0.2609
7	0.3571	0.4643	0.1786
8	0.4138	0.3448	0.2414

2. The Size Assortment Heuristic

Although finding an optimal set of m bins for n stores can be accomplished by evaluating the objective function $f(\mathcal{B})$ for all possible partitions $\mathcal{B}$ of the n stores into m groups, the number of such partitions is the Stirling number $S(n,m)$ of the second kind [6], which grows exponentially (asymptotically like $m^n/m!$). Indeed, even for $n = 16$ stores and $m = 5$ bins, the number of partitions is 1,096,190,550. Consequently, checking all possible partitions is not feasible. In general, this optimization problem (like most clustering problems in higher dimensions) is NP-hard [2] and so a reasonable approach is to seek heuristic solutions that can be computed in a modest amount of time. The remainder of this section describes the components of a multistart, interchange heuristic combined with a variable neighborhood search.

2.1 Bin-Center Algorithm

First we describe an important part of a heuristic tailored to this particular problem. Let us suppose that we are given an initial set of m bins $\mathcal{B} = \{B_1, \ldots, B_m\}$. (Various techniques for generating an initial set of bins will be discussed in Section 3.) From this set of m bins, a corresponding set of m centers $\{c_1, \ldots, c_m\}$ can be computed using (1). Next we can assign each store i to its nearest center c_j with respect to the distance defined in (2). This assignment phase in general results in a new set of bins $\mathcal{B}$ and the entire process can thus be repeated. Such an alternating sequence of computing centers from bins and assigning stores to bins (based on the distance to the nearest center) is repeated until there is no change in the bins from the previous iteration. This *Bin-Center (BC) algorithm* follows an alternating strategy familiar in location-allocation approaches in the location literature [7, 11] as well as in the HMEANS clustering algorithm [8, 14].

The feasible solution (set of bins) resulting from applying the BC algorithm is not guaranteed to be a globally optimal solution, or even a locally optimal solution. Accordingly, a more involved heuristic that employs the BC algorithm as an integral part will be described next.

2.2 Local Improvement Phase

Once the BC algorithm has terminated, finding no further changes to the current set of bins, we can invoke a *local improvement phase* [3, 7, 8, 11]. This phase will produce a locally optimal solution, one that cannot be further improved by a simple interchange.

To describe this local improvement phase, let the current solution be defined by an *allocation vector* $X = (x_1, x_2, \ldots, x_n)$ relative to the n stores. That is, $x_i = j$ means that store i is currently allocated to bin B_j, where $1 \leq x_i \leq m$. The misallocation cost associated with the allocation vector X is denoted $f(X)$. Also, define a ***neighbor*** of the allocation vector X to be any allocation vector Y that differs from X in exactly one component. The *neighborhood* $N(X)$ of X, which consists of all neighbors of X, contains $n(m-1)$ vectors. With this notation, the local improvement phase is defined by the pseudocode in Figure 1. In our implementation, the neighborhood of X is explored in a natural sequential order, changing the allocation x_i of store i $(i = 1, \ldots, n)$ in turn to each $j = 1, \ldots, m$ $(j \neq x_i)$. For each neighbor Y of X, we first invoke the BC algorithm. If the resulting vector has a smaller objective function value than X, then it replaces X and the process continues anew. The local improvement phase continues until there is no neighbor of the current solution that can be used to improve the misallocation cost. To verify that we have a locally optimal solution X, we must consider all $n(m-1)$ neighbors of X, which can be time consuming if n is large.

2.3 Multistart Algorithm

Starting from a given initial solution X^0, we can use the BC algorithm followed by the local improvement phase to obtain a local solution $Y = g(X^0)$,

```
local_improvement(X)
  repeat
     select Y ∈ N(X)
     Ȳ ← BC_algorithm(Y)
     if f(Ȳ) < f(X) then X ← Ȳ
  until no change in N(X)
```

Figure 1. Pseudocode for the local improvement phase

```
multistart
  for i = 1, ..., imax
     generate X^0
     X ← BC_algorithm(X^0)
     Y ← local_improvement(X)
     if f(Y) < f(Y*) then Y* ← Y
  endfor
  output the solution Y*
```

Figure 2. Pseudocode for the multistart heuristic

a function of the initial solution X^0. Moreover, as we vary the initial solution X^0, the final solution $Y = g(X^0)$ can vary also. In order to explore various regions of the (vast) solution space of possible allocation vectors, and to avoid getting trapped in inferior local solutions, we systematically vary the starting configuration X^0 and then keep track of the best solution Y^* (in terms of objective function value) obtained over $imax$ iterations. Pseudocode for this *multistart* heuristic, similar to that used in [3, 9, 11], is given in Figure 2, where it is assumed that $f(Y^*)$ has been initialized appropriately.

2.4 Postprocessing Phase

Upon termination of the multistart algorithm we obtain a locally optimal solution Y^*, which represents the best solution found in $imax$ runs. Thus, no improvement in the objective function value is possible by changing a single component of the allocation vector Y^*. That is, the neighborhood of Y^* contains no improving allocation vectors. It may however be useful to explore a "higher order" neighborhood of the solution Y^*. Consequently, let the k-th order neighborhood $N^k(X)$ of an allocation vector X consist of all allocation vectors that differ from X in exactly k components. Since the size of $N^k(X)$ is quite large it is not feasible to explore the entire k-th order neighborhood of the current solution X. As in [3, 7, 11] we randomly sample a fixed number of times ($rmax$) from each higher order neighborhood. Each such sampled Y is then explored for improvement using the BC algorithm. Once a given neighborhood is explored without an improvement, the next higher order neighborhood is explored up to $kmax$ neighborhoods. If at any point an improvement is detected, we restart the process by resetting the value k to 1. This postprocessing phase, which carries out a *variable neighborhood search* [10], is described by the pseudocode in Figure 3.

```
postprocessing(X)
for k = 2,...,kmax
    for r = 1,...,rmax
        randomly select Y ∈ N^k(X)
        Ȳ ← BC_algorithm(Y)
        if f(Ȳ) < f(X) then
            X ← Ȳ, k ← 1, r ← 1
    endfor
endfor
output the solution X
```

Figure 3. Pseudocode for the postprocessing phase

3. Initial Solutions

Application of the heuristic algorithm presupposes that we have in hand an initial allocation of stores to bins. Several different techniques for generating an initial solution were investigated. The first method for initially allocating n stores to m bins is the *random binning* method. That is, taking each store i in turn, we randomly and independently select one of the m bins and allocate store i to that bin. This method has the property that on average all bins will have the same size (contain approximately n/m stores).

A second method of constructing initial solutions is based on solving s one-dimensional versions of the original store allocation problem. Specifically, let us fix the size $k \in \{1, \ldots, s\}$ and consider the one-dimensional problem P_k defined by the quantities $\{p_i(k) : i = 1, \ldots, n\}$ and the weights $\{V_i : i = 1, \ldots, n\}$. That is, we wish to determine m bins B_j that minimize

$$f_k(\mathcal{B}) = \sum_j \sum_{i \in B_j} V_i \, |p_i(k) - c_j(k)|, \tag{3}$$

where $c_j(k)$ is the one-dimensional centroid (1) of the demands in bin B_j for size k. If we assume that the n stores are re-ordered so that $p_1(k) \leq p_2(k) \leq \cdots \leq p_n(k)$, then there is an optimal solution for (3) in which each bin consists of consecutively numbered stores [12, 17]. This "string property" can then be exploited to develop a dynamic programming solution algorithm. Namely, let

$F_b(r)$ be the minimum value of (3) achievable for stores $1, \ldots, r$ and using exactly b bins. By conditioning on the smallest store i in the last bin B_b we obtain the recursion

$$F_b(r) = \min_{b \leq i \leq r} \{F_{b-1}(i-1) + c_{ir}\}, \tag{4}$$

where c_{ij} represents the cost of grouping together stores $i, i+1, \ldots, j$ into a single bin. Using the initial conditions $F_1(r) = c_{1r}$, one can apply (4) to successively calculate $\{F_b(r) : r = b, \ldots, n\}$ for $b = 2, \ldots, m$ and thus the required $F_m(n)$.

By use of the above recursion, we can obtain optimal one-dimensional partitionings of $\{1, \ldots, n\}$ for all s different sizes in time $O(n^2 ms)$ plus the time to calculate all c_{ij}, which can be done in $O(n^2 s)$ time. Each of these s partitionings can be used as an initial binning for the multistart heuristic described in Section 2. We term this method *intelligent binning*, in contrast to the random binning method described earlier.

For the 8 store example in Table 1a, sorting the stores in nondecreasing order by the values $\{p_i(1)\}$ places stores in the order $1, 3, 2, 5, 7, 6, 8, 4$. Applying the recursion (4) then produces the optimal bins $\{1, 3\}, \{2, 5\}, \{7, 6, 8, 4\}$ for size 1. Similar calculations yield the optimal bins $\{4, 8, 6, 1, 3\}, \{7, 5\}, \{2\}$ for size 2 and the optimal bins $\{2, 7\}, \{5, 8, 4, 6\}, \{1, 3\}$ for size 3. When these three intelligent starting solutions are used to initiate the multistart heuristic we obtain total misallocation costs of 17.949, 14.884, and 14.884 respectively (without postprocessing). Thus in two of the three cases, we obtain the optimal solution to the original 3-dimensional size assortment problem. (When postprocessing is applied to the solution with objective function value 17.949, then the optimal solution is also obtained.) Computational results on larger test examples will be reported in Section 4.

Since each one-dimensional solution focuses only on a single size k, it seems natural to investigate combining the s individual solutions into one (or several) "consensus" solutions for initiating the multistart heuristic. Let π_k denote the optimal binning obtained by solving problem P_k. Hence π_k is an m-partition of $\{1, \ldots, n\}$ into m nonempty subsets or *blocks*. We therefore want to obtain a consensus m-partition π of $\{1, \ldots, n\}$ from $\{\pi_k : k = 1, \ldots, s\}$. In general, consensus problems of this type are known to be NP-hard [4] so we will investigate some simple, computationally efficient techniques to deliver partitions that provide near-consensus solutions.

To this end, rather than seeking the partitions directly, we first search for a set of m "leaders" or "representatives" that in some sense will represent the m blocks of π. More precisely, suppose that we identify a set of m representatives $\{r_1, \ldots, r_m\} \subseteq \{1, \ldots, n\}$. By using the l_1 distance measure, as in (2), we can allocate each store i to a closest representative r_j, creating a collection of m bins and thus an m-partition π. Ideally, we would like the representatives

r_j to be consistent with each of the given s partitions π_k. Namely, each of $\{r_1, \ldots, r_m\}$ occurs in a separate block of π_k, for $k = 1, \ldots, s$. We call such a set $\{r_1, \ldots, r_m\}$ a *simultaneous system of distinct representatives* (s-SDR) for the π_k.

When $s = 2$ it is not difficult to obtain a 2-SDR, or to determine that none exists, by means of bipartite matching. Namely, suppose that π_1 has blocks $A_1, \ldots, A_m$ and that π_2 has blocks $B_1, \ldots, B_m$. Define the bipartite graph G having vertex set $\{A_1, \ldots, A_m\} \cup \{B_1, \ldots, B_m\}$ and edges (A_u, B_v) whenever $A_u \cap B_v \neq \emptyset$. Then a perfect matching M of G provides the required 2-SDR. That is, for each edge $(A_u, B_v) \in M$ select $r_j \in A_u \cap B_v$. On the other hand, if no perfect matching exists, then there can be no 2-SDR. To remedy this situation, we can more generally seek an approximate 2-SDR. Namely, let G' now be the complete bipartite graph with partite sets $\{A_1, \ldots, A_m\}$ and $\{B_1, \ldots, B_m\}$. Define the weight of edge (A_u, B_v) to be the minimum distance $||p_i - p_j||_1$ between any two profiles p_i, p_j where $i \in A_u$, $j \in B_v$. We then seek a minimum weight perfect matching M' in G'. For each matching edge $(A_u, B_v) \in M'$, we can select a representative $r_j \in A_u$ that achieves the minimum distance between A_u and B_v.

Unfortunately, for $s \geq 3$ the s-SDR problem becomes NP-hard, based on a reduction from the 3-dimensional matching problem [5]. Consequently, one approach to finding a near-consensus partition for $\pi_1, \pi_2, \ldots, \pi_s$ is to apply the minimum weight perfect matching approach to successive pairs of partitions. Specifically, let π_{12} be a partition derived from π_1, π_2 by first obtaining a set of m representatives (via a minimum weight perfect matching) and then assigning each store i to a nearest such representative. Next, apply the same procedure to π_{12} and π_3 to obtain π_{123}, and so forth, eventually obtaining a near-consensus partition $\pi_{12\ldots n}$.

A somewhat more direct, and satisfying, approach directly models the interactions between all the partitions $\pi_1, \pi_2, \ldots, \pi_s$. We will employ the *conflict graph* CG, having vertices that correspond to stores and edges that correspond to "conflicts." To this end, let the *label* on vertex $v_i \in V$ be $i_1 i_2 \ldots i_s$ where $i_k \in \{1, \ldots, m\}$ is the number of the (unique) block that contains store i in partition π_k. An edge $(i, j) \in E$ exists for all $i < j$ and has the associated weight w_{ij} indicating the number of components k for which $i_k = j_k$. We seek then a minimum weight clique consisting of exactly m vertices of CG. Since this problem is NP-hard, we describe a simple heuristic to obtain an m-clique of low total weight, which in turn provides a set of m representatives. If the total clique weight is 0, then we have in fact obtained an s-SDR.

The heuristic algorithm given Figure 4 when applied to the weighted graph $CG = (V, E, w)$ first sorts the vertices v_i by their *weighted degree* $w_i = \sum_j w_{ij}$. Starting with the first vertex, it successively appends a new vertex v_j to the current set S of selected vertices. Vertex v_j is chosen to have minimum

```
min_clique(V, E, w)
reorder the vertices v_i so that w_1 ≤ w_2 ≤ ··· ≤ w_n
S ← {v_1}
while |S| < m
    select v_j ∉ S to minimize Σ_{v_i∈S} w_{v_i v_j}
    S ← S ∪ {v_j}
endwhile
```

Figure 4. Pseudocode for the minimum weight clique heuristic

contribution to the total weight of the clique induced by $S \cup \{v_j\}$; in the case of ties, the lowest indexed such vertex is appended. This process continues until $|S| = m$. In practice the algorithm *min_clique* can be implemented by a suitable modification of Prim's $O(n^2)$ minimum spanning tree algorithm [16], which is particularly appropriate since CG is dense. Variations on this basic algorithm are possible. For example we could randomly sort all vertices having the same weighted degree rather than break ties based on indices. Or we could simply sort all vertices randomly, thus ignoring their weighted degree.

4. Computational Results

We report on computational experiments in which we used synthetic test problems of varying sizes as well as some real-world problems. Our objective is to see how well the heuristic approach performs, possibly augmented by the postprocessing phase. We also wish to compare random binning (which produces a diverse set of initial solutions) with intelligent binning choices (based on solving one-dimensional problems). By injecting randomness into the sorting phase of *min_clique*, we can combine both intelligence with diversification to produce a variety of consensus solutions.

Preliminary computational studies revealed that, especially for larger problems, the local improvement phase, which thoroughly searches the neighborhood of each successive solution X, can be time consuming. In particular, the local improvement phase takes $n(m-1)$ steps simply to verify local optimality at termination. In our empirical testing we observed that if an improvement is found, it is generally found quite early in the exploration of the neighborhood $N(X)$. Therefore, we propose exploring only a certain fraction (say 10%) of the $n(m-1)$ neighbors at each step. Throughout the computational results reported here, we have adopted this strategy of purposely limiting the exploration of each $N(X)$.

The first set of test problems involved $n = 30, 90, 150$ stores having $s = 4$ sizes and $m = 6$ bins. To generate these test problems, we began with $m = 6$ beta distributions, each representing a distinctive shape, obtained by varying the beta parameters α, β. Each such beta(α, β) distribution was discretized to produce a "base" probability distribution $p_{\alpha\beta}$ over the s sizes. Then we perturbed each of the base probabilities $\{p_{\alpha\beta}(k) : k = 1, \ldots, s\}$ by a random multiplicative factor drawn uniformly from $[1-\epsilon, 1+\epsilon]$. After normalization this produced a set of "sibling" profiles $p_i(k)$ associated with the given base distribution $p_{\alpha\beta}$. The volume V_i was randomly generated to yield the demands $\{v_{ik}\}$ for each store i. In our computational study, we varied the parameter $\epsilon \in \{0.15, 0.20, 0.25\}$. Five replications (test data sets) were generated for each pair (n, ϵ). The test data sets are available at the web site [13], together with the best known solutions.

Results for this suite of test examples are summarized in Table 2, in which we compare the use of two types of initial starting solutions X_0 within the multistart heuristic: random binning and one-dimensional binning. For the former, we generated $imax = 30$ random starting solutions, while for the latter we simply used the $s = 4$ available one-dimensional solutions. Table 2 shows the overall percent of time the best known solution was achieved by each procedure (RANDOM or 1-D) after the local improvement phase was executed. In fact for every one of the 45 test problems, the multistart heuristic (*without* postprocessing) was able to determine the best known binning by using at least one of the $imax = 30$ random starts or one of the $s = 4$ one-dimensional starts. As observed in Table 2, the 1-D starts are much more effective, obtaining the best known solutions in 50–80% of the cases, whereas the random starts are succcessful in only 19–45% of the cases. It is interesting to note that for random starts this "yield" percent did increase for larger values of n. Even for the largest test problems ($n = 150$), the heuristic was fast, requiring only 2–4 seconds to execute all $imax = 30$ random starts on a desktop computer running at 533 MHz.

Table 2. We compare random and 1-D starts on generated problems, indicating in the third and fourth columns, respectively, the percentage of times the best known solution was obtained.

n	ϵ	*RANDOM*	*1-D*
30	0.15	18.7	70.0
30	0.20	20.0	55.0
30	0.25	20.7	50.0
90	0.15	21.3	70.0
90	0.20	22.0	60.0
90	0.25	24.0	70.0
150	0.15	45.3	80.0
150	0.20	32.7	60.0
150	0.25	41.3	55.0

We also applied the heuristic approach to a pair of larger, proprietary problems supplied by an industrial client. In both examples, postprocessing was used with the parameters $kmax = 10$ and $rmax = 100$. The first industrial problem had $n = 401$ stores, $s = 3$ sizes, and $m = 9$ bins. When we compared random and 1-D starts, a better solution was found using the multistart heuristic with $imax = 30$ random starts, followed by postprocessing. Postprocessing decreased the misallocation cost (objective function value) by only 0.05%. Using the multistart heuristic with the $s = 3$ one-dimensional starts and postprocessing produced a solution with cost 0.5% higher than the best random solution. On the other hand, the time to find this solution (9 seconds in total) was only one-fifth of that required to find the best random solution.

We also investigated the use of consensus solutions to initiate the multistart heuristic. In particular we generated 10 consensus solutions (from the three 1-D initial solutions) by applying the *min_clique* algorithm. The best solution was obtained by randomly ordering all vertices v_i rather than ordering them by their weighted degree w_i. This produced a solution with clique weight 1 and cost 0.09% lower than the best random solution obtained in $imax = 30$ runs.

The second industrial problem had $n = 820$ stores, $s = 6$ sizes, and $m = 9$ bins. The best solution was found using the multistart heuristic with $imax =$ 30 random starts, followed by postprocessing. Postprocessing decreased the objective function value by only 0.04%. Using the multistart heuristic with the $s = 6$ one-dimensional starts and postprocessing produced a solution with cost 0.5% higher than the best random solution in 192 seconds, one-fourth of the time needed to obtain the best random solution.

Ten consensus solutions were generated from the six 1-D initial solutions using the *min_clique* algorithm, ordering vertices by their weighted degree. In every case we got a clique with total weight 1, and found a solution (after postprocessing) with cost 0.03% higher than the best random solution. When vertices were randomly ordered within the *min_clique* algorithm, we obtained a solution with clique weight 3 and cost 0.3% higher than the best random solution.

In conclusion, the preliminary computational tests conducted here showed that the overall multistart heuristic was effective in solving all synthetically generated test problems. Also for two larger, real-world problems the use of random (diverse) starts consistently gave high quality solutions, when augmented by postprocessing. (In fact our solutions were 7–10% better in objective value compared with the solutions being used by our industrial client.) The intelligent 1-D starts were almost as good as the random starts and were computationally less demanding. Since the number of such 1-D solutions is limited by the problem parameter s, we also investigated the generation of consensus solutions from these 1-D solutions to initiate the multistart heuristic. Using the best solution from 10 such consensus starts provided a slight improvement over using 1-D starts alone, though it did not show a clear advantage over using 30 random starts. Our tentative conclusion is that generating diverse solutions (random starts) seems to work quite well for the size assortment problem. Additional computational testing on large, structured test problems (combined with exploring the effects of varying the algorithm parameters) is certainly warranted to verify this conclusion. We believe that the use of 1-D solutions and the derived consensus solutions seems promising and may prove to be useful in other contexts. This too is an area for further research.

References

[1] I. Bongartz, P. H. Calamai, and A. R. Conn, A projection method for l_p norm location-allocation problems, *Mathematical Programming* 66 (1994), 283–312.

[2] P. Brucker, On the complexity of clustering problems, in *Optimization and Operations Research*, M. Beckmann and H. P. Kunzi (eds.), Lecture Notes in Economics and Mathematical Systems 157, Springer, Berlin, 1978, pp. 45–54.

[3] S. A. Canuto, M. G. C. Resende, and C. C. Ribeiro, Local search with perturbations for the prize-collecting Steiner tree problem in graphs, *Networks* 38 (2001), 50–58.

[4] W. H. E. Day and F. R. McMorris, *Axiomatic Consensus Theory in Group Choice and Biomathematics*, SIAM, Philadelphia, 2003.

[5] M. R. Garey and D. S. Johnson, *Computers and Intractability: A Guide to the Theory of NP-Completeness*, W. H. Freeman, San Francisco, 1979.

[6] R. P. Grimaldi, *Discrete and Combinatorial Mathematics*, 5th edition, Pearson Addison Wesley, Boston, 2004.

[7] P. Hansen and N. Mladenović, Variable neighborhood search for the p-median, *Location Science* 5 (1997), 207–226.

[8] P. Hansen and N. Mladenović, J-MEANS: a new local search heuristic for minimum sum of squares clustering, *Pattern Recognition* 34 (2001), 405–413.

[9] P. Hansen, N. Mladenović, and E. Taillard, Heuristic solution of the multisource Weber problem as a p-median problem, *Operations Research Letters* 22 (1998), 55–62.

[10] N. Mladenović and P. Hansen, Variable neighborhood search, *Computers & Operations Research* 24 (1997), 1097–1100.

[11] N. Mladenović, M. Labbé, and P. Hansen, Solving the p-center problem with tabu search and variable neighborhood search, *Networks* 42 (2003), 48–64.

[12] M. R. Rao, Cluster analysis and mathematical programming, *Journal of the American Statistical Association* 66 (1971), 622–626.

[13] D. Shier, http://www.math.clemson.edu/faculty/Shier/Shier, Test problems for the size assortment problem, July 2004.

[14] H. Späth, *Cluster Analysis Algorithms for Data Reduction and Classification of Objects*, Ellis Horwood, Chichester, 1980.

[15] Sun Microsystems, http://www.sun.com/br/retail_415/feature_merch.html, Merchandise optimization drives the sales, 2003.

[16] R. E. Tarjan, *Data Structures and Network Algorithms*, SIAM, Philadelphia, 1983.

[17] H. D. Vinod, Integer programming and the theory of grouping, *Journal of the American Statistical Association* 64 (1969), 506–519.

HEURISTIC METHODS FOR SOLVING EUCLIDEAN NON-UNIFORM STEINER TREE PROBLEMS

Ian Frommer,[1] Bruce Golden,[2] and Guruprasad Pundoor[2]

[1] *Department of Mathematics, University of Maryland*
College Park, MD 20742

[2] *R.H. Smith School of Business, University of Maryland*
College Park, MD 20742

Abstract We consider a variation of the Euclidean Steiner Tree Problem in which the space underlying the set of nodes has a specified non-uniform cost structure. This problem is significant in many practical situations, such as laying cable networks, where the cost for laying a cable can be highly dependent on the location and the nature of the area through which it is to be laid. We empirically test the performance of a genetic-algorithm-based procedure on a variety of test cases of this problem. We also consider the impact on solutions of charging an additional fee per Steiner node. This can be important when Steiner nodes represent junctions that require the installation of additional hardware. In addition, we present a novel way of visualizing the performance and robustness of the genetic algorithm.

1. Introduction

The minimal spanning tree (MST) problem deals with connecting a given set of nodes in a graph in a minimal cost way. In a Steiner tree, we are allowed to use additional nodes, known as Steiner nodes, if doing so reduces the total cost. Steiner tree problems find applications in various areas such as the design of communication networks, printed circuits, and the routing of transmission lines [1, 2, 4]. There are many versions of the Steiner tree problem. In the Steiner problem in graphs, a graph $G = (V, E)$ is given, where V is a set of nodes and E is a set of edges connecting nodes in V. The goal is to obtain a least-cost-tree configuration which contains all of the nodes in a specified subset of V. Nodes in V that are not in the given subset may also be used. In the Euclidean Steiner problem, a given set of nodes is to be connected in Euclidean space. Another version involves rectilinear configurations, in which the arcs have to be either vertical or horizontal. A modified version of this involves nodes located on a hexagonal grid [2, 4].

In general, Steiner tree problems have been shown to be NP-hard. So, exact techniques proposed to find optimal solutions have exponential computational complexity in the worst case [5, 6], although special cases are solvable in polynomial time. For the NP-hard cases, heuristic approaches have been proposed to find near-optimal solutions. Gröpl et al. [7] reviews several mostly greedy algorithms that have been applied to the Steiner problem in graphs. Numerous randomized algorithms exist as well. Barreiros [8] uses a genetic algorithm (GA) to solve the Euclidean Steiner problem. Ribeiro and de Souza [9] use tabu-search-based techniques for solving Steiner tree problems on graphs, where a candidate set of Steiner nodes is prespecified. For the same problem with directed arcs, simulated-annealing-based procedures have been proposed in [10]. Julstrom [11, 12] applies various GA's, including one using an edge-set encoding, to the rectilinear Steiner tree problem. Coulston [4] applies a genetic algorithm using a full-components-based encoding to the non-uniform problem. Coulston's work is closest to the work being presented here, but he considers any path between two nodes to be an edge, while we restrict edges to be straight-line connections between nodes. This restriction is relevant to some applications, such as circuit design, in which components are connected by straight-line segments of rigid wire. Another difference is our consideration of the impact of charging additional penalties for Steiner nodes.

We consider the problem of finding near-optimal, non-directed Steiner trees on a non-uniform grid. Each location in the grid has an associated cost and a given set of nodes has to be connected in the form of a spanning tree. This kind of problem is of relevance to many practical situations, such as laying cable networks. There may be regions where laying cable is prohibitively expensive, such as a prime location in a metropolitan area, or other regions where it is cheaper. So the objective is to find a set of additional nodes in the grid that can be used to connect all the given nodes at a minimum cost.

In order to represent the cost structure of the space underlying the nodes, it is necessary to use some type of grid. A hexagonal grid is a reasonable choice since hexagons are the highest degree regular polygon that can tile the plane, and such a tiling has the desirable property that distances between centers of adjacent cells are equal [4]. While the size of the cells may affect that quality of the solution, for the purposes of this paper, we focus on finding the lowest cost Steiner tree for a cost structure using a prespecified cell size. Note the use of a grid cost structure will also simplify the search for optimal solutions because it reduces an otherwise uncountable search space to one of a finite size. In effect, this approach reduces the non-uniform variant of the Euclidean Steiner problem we are studying, to a Steiner problem in a graph. The two-dimensional Euclidean space is divided into hexagonal cells as shown in Fig.1. Each cell has a cost associated with it and may contain at most one of the nodes in the graph. Two nodes can be connected directly only if a straight line of cells can be drawn between the cells containing the two nodes. For example, in Fig.1, cells A and B can be connected, but not cells A and C.

We are given a set of nodes, called terminal nodes, which have to be connected in the cheapest way possible. As in the usual Steiner problem, additional nodes not in the terminal node set may be used to connect the terminal nodes. These additional nodes are the Steiner nodes. When an edge is drawn connecting two cells such as A and B, the cost of the edge is calculated as the sum of the costs associated with all the intermediate cells. The cost of the tree is calculated as the sum of the costs corresponding to all the edges plus the costs corresponding to each node in the tree. We may also charge an additional fee for each Steiner node.

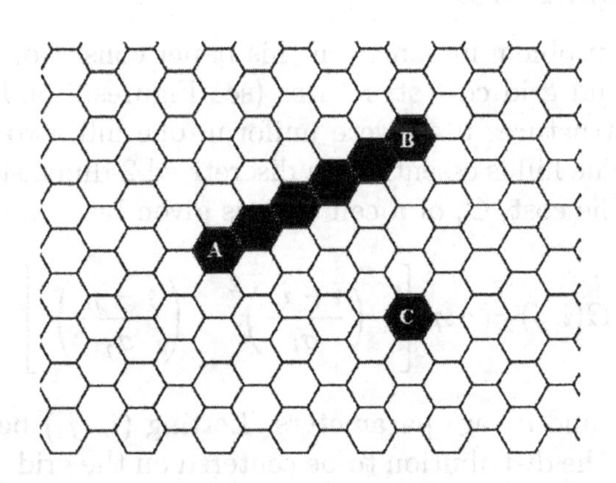

Figure 1. Hexagonal tiling of a 2-dimensional space.

2. Genetic Algorithm: Queen Bee Selection with Spatial-Horizontal Crossover

2.1 The Algorithm

The optimal solution to a Steiner tree problem is known as the Steiner minimal tree (SMT). We implement a straightforward genetic algorithm (GA) to find a Steiner tree (ST) whose cost is as close as possible to that of the SMT. This GA was the top performer out of several variants we tested, some of which were similar to ones used to solve the uniform problem. The algorithm is as follows:

1. **Input:** terminal node set, grid cost structure
2. Generate **Initial Population** randomly
   Steps 3–7 repeated for TMAX iterations
   3. Find **Fitness** (= ST cost) of each individual
   4. **Queen Bee Selection** to select parents
   5. **Spatial-Horizontal Crossover** on parents to produce offspring
   6. **Mutation 1** – add Steiner nodes at arc crossings
   7. **Mutation 2** – randomly move Steiner nodes
8. **Output**: final Steiner tree, time series of best Steiner tree costs, MST on the terminal node set (for rough comparison), and total run time.

2.2 Explanation

Input. The problem instances in this paper consist of various terminal node sets and grid cost structures (see Figures 2 and 3).[1] Some of the grid cost structures used were uniform, one hill, two hills, one pit, and two pits. The hill is essentially a discretized 2-dimensional Gaussian distribution. The cost, C, of a cell (i, j) is given by

$$C(i,j) = exp\left[-\left(\frac{i - i_o}{\sigma_i}\right)^2 - \left(\frac{j - j_o}{\sigma_j}\right)^2 \right]$$

where i_o, j_o, σ_i, and σ_j are parameters. Letting (i_o, j_o) be the center of the grid causes the distribution to be centered on the grid. The standard deviations, σ_i and σ_j control the shape of the hill. Variations on this basic formula are used to create grid cost structures with two hills, one pit (or negative hill), and two pits.

[1]To our knowledge, besides [4], there are no other test cases for this non-uniform problem. Again, since [4] uses a different edge definition, any comparison would be indirect.

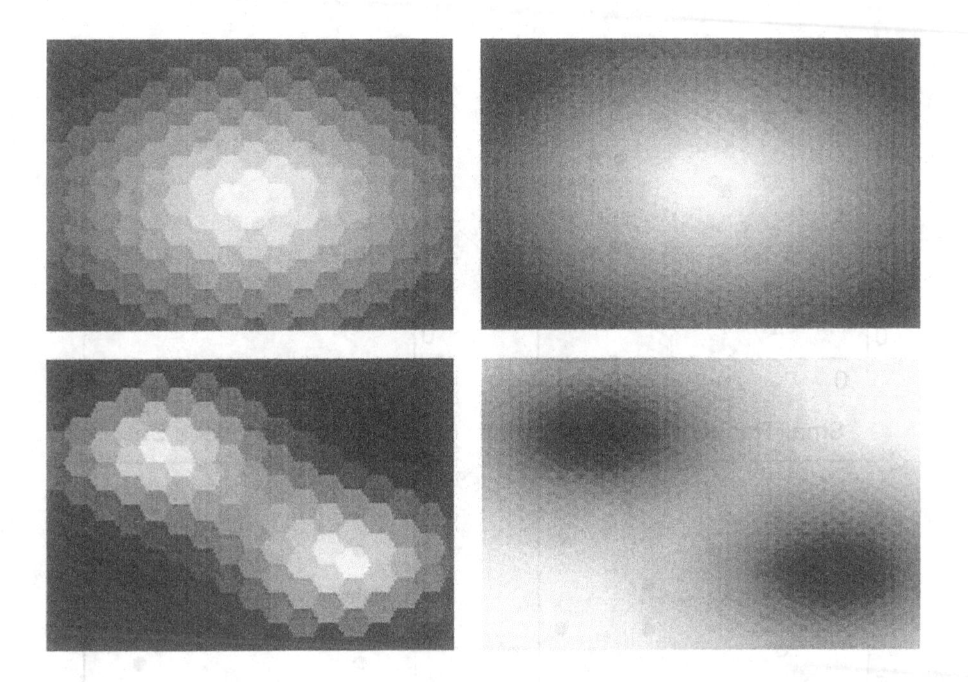

Figure 2. Sample hexagonal grid cost structures. The lighter the shading, the more costly the cell. Clockwise from upper left: small hill, large hill, large 2-pit, small 2-hill.

We generated random terminal node sets and also constructed terminal node sets of specific structures, such as a ring-like structure. Recall from Sect. 1 that due to the hexagonal grid structure, it is not always possible to connect any two nodes. Hence, given a set of terminal nodes it may not necessarily be possible to connect them in a tree. Since we wanted to use the minimal spanning tree as a rough comparison for our GA Steiner tree, we generated our terminal node sets so that all of the nodes in them could be connected by a tree. In Table 1, we define the set of problems that we consider in this paper.

Initial Population. We use a population size of 40. Each individual is represented by a fixed-length chromosome that can hold up to $1.1N$ Steiner node locations where N is the number of terminal nodes. The locations are randomly generated from a uniform distribution of grid locations. An example of a chromosome for $N = 10$ and a grid size of 20 by 20 is:

$$individual_k = \{(18,5),(11,17),(1,4),(13,2),(3,10)$$

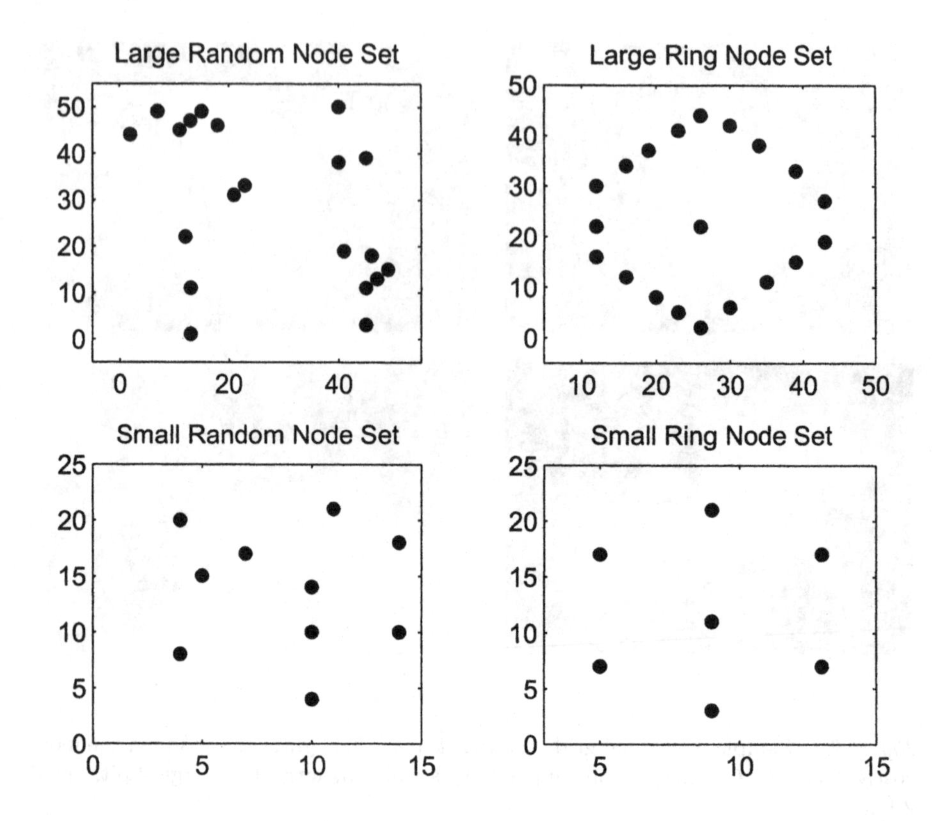

Figure 3. The node sets.

Table 1. Problem Instances

Problem #	*Grid Size*	*Grid Type*	*Number of Terminal Nodes*	*Terminal Node Configuration*
1	50 x 50	Hill	20	Random
2	50 x 50	2-Pit	20	Random
3	50 x 50	Hill	20	Ring
4	50 x 50	2-Pit	20	Ring
5	21 x 17	Hill	10	Random
6	21 x 17	2-Hill	10	Random
7	21 x 17	Hill	7	Ring

$$(2, 14), (7, 9), (3, 9), (16, 19), (5, 5), (9, 17)\},$$

where each of the 11 ($=1.1N$) pairs represents a Steiner node location. Checks are performed to ensure that the Steiner node locations do not coincide with the terminal node locations. No information from the original problem other than the grid size, the number of terminal nodes and their locations, is used to construct the initial population; the initial population is otherwise random.

Fitness. For each individual in the population, the algorithm determines the Steiner tree as follows. Given the complete graph over the terminal nodes and the Steiner nodes encoded in a particular individual, a MST is found. Degree-1 Steiner nodes and their incident arcs are removed as described in Sect. 3, to yield the individual's Steiner tree. The fitness of the individual is the cost of its Steiner tree. This cost includes the cost of the cells containing tree edges and nodes, and possibly an additional charge for each Steiner node used (see Sect. 7). This method for finding a Steiner tree is quick, though not necessarily optimal for the individual's set of Steiner nodes.

Queen Bee Selection. The fittest individual (*the Queen Bee*) mates with the remainder of the population [13]. Out of the resulting set of offspring, the 40 fittest individuals are chosen to replace the current population. The mating procedure used is the spatial-horizontal crossover described below.

Spatial-Horizontal Crossover. This operator produces two offspring from two parents by first splitting the grid into a top and bottom (see Fig. 4). The exact location of the horizontal line is the vertical midpoint of the grid plus a normal random variable (of mean 0 and positive standard deviation). The Steiner nodes from Parent 1 that are below the horizontal line are combined with the Steiner nodes from Parent 2 that are above the horizontal line to form Offspring 1. Likewise, Offspring 2 is formed from the Steiner nodes of Parent 1 that are above the horizontal line combined with the Steiner nodes from Parent 2 that are below the horizontal line. The Queen Bee is always one of the two parents.

Mutation 1. It is possible that some of the arcs in the Steiner tree found for a given individual, may cross each other. It is advantageous to add Steiner nodes at these crossing locations. Within the algorithm, these new Steiner nodes replace Steiner nodes in the individual's chromosome that are unused in the current tree. In practice, the percentage of Steiner nodes in an individual that are used in a tree tends to be small. Thus, there are typically many free locations into which a new Steiner node may be added.

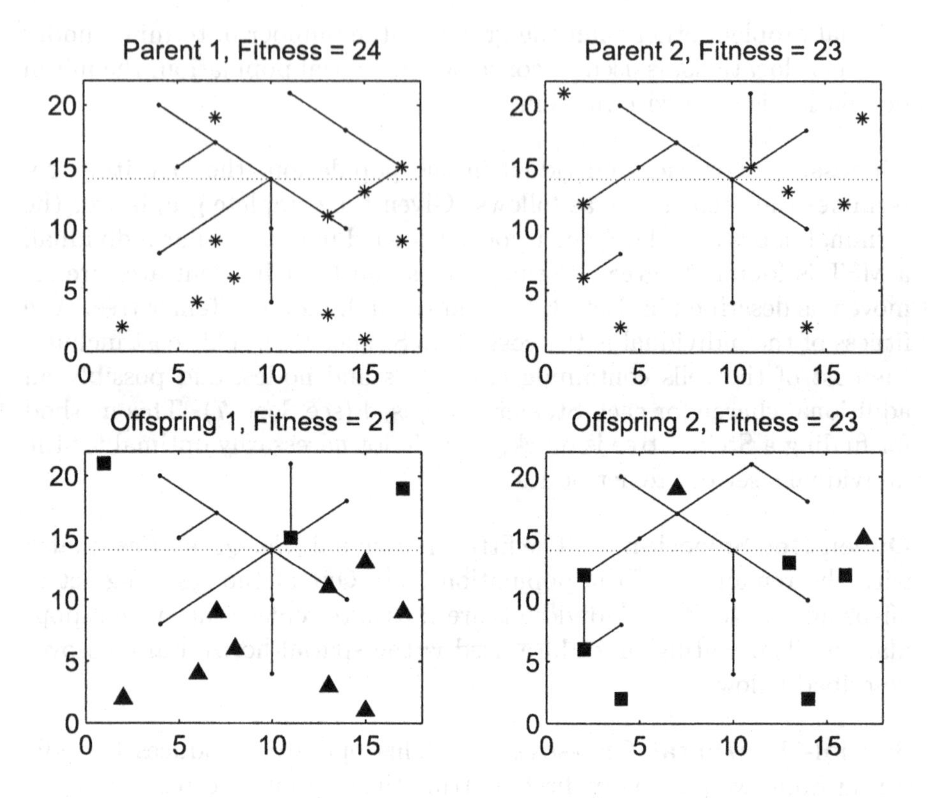

Figure 4. Horizontal crossover. The small dots in each plot represent the terminal nodes. In the top two plots, stars represent the Steiner nodes for the two parents. In the bottom two plots, triangles and squares represent the Steiner nodes in each offspring that came from Parents 1 and 2, respectively. The horizontal crossover location is indicated by the dotted line. The solid line segments indicate the Steiner tree arcs.

Mutation 2. This operator randomly moves Steiner nodes belonging to a given individual. The two fittest individuals in the population are not subject to this mutation. A randomly chosen Steiner node is moved to a random location with probability 0.20. This helps prevent solutions from stagnating at local minima.

Output. Steps 4 through 8 are repeated TMAX times. Testing indicated a value of 70 for TMAX was reasonable. At the conclusion of a run, the following are reported:

a) the coordinates of the Steiner nodes in the best final Steiner tree,

b) a record of the Steiner tree cost of the fittest individual in the population after each iteration, and

c) the time it took for the algorithm to run.

In addition, the MST over the terminal nodes and the best final Steiner tree are displayed along with the terminal nodes and Steiner nodes over a color or gray-scale image of the grid cost structure (e.g., see Figures 5 through 7). The MST on the terminal nodes, found by Prim's algorithm, serves as a very rough comparison for the best final Steiner tree.

3. Improvement Procedures

The GA utilizes additional procedures in order to improve each Steiner tree and to run more efficiently. These procedures remove degree-1 Steiner nodes, suggest potential Steiner nodes from arc intersections (see Sect. 4), and reduce the amount of time spent finding potential arcs.

The degree of a node is calculated as the number of arcs having an end-point at that node. If there are any degree-1 Steiner nodes in a Steiner tree, then removing those nodes will lead to an improvement in the solution value. Removing such Steiner nodes might lead to the degree at some other Steiner node dropping to one. Those Steiner nodes can also be removed from the tree. Hence, the GA employs a procedure to iteratively remove degree-1 Steiner nodes while finding the Steiner tree. Note that these Steiner nodes are not needed to connect the terminal nodes.

As was mentioned previously, if there are any intersecting arcs in the Steiner tree, the solution value might be improved by converting the intersection node into a new Steiner node. This is because one of the four arcs originating at that intersection can be deleted from the tree if that node is converted into a Steiner node. This is implemented in the GA as a mutation as described in Sect. 4. This procedure may not be useful if the fee for additional Steiner nodes is very high.

In a straightforward implementation of the tree-finding routine, one has to compute all the possible arcs during each iteration. This can be a very time consuming task. One way to reduce the run time is to use an arc saving approach in which we store the arc set at the end of each iteration. This way, during the next iteration, if a node gets added, one has to compute only those arcs that are new. This procedure leads to a significant reduction in the run time.

4. Progressive Addition of Steiner Nodes

Since the formulation of this problem is unique, it is not possible to directly compare our results with previous research. As a basis for comparison, we developed a simple enumerative algorithm named *progressive addition* (PA). The idea behind this approach is to add a Steiner node at

Table 2. Final Steiner Tree Costs for Each Algorithm

	Large Problems (50 x 50)				*Small Problems (21 x 17)*		
Algorithm	*1*	*2*	*3*	*4*	*5*	*6*	*7*
GA	29.35	39.81	31.60	29.08	7.82	3.50	8.41
PA	29.37	40.74	30.92	28.50	7.68	3.44	8.82
MST	56.24	57.81	34.67	33.12	8.61	6.38	8.82

each stage that gives the best Steiner tree. So Steiner nodes are added in a sequential manner, one at a time. To begin with, each non-terminal node in the hexagonal grid is tried as a potential Steiner node. The node that gives a tree with the lowest cost is made permanent. In the next stage, a second Steiner node is chosen that gives the minimum value when used along with the terminal nodes and the first Steiner node. This procedure is repeated until there is no further reduction in the tree cost upon addition of new Steiner nodes.

This procedure is very time consuming since each potential Steiner node in the grid is considered during each iteration. Similar to the arc saving approach, we can implement an arc exchange approach here. When one Steiner node replaces another, remove all the arcs corresponding to the removed node and add only those arcs that originate from the newly added node. But even with this improvement, the procedure is very computationally burdensome.

5. Computational Experiments

We ran each algorithm (GA and PA) on numerous test cases using the grid and node sets described in Sect. 2.2. We report the results on seven of them in Table 2. In each case, the GA was run seven times for 70 iterations each time and the mean of the best fitness found in each of the runs is reported. Seven runs is generally not sufficient to provide good statistics. But additional experiments indicated that the final fitness distribution for the GA is fairly tight about the mean. Overall, the results indicate that the performances of the GA and PA are about even and both are much better than the MST. The MST performs so poorly because of the restriction on node connections imposed by the hexagonal grid. Steiner nodes provide a way around this restriction.

We present examples of the final minimal spanning tree and best final Steiner trees in Figures 5 through 7. Figure 5 illustrates how the GA nicely finds a tree which avoids the high cost central region. This can also be seen in Fig. 6. Note that in Fig. 5, the right-most Steiner node

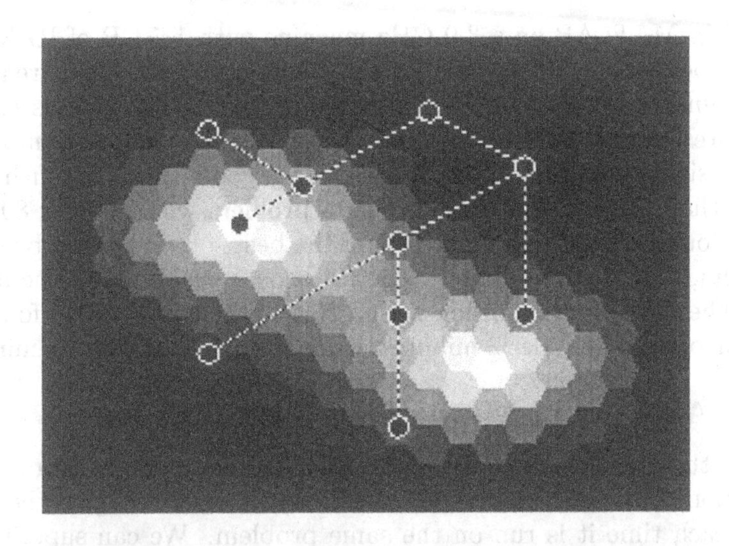

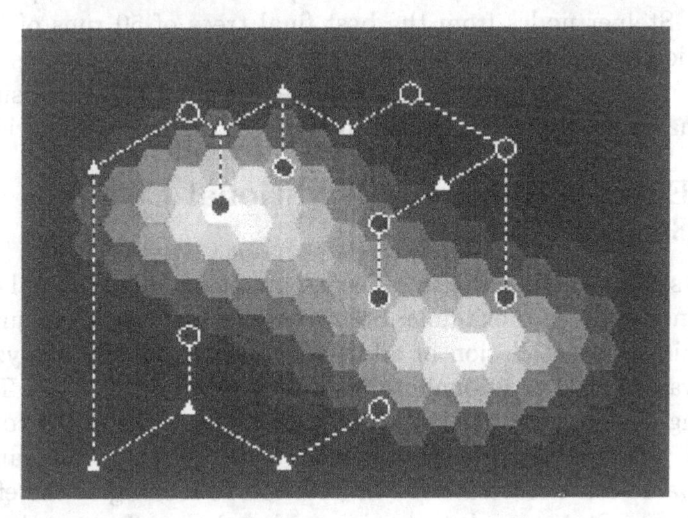

Figure 5. MST (top, cost=6.38) and best final GA Steiner tree (bottom, cost=3.38) for Problem # 6. Black circles indicate terminal nodes. White triangles indicate Steiner nodes. The lighter the shading, the more costly the cell.

is superfluous; it can be removed without affecting the cost of the tree. In Sect. 7, we consider charging an additional fee for Steiner nodes that will help prevent the occurrence of superfluous ones. In Fig. 7, Steiner nodes can be seen to cluster in the two low cost pits.

In terms of run times, PA tends to be quicker than the GA for the small problems (21 x 17), but is often significantly slower for the large problems (50 x 50) due to its combinatorial complexity. For example, running in MATLAB on a 3.0 GHz machine with 1.5 GB of RAM, PA usually took under 10 seconds on the small problems, compared with 2 to 3 minutes for the GA. On the large problems, PA took as long as 20 minutes, compared with 10 minutes for the GA. In addition, as the problem size grows, the GA finds reasonably good solutions much more quickly than the PA. For example, on a problem with an 80 x 80 grid, the GA found a solution within 10% of the best solution found by either algorithm, in under 10 minutes. The PA took over 30 minutes to do the same. The GA found a solution within 20% of the best solution found by either algorithm in under 2 minutes, while the PA took over 18 minutes.

6. Additional Methods of Visualization

Since the GA has randomness built in at many stages (e.g., initial population, crossover, mutation, selection), it may not give the same results each time it is run on the same problem. We can superimpose all of the Steiner nodes from the best final trees of 50 runs of the GA on a particular problem, on top of the grid. This image gives us an idea of where most of the useful Steiner nodes reside. Not surprisingly, it reveals that most Steiner nodes lie in low cost regions of the grid.

7. Effect of Imposing Additional Costs on Steiner Nodes

In this section, we analyze the effect of imposing an additional cost for each Steiner node. For example, Steiner nodes may represent junctions that require the installation of additional hardware. We analyzed the effect of varying the additional Steiner cost on Problem #1 (see Table 1) by varying the cost per Steiner node from a low value of 0.1 to a high value of 1. We tried ten different values of Steiner costs in this range and for each value ran ten iterations. In addition to looking at the effect on the number of Steiner nodes, we also analyzed the effect on the search space. Table 3 gives the results obtained.

It can be observed that the cost of the final tree increases with an increase in the additional cost for Steiner nodes. Also, we note that the average number of Steiner nodes drops as we increase the Steiner cost. Both of these are straightforward effects. What is more interesting is the effect of the Steiner cost on the search space. Table 3 shows that the standard deviations of the final tree cost and final number of Steiner nodes both drop as we increase the penalty for adding Steiner nodes. This indicates that the final results tend to be clustered together more closely when the penalty is higher. This is because, in the case of low penalties, even those Steiner nodes that lead to a slight reduction

Table 3. Effect of imposing additional costs on Steiner nodes

Steiner cost	*Mean ST cost*	*Stdev for ST cost*	*Mean number of Steiner nodes*	*Stdev for number of Steiner nodes*
0.1	30.4794	0.8832	7.8	1.3984
0.2	31.2877	0.5613	7.4	0.9661
0.3	32.1625	0.7036	6.3	1.3375
0.4	33.0306	0.8068	5.7	1.0593
0.5	33.5788	0.6456	5.9	0.8756
0.6	34.5048	0.5213	5.0	1.1547
0.7	34.8521	0.5724	4.8	0.7888
0.8	35.1604	0.2907	4.8	0.6325
0.9	36.2098	0.4160	4.1	0.3162
1.0	36.4957	0.4023	4.1	0.5676

in the MST cost get added since the slight reduction may more than compensate for the increase in cost due to the additional Steiner node. On the other hand, when the penalty is high, each Steiner node gets added only if it results in a significant reduction in the MST cost. So, in effect, the search space becomes smaller when we increase the penalty for Steiner nodes.

8. Conclusions

In conclusion, we have considered a variation of the Euclidean Steiner tree problem with non-uniform underlying cost structure. Using a genetic algorithm with fairly simple operators we have found solutions significantly better than those of the minimal spanning tree and comparable to those of our enumerative algorithm, progressive addition. We point out, however, that the GA scales better with increasing problem size than the PA. The solution trees are able to correctly avoid high cost areas while finding low cost regions. We have also indicated a novel way of visualizing the set of solutions. And we have shown the impact on the solutions of charging additional fees for the Steiner nodes.

Ongoing work includes developing standardized test cases using node sets from Beasley's OR library [14] and grids similar to the ones used here. We plan to solve the smaller cases exactly using a scheme such as the Dreyfus-Wagner algorithm [15]. At the same time, we aim to improve upon the results of our current algorithms. In particular, we plan to employ smarter (but still simple) operators in the GA. One possibility is to find the exact solution for some individuals, such as the queen bee, rather than approximating them by the MST. Another potential research direction is the rectilinear version of the non-uniform problem, which easily lends itself to discretization using a square grid.

References

[1] Winter, P.: The Steiner problem in networks: A survey. Networks **17** (1987) 129–167.

[2] Thurber, P.A., Xue, G.: Computing hexagonal Steiner trees using PCX. In: International Conference on Electronics, Circuits and Systems. (1999) 381–384.

[3] Coulston, C.: Constructing exact octagonal Steiner minimal trees. In: Proceedings of the 13th ACM Great Lakes Symposium on VLSI 2003, Washington, DC, USA, ACM (2003) 1–6.

[4] Coulston, C.: Steiner minimal trees in a hexagonally partitioned space. International Journal of Smart Engineering System Design **5** (2003) 1–6.

[5] Ganley, J.L., Cohoon, J.P.: A faster dynamic programming algorithm for exact rectilinear Steiner minimal trees. In: Proceedings of the Fourth Great Lakes Symposium on VLSI, University of Notre Dame, Notre Dame, IN, USA (1994) 238–241.

[6] Hakimi, S.L.: Steiner's problem in graphs and its implications. Networks **1** (1971) 113–133.

[7] Gröpl, C., Hougardy, S., Nierhoff, T., Prömel, H.J.: Approximation algorithms for the Steiner tree problem in graphs. In: Steiner Trees in Industry. Kluwer Academic Publishers, Dordrecht, The Netherlands (2001).

[8] Barreiros, J.: An hierarchic genetic algorithm for computing (near) optimal euclidean Steiner trees. In: Proceedings of the Genetic and Evolutionary Computation - GECCO 2003, Chicago, IL, USA, Springer-Verlag (July 12-16, 2003) 56–65.

[9] Ribeiro, C.C., de Souza, M.C.: Tabu search for the Steiner problem in graphs. Networks **36** (2000) 138–146.

[10] Osborne, L.J., Gillett, B.E.: A comparison of two simulated annealing algorithms applied to the directed Steiner problem on networks. ORSA Journal on Computing **3** (1991) 213–225.

[11] Julstrom, B.A.: A scalable genetic algorithm for the rectilinear Steiner problem. In: Proceedings of the 2002 Congress on Evolutionary Computation, IEEE, New York (2002) 1169–1173.

[12] Julstrom, B.A.: A hybrid evolutionary algorithm for the rectilinear Steiner problem. In: Proceedings of the Genetic and Evolutionary Computation - GECCO 2003, Chicago, IL, USA, Springer-Verlag (July 12-16, 2003) 49–55.

[13] Jung, S.H.: Queen-bee evolution in genetic algorithms. IEEE Electronic Letters **39** (2003) 575–576.

[14] Beasley, J.E.: OR-library: distributing test problems by electronic mail. Journal of the Operational Research Society **41** (1990) 1069–1072.

[15] Prömel, H.J., Steger, A.: The Steiner Tree Problem: a tour through graphs, algorithms, and complexity. Vieweg, Braunschweig/Wiesbaden, Germany (2002).

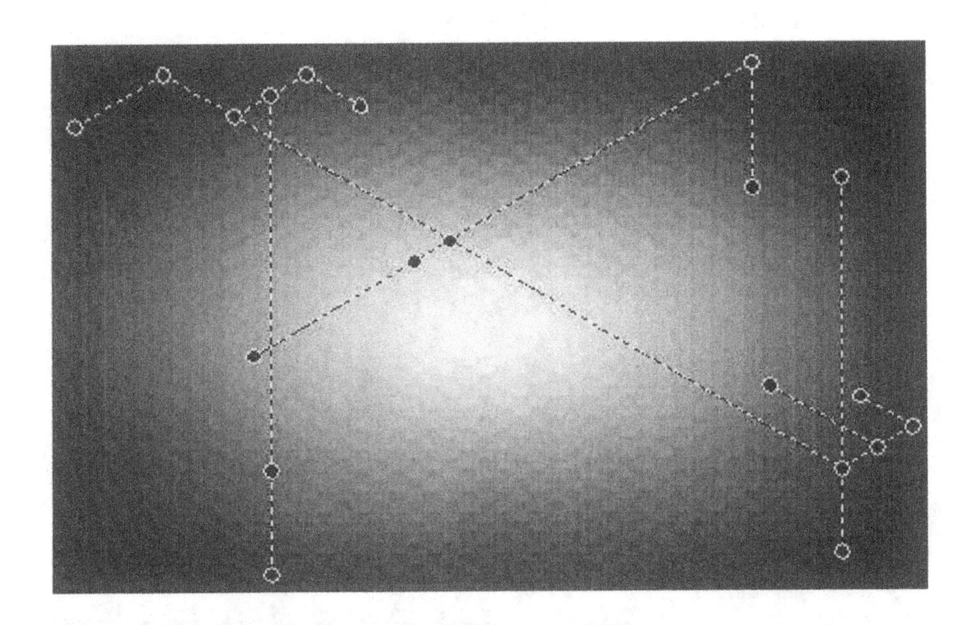

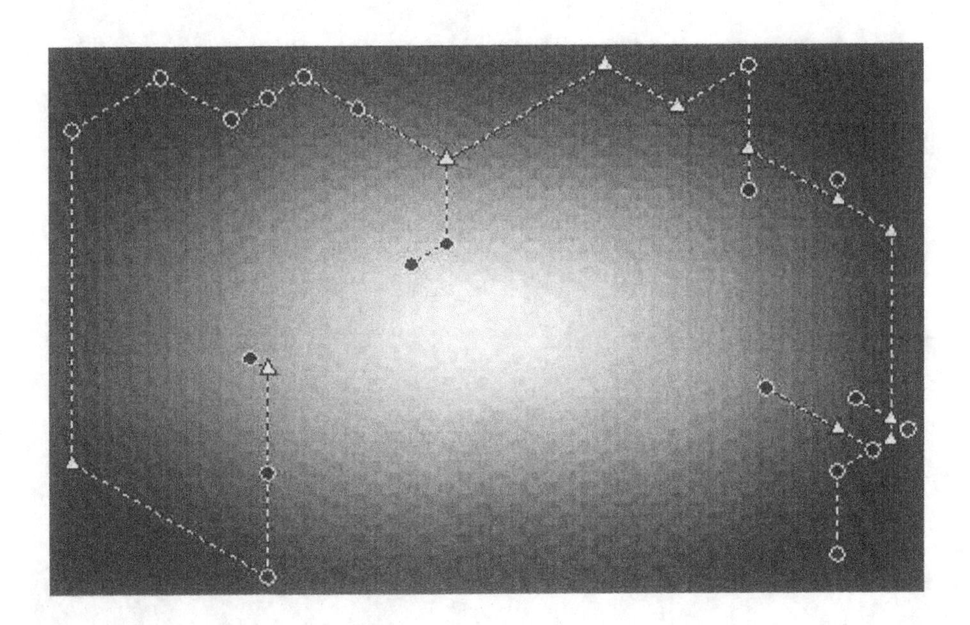

Figure 6. MST (top, cost=56.24) and best final GA Steiner tree (bottom, cost=29.01) for Problem # 1. Black circles indicate terminal nodes. White triangles indicate Steiner nodes. The lighter the shading, the more costly the cell.

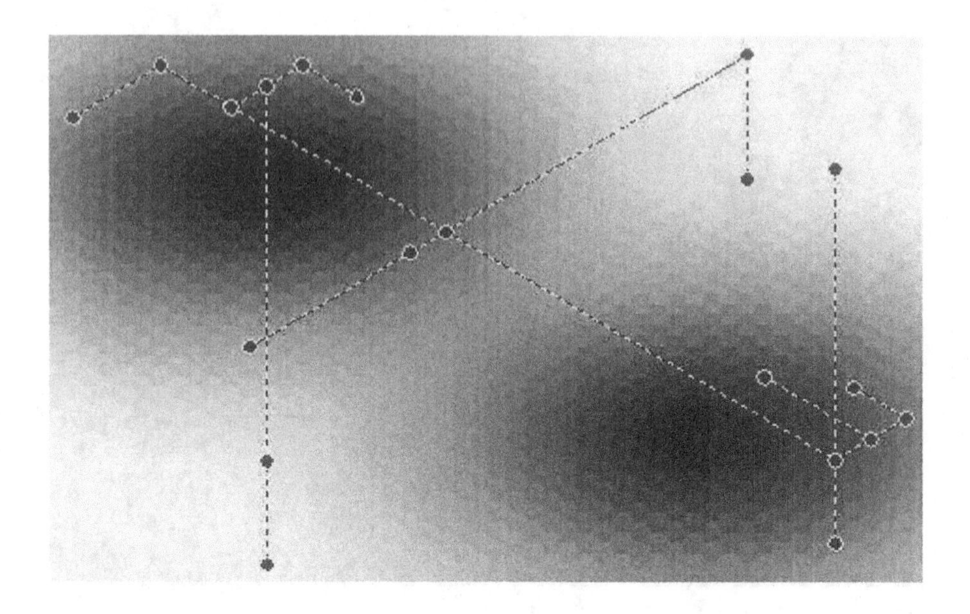

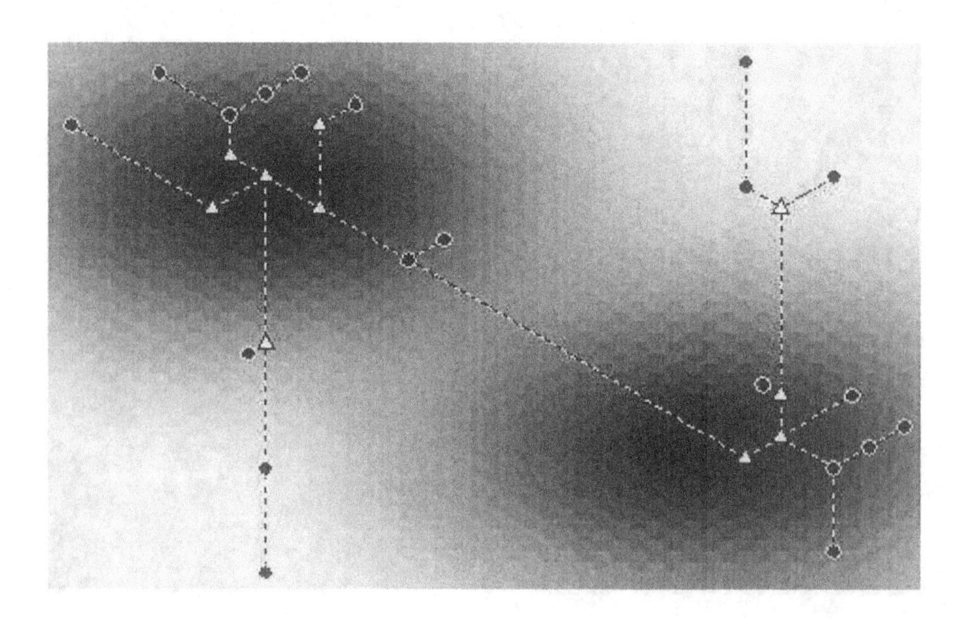

Figure 7. MST (top, cost=57.81) and best final GA Steiner tree (bottom, cost=39.41) for Problem # 2. Black circles indicate terminal nodes. White triangles indicate Steiner nodes. The lighter the shading, the more costly the cell.

MODELING AND SOLVING A SELECTION AND ASSIGNMENT PROBLEM

Manuel Laguna and Terry Wubbena
Leeds School of Business, University of Colorado, Boulder, CO 80301 and OptTek Systems, 1919 7th Street, Boulder CO 80304

Abstract: In this paper, we first provide an MIP formulation of a selection/assignment problem and then we discuss a solution method based both on the use of a commercial general-purpose scatter-search and a simple implementation of tabu search. This optimization problem is related to a research project supported by the Office of Naval Research where sailors need to be selected to perform a set of jobs that require specific skill levels. The results of our computational experiments indicate the usefulness of the software system for workforce planning that we have developed.

Key words: assignment/selection problem; bin packing; metaheuristic optimization.

1. INTRODUCTION

The optimization problem that we address in this paper is part of a workforce planning problem in the U.S. Navy. The problem may be described as follows. A set of jobs $J = \{1, \ldots, m\}$ must be completed during a fixed planning period (e.g., a week). Each job j requires d_j hours from a single sailor during the planning period. There is a set $I = \{1, \ldots, n\}$ of available sailors. The availability of sailor i during the planning period is s_i hours. For the purpose of efficiency, a sailor must perform a minimum number of hours (h_{min}) of any job to which he/she is assigned, but at the same time, no sailor may be assigned to more than j_{max} jobs during the planning period. Sailors have different skills and therefore there is a set A_i, associated with each sailor, which consists of the jobs that sailor i is qualified to perform. No more than t sailors may be assigned during the planning period. In other words, at most t sailors may be chosen from the set I of n sailors and the subset of selected sailors must be capable of completing all the jobs. The objective is to find a feasible solution that optimizes a given objective function. The problem falls within the general area of manpower planning and scheduling. Burke, at al. (2001) and Ernst, et al. (2004) are two recent references in the subject.

In our model, we assume that there is a cost c_{ij} of assigning sailor i to job j and formulate the optimization problem as follows:

$$x_{ij} = \begin{cases} 1 & \text{if sailor } i \text{ is assigned to job } j \\ 0 & \text{otherwise} \end{cases}$$

$$y_i = \begin{cases} 1 & \text{if sailor } i \text{ is selected} \\ 0 & \text{otherwise} \end{cases}$$

z_{ij} = number of hours that sailor i is assigned to perform job j

Q_j = set of sailors that are qualified to perform job j

$$\text{Minimize} \sum_{i \in I} \sum_{j \in A_i} c_{ij} x_{ij} \tag{1}$$

Subject to

$$\sum_{j \in A_i} z_{ij} \leq s_i y_i \qquad \forall i \in I \tag{2}$$

$$\sum_{i \in Q_j} z_{ij} \geq d_j \qquad \forall j \in J \tag{3}$$

$$\sum_{j \in A_i} x_{ij} \leq j_{\max} y_i \qquad \forall i \in I \tag{4}$$

$$h_{\min} x_{ij} \leq z_{ij} \leq s_i x_{ij} \qquad \forall i \in I, j \in A_i \tag{5}$$

$$\sum_{i \in I} y_i \leq t \tag{6}$$

$$x_{ij} \in \{0, 1\} \qquad \forall i \in I, j \in A_i$$

$$y_i \in \{0, 1\} \qquad \forall i \in I$$

$$z_{ij} \geq 0 \qquad \forall\, i \in I, j \in A_i$$

In the model above, the objective function (1) minimizes the total assignment cost. Constraint set (2) limits the number of hours for each selected sailor. If a sailor is not chosen, then this constraint does not allow any assignment of hours. Constraint set (3) enforces the job requirements, as specified by the number of hours needed during the planning period. Constraint set (4) limits the number of jobs that a chosen sailor is allowed to perform. Constraint set (5) enforces the requirement that a sailor may not perform a job for less than a minimum number of hours. At the same time, this constraint set does not allow the assignment of hours if the sailor has not been assigned to a given job. Finally, constraint set (6) limits the number of sailors chosen during the current planning period.

The same model may be used to optimize a different objective function, for instance, the minimization of the total assignment cost on an hourly basis or the maximization of an aggregate preference value. However, as mentioned above, we assume that the decision maker wants to minimize the total assignment cost as calculated in (1).

The most common mathematical model used to assign jobs to resources (e.g., machines, agents or workers) is the formulation of the *generalized assignment problem* (GAP). The GAP is a well-known combinatorial optimization problem and it would be advantageous to make use of solution procedures and strategies that the extensive GAP literature has to offer (Cattrysse and van Wassenhove, 1992). There are, however, significant differences between our assignment problem and the GAP. First of all, only a subset of the sailors in the pool may be assigned to jobs during the planning period. This adds a layer of decision-making that the GAP does not have. Another difference is that although resources may work on more than one job, splitting of jobs is not allowed in the GAP. Since in our problem some of the demand values (d_j) may be greater than the capacity values (s_i), we must consider job splitting in the solution process.

The problem at hand is also related to the capacitated facility location problem (CFLP) as well as the capacitated p-median problem (Aardal, 1998; Klose, 1999). In fact, our location-allocation problem reduces to a CFLP if the complicating constraints (4)-(6) are relaxed in a Lagrangean manner.

In the remainder of this paper, we describe a metaheuristic procedure for the solution of the selection/assignment problem represented by (1)-(6). The development of the metaheuristic procedure was triggered by the realization that commercial off-the-shelf MIP solvers would be incapable of solving even fairly small instances of the problem at hand. In particular, we attempted the solution of a problem with n = 20, m = 20 and t = 10 using Cplex 8.1 (with default parameter settings) on a Pentium 4 machine at 2.53

GHz and after 6.7 hours of CPU time the branch-and-bound search terminated in an out-of-memory error and an optimality gap of 71.41%. (The problem was generated using the data generator described at the beginning of section 3.) In order to avoid the out-of-memory error, we changed the node selection strategy in Cplex from its default value of 1 (best-bound search) to the value of 0 (depth-first search). We attempted the solution of the small problem with the new parameter setting and after 8 hours of CPU time, the best solution was worst than the solution found when the process ran out of memory in our first attempt.

2. SOLUTION APPROACH

The approach that we present in this section decomposes our problem into two interacting phases: selection and assignment. The selection phase consists of instantiating the y-variables in our model. The assignment phase consists of assigning work to the sailors chosen in the selection phase.

2.1 Selection Phase

For the selection phase we make use of the commercial implementation of scatter search known as *OptQuest* (Laguna and Martí 2002 and 2003). The optimization problem is simply stated as:

Minimize $f(y)$

Subject to

$$\sum_{i \in I} y_i \leq t \tag{6}$$

$$\sum_{i \in I} s_i y_i \leq \sum_{j \in J} d_j \tag{7}$$

$$y_i \in \{0, 1\} \qquad \forall\, i \in I$$

Since *OptQuest* is a general-purpose optimizer, the objective function does not need to be specified in mathematical form, such as a linear or nonlinear relationship of the decision variables. *OptQuest* is capable of searching for optimal values of the decision variables without knowing the structure of the objective function. In other words, *OptQuest* treats the evaluation of the objective function as a black-box process that maps the decision variables into an objective function value (i.e., $y \rightarrow f(y)$). Constraint (7) is added to eliminate instantiations of the y-variables that do not include enough total capacity. This constraint, however, does not guarantee that a feasible assignment will be found in the assignment phase because the sailors can perform only a subset of the jobs.

The evaluation of $f(y)$ entails the assignment of jobs to the sailors selected by the *OptQuest* engine. Therefore, the black-box evaluation of $f(y)$ is the assignment phase of our procedure. The assignment phase is launched for each new set of values for the y-variables. The number of time the selection phase is executed during the search process is controlled by the input parameter *SelectIter*.

2.2 Assignment Phase

For a given set of sailors, the assignment phase consists of solving the optimization problem defined by expressions (1) to (5) with a reduced set of variables and constraints. The reduced set includes all x- and z-variables associated with each sailor $i \in Y$, where $Y = \{i : y_i = 1\}$, as specified in the selection phase. Solving the resulting assignment problem optimally with a commercial solver such as Cplex turns out to be impractical. The main reason for this is that the LP relaxation of the resulting problem is weak, resulting in a long branch-and-bound search.

The problem remains difficult even after fixing the values of the y-variables due to the nature of constraint set (5). This set of constraints specifies that any sailor i is assigned to a job j must work at least $h_{\min}$ hours on the assigned job. This constraint creates a minimum "block of hours" that must be allocated to a sailor if the sailor is to perform a given job. Hence, if a job is split to accommodate its demand, the resulting parts of a job must not be smaller than $h_{\min}$ hours. With this in mind, our approach divides each job j with demand d_j into $\left\lfloor \frac{d_j}{h_{\min}} \right\rfloor$ items. There are $\left\lfloor \frac{d_j}{h_{\min}} \right\rfloor - 1$ items of size $h_{\min}$ and 1 item of size $d_j - h_{\min}\left(\left\lfloor \frac{d_j}{h_{\min}} \right\rfloor - 1\right)$. We denote the size (or weight) of the kth item of job j by w_{jk}. We let K_j be the set of items associated with job j.

Once the jobs have been divided into items, the problem becomes to assign items to sailors in such a way that capacity constraints (2) and (4) are not violated. An optimal assignment of items to sailors may be found by solving the following integer program:

$$u_{ijk} = \begin{cases} 1 & \text{if worker } i \text{ assigned to item } k \text{ of job } j \\ 0 & \text{otherwise} \end{cases}$$

$$\text{Minimize } \sum_{i \in I} \sum_{j \in A_i} c_{ij} x_{ij} \tag{1}$$

Subject to

$$\sum_{i \in Y \cap Q_j} u_{ijk} = 1 \qquad \forall j \in J, k \in K_j \tag{8}$$

$$\sum_{j \in A_i} \sum_{k \in K_j} w_{jk} u_{ijk} \leq s_i \qquad \forall i \in Y \tag{9}$$

$$\sum_{k \in K_j} u_{ijk} \leq \left\lfloor \frac{d_j}{h_{\min}} \right\rfloor x_{ij} \qquad \forall i \in Y, j \in A_i \tag{10}$$

$$\sum_{j \in A_i} x_{ij} \leq j_{\max} \qquad \forall i \in Y \tag{11}$$

$$x_{ij} \in \{0,1\} \qquad \forall i \in Y, j \in A_i$$

$$u_{ijk} \in \{0,1\} \qquad \forall i \in Y, j \in A_i, k \in K_j$$

The IP above may be infeasible for certain values of the y-variables as specified in the selection phase. For instance, if $Y \cap Q_j = \varnothing$ then the assignment problem (8)-(11) is infeasible. If the assignment problem is infeasible, the assignment phase returns $f(y) = M$, where M is a large positive number.

The problem of assigning items to sailors may be viewed as a special case of the well-known *bin packing problem*, where the goal is to accommodate a set of items of different weights into a set of bins with a fixed capacity. Other versions of this problem include minimizing the number of bins used to fit all items or minimizing the capacity of the largest bin (Coffman, Garey and Johnson, 1996). If we equate sailors with bins, the special features of our bin packing problem are:

1. There is a cost associated with assigning items to specific bins. Therefore, optimality is not simply defined by successfully packing all items into the available bins.
2. Not all the assignments of items to bins are possible.
3. The capacity of a bin is defined by both total weight and a function of the number of items. (For instance, if two items were originated by the same job then they count as one instead of two different items.).

Considering the characteristics of our special bin packing problem, we search for a high-quality assignment with a construction and improvement procedure. The construction is greedy in nature and the improvement employs a simple tabu-search memory structure. For a complete description of the tabu search methodology and its applications see Glover and Laguna (1997).

2.2.1 Construction

Given the set of selected sailors Y, we know that there is a total capacity of $H = \sum_{i \in Y} s_i$ hours, while the total demand is $D = \sum_j d_j$ hours. We define the target relative load for each sailor as $r = \frac{D}{H}$.

Before initiating the construction process, we order the items in non-increasing ω values, where $\omega_{jk} = \frac{w_{jk}}{|Q_j|^2}$ for $j \in J$ and $k \in K_j$. This ordering gives preference to the heaviest items that have the least amount of flexibility (that is, those items associated with jobs that not many sailors can perform). We then construct a solution in the following way:

1. Select the next item in the ordered list. Let this item be the kth item of the jth job.
2. Let l_i be the current load of bin i, where the load is the sum of the weights of the items already assigned to the bin. Calculate the updated load $l'_i = l_i + w_{jk}$.
3. Build the list of candidate bins (CL) to assign the item identified in step 1. We first try to build the list with bins whose updated load l'_i is not greater than s_i - $h_{\min}$. That is, $CL = \{i : l'_i \leq s_i - h_{\min}, i \in Y \cap Q_j\}$. If $CL = \varnothing$, then we try to build the candidate list with those bins for which the updated load does not exceed the capacity but is at least as large as the target load. That is, $CL = \{i : r^*s_i \leq l'_i \leq s_i, i \in Y \cap Q_j\}$. If $CL = \varnothing$,

then we try to build the candidate list with those bins that remain feasible after the load is updated. That is, $CL = \{i : l'_i \leq s_i, i \in Y \cap Q_j\}$. If the candidate list is still empty, it means that we will have to make an infeasible assignment. The candidate list then consists of all bins with minimum infeasibility (as measured by the number of hours that the load exceeds the available capacity).

4. Select the bin from *CL* that minimizes the cost of assigning the item to it. If we have reached the end of the ordered list of items then we stop. Otherwise we go to 1.

This construction procedure does not guarantee the construction of a feasible solution with respect to constraint set (2). Feasibility with respect to constraint set (4) is managed by the procedure by not allowing the assignment of items to bins that will exceeds the total number of jobs j_{max}. The greedy construction can be modified to accommodate other strategies, such as those based on semi-greedy rules. An instance of this family of approaches is the one known as GRASP, where the bin to be selected would be randomly chosen from a reduced candidate list. See Feo and Resende (1995) for a detailed description of GRASP. In our experimentation we use the deterministic construction procedure described above.

2.2.2 Improvement

The improvement method starts from an initial assignment, which may or may not be feasible with respect to (2), and performs exchanges in search for an improved outcome. Two types of exchanges are performed: swaps and inserts. A swap is an exchange of the bin assignment of two items. An insert is a move of an item from one bin to another. These exchange mechanisms result in a neighborhood that is a special case of what Osman (1995) refers to as λ-neighborhood. If the initial solution yielded by the construction procedure is infeasible, then the improvement procedure first focuses on finding a feasible assignment in the following way:

1. Identify all the infeasible bins. An infeasible bin is one for which its load is larger than its capacity.
2. Evaluate the change in the current infeasibility value that results from swapping an item from an infeasible bin to a feasible one. Only swaps for which the heavier item is moved to the feasible bin are considered.
3. Evaluate the change in the current infeasibility value that results from moving items currently assigned to infeasible bins to feasible ones.
4. Perform the exchange that minimizes the resulting infeasibility and update the tabu structures and the best solution found.

In steps 2 and 3 above, the procedure verifies that an exchange is possible in terms of not violating constraints (4) and (5). It also makes sure that when moving an item to a bin, the corresponding sailor is qualified to perform the job that originated the item.

The tabu structure that we employ simply records the time (i.e., iteration number) that the bins participating in the current move are allowed to exchange items again. The evaluation of exchanges in steps 2 and 3 above is done only for those bin-pairs that are not tabu-active. Note that at this stage of the improvement procedure it is not necessary to evaluate the swaps of items with the same weight, given that this move does not change the infeasibility of the current solution.

If *AssignIter* exchanges are performed without improving the infeasibility of the best solution, then the procedure stops and it returns $f(y) = M$ to the selection phase. However, if a feasible solution is found, the search focuses on finding a solution with an improved assignment cost. The search is modified as follows:

1. Identify all pairs of bins.
2. Evaluate the change in the current infeasibility value and the change in the assignment cost that results from swapping items in all the bin-pairs identified in step 1. The swaps include items with the same weight.
3. Evaluate the change in the current infeasibility value and the assignment cost that results from eliminating one item from a bin and inserting it in another one, considering all the bin-pairs identified in step 1.
4. If the current assignment is infeasible, then perform the exchange that minimizes the infeasibility of the resulting assignment. Otherwise, perform the exchange that minimizes the total assignment cost. Update the tabu structures and the best solution found.

Note that the swaps in step 2 include those that exchange the bin assignment of two items with the same weight. While this is a "null exchange" in typical bin packing procedures, in our case, the assignment cost may improve as a result of such an exchange. For example, suppose that job 1 is divided into 3 items with corresponding weights of 14, 10 and 10. Also suppose that items 1 and 2 of job 1 are currently assigned to bin 6, while item 3 is assigned to bin 8. Finally, suppose that an item with weight 10 from another job is assigned to bin 6 and that a swap is being considered that will exchange this item with item 3 of job 1 that is currently assigned to bin 8. Clearly, this exchange does not modify the load of neither bin 6 nor bin 8, because the swapping items have the same weight; however, the swap "consolidates" all items of job 1 into one bin (number 6). In terms of the objective function value, the swap saves the assignment cost of sailor 8 to job 1.

The improvement procedure returns the objective function value of the best solution found during the search. The search stops after *AssignIter* exchanges without improvement. The tabu-search memory structure utilized by the improvement procedure consists of a two dimensional $n \times n$ array labeled *TabuTime.* The (i, i') element of the array indicates the iteration number at which bin i is allowed to exchange items with bin i'. The array is initialized with a value of zero for all elements. After a swap or an insert involving bins i and i' in iteration *iter*, the *TabuTime* array is updated as follows:

$$TabuTime(i, i') = iter + U(TabuMin, TabuMax)$$

where

$$TabuMin = \min\left(5, \sqrt{n}\right)$$

$$TabuMax = \max\left(10, \sqrt{n}\right)$$

and $U(a, b)$ is a discrete uniform probability distribution with parameters a and b.

At iteration *iter*, if the following expression is true, then bin i is allowed to exchange items with bin i':

$$TabuTime(i, i') < iter$$

Most tabu search implementations incorporate what is known as the aspiration level criteria. Due to the computational burden associated with implementing even the simplest aspiration level version, our implementation operates without it.

3. COMPUTATIONAL EXPERIMENTS

In order to test the merit of our metaheuristic procedure, we generated artificial problem instances. Given the values of n, m, and t the problem instances were generated with the following characteristics:

$s_i = U(50, 70)$
$j_{max} = U(3, 5)$

$h_{\min} = U(10, 15)$
Category of sailor $i = U(0, 2)$
$P(i \in Q_j) = 0.25 * (1 + \text{Category of sailor } i)$

$$d_j = \max\left(h_{\min}, U\left(\frac{\bar{s}*t}{2*m}, \frac{1.5*\bar{s}*t}{m}\right)\right)$$

where $\bar{s} = \frac{\sum_i s_i}{n}$ and $\frac{\sum_j d_j}{\bar{s}*t} \leq \alpha$.

$c_{ij} = |A_i| + d_j + U(10, 20)$

The generator establishes a relationship between the flexibility of a sailor and his/her assignment cost. That is, sailors that are able to perform more jobs have larger assignment costs. Also, the range for $h_{\min}$ relative to s_i makes the resulting problem instances difficult, because the "items" in the bin packing problem associated with the Assignment Phase of the procedure are relatively large with respect to the capacity of the bins. We referred to these problems as "unstructured". Slightly easier (from the point of view of a branch-and-bound approach) problems can be constructed by using a small value of $h_{\min}$, for example 4, and making the demand values d_j a multiple of $h_{\min}$. We refer to these problems as "structured". We address the performance of the proposed procedure in both cases.

Note that the problem generator uses α as the limit for the expected relative load of each sailor. For our first experiment, 10 unstructured problems were generated with the following parameter values:

$n = 20$
$m = 20$
$t = 10$
$\alpha = 0.97$

Instead of generating a different capacity value (s_i) for each sailor, our set of problems has a single value $s_i = \bar{s}$. This makes constraint (7) of the selection problem redundant. We use the set of problems to compare the solutions obtained with our metaheuristic procedure and those obtained solving the MIP formulation (1)-(6) with Cplex 8.1. All experiments were performed on a Pentium 4 machine at 2.53 GHz. Cplex was terminated after 1 hour of CPU time and all default parameter values were used. The parameter values used for the stopping rules within the metaheuristic were:

SelectIter = 5000
AssignIter = 500

Although these parameter values were set after limited experimentation, they were not customized to make the procedure perform best on the set of test problems presented below. The parameters were set using problem instances that are not part of the 10 that appear in our computational results summarized in Table 1.

Table 1. Summary of results for unstructured problems

	Cplex		Metaheuristic	
Prob.	Upper Bound	Optimality Gap	Best Solution	Optimality Gap
1	1215	27.27%	1106	20.11%
2	1039	11.52%	937	1.89%
3	1151	14.79%	1090	10.02%
4	1228	22.63%	1143	16.88%
5	1093	15.01%	1046	11.19%
6	1083	11.12%	1019	5.54%
7	995	8.35%	1008	9.53%
8	1009	14.25%	1003	13.74%
9	890	4.46%	867	1.92%
10	913	5.29%	904	4.35%

Table 1 shows the best solutions found by both Cplex (Upper Bound) and our metaheuristic procedure (Best Solution). This table also shows the optimality gap calculated against the best lower bound found by the Cplex branch-and-bound process. The total search time for Cplex was 1 CPU hour, while the total metaheuristic search time averaged 305.5 seconds. The best upper bound values were found on an average of 2091.5 seconds. The best metaheuristic solutions were found on an average of 156.8 seconds. The table indicates that our procedure has merit, both in terms of the quality of the solutions found (when compared to those produced by Cplex) and the CPU time needed to find them. Only in one instance (Problem 7) Cplex was able to find a better solution than the one found by our approach. In addition, the metaheuristic was, on the average, at least one order of magnitude faster than Cplex at arriving to the best solutions.

As mentioned above, the problem instances in Table 1 are difficult because not only the h_{min} value is large relative to the s_i values but also the α-value is close to 1. This combination complicates the search for feasible assignments and weakens the LP relaxation, which cripples Cplex's ability to find and confirm optimal solutions. For our next experiment, we construct a set of problems for which h_{min} is set to 4 and the d_j values are generated as before and then adjusted as follows:

$$d_j = d_j - \mathbf{mod}(d_j, 4)$$

where **mod**(x,y) returns the remainder of x/y. The problem generator parameters are set as before: $n = m = 20$, $t = 10$ and $\alpha = 0.97$. Table 2 shows the results of solving the MIP formulation with Cplex and applying the proposed metaheuristic.

Table 2. Summary of results for structured problems

	Cplex		Metaheuristic	
Prob.	Upper Bound	Optimality Gap	Best Solution	Optimality Gap
1	855	0.01%	922	7.28%
2	899	0.01%	936	3.96%
3	980	0.94%	1131	14.17%
4	935	0.01%	970	3.62%
5	961	1.62%	996	5.08%
6	968	0.01%	1029	5.94%
7	904	0.01%	989	8.60%
8	994	4.88%	1049	9.87%
9	931	6.00%	937	6.60%
10	866	0.01%	887	2.38%

The results in Table 2 indicate that the structured problems make the solution of the MIP formulation with a branch-and-bound code such as Cplex a viable alternative. Cplex is capable of solving six out of the ten problems to optimality (with the default value of a 0.01% gap). The average optimality gap for the metaheuristic is a respectable 6.75%, but the procedure is unable to find any of the known optimal solutions.

4. CONCLUSIONS AND FUTURE WORK

We have described the development of a solution procedure for a problem in the area of manpower planning and scheduling. The solution procedure takes advantage of existing commercial software by coupling it with a specialized heuristic. The merit of the proposed approach is established with comparisons to upper bounds found by a truncated branch-and-bound search.

Additional experiments are necessary to confirm the usefulness of our approach. In particular, we seek to solve instances of a larger size that are typical in the setting that trigger the investigation of this problem. A typical problem size in the real setting consists of $n \sim 200$, $m \sim 200$ and $t \sim 100$.

To deal with these large problems, a specialized heuristic for the selection phase of our approach must be developed. The heuristic must use context information to strategically select sets of sailors for the application of the assignment phase. The specialized heuristic should replace the generic *OptQuest* optimizer and the assignment phase should not be treated as a black-box evaluator. Additionally, the assignment phase should be examined, given that the current partition of items is quite arbitrary. Specifically, we currently are unable to show that our partition can lead to optimal solutions.

Another avenue for investigation consists of the application of constraint programming. The objective functions discussed in this paper are somewhat artificial, given that the Navy would be hard-pressed to find accurate cost or preference information to be able to address the current situation as an optimization problem. Therefore it is likely that a constraint programming approach that simply seeks to find "interesting" and feasible sailor selections and assignments may be more applicable in practice.

REFERENCES

Aardal, K. (1998) Capacitated Facility Location: Separation Algorithm and Computational Experience, *Mathematical Programming*, vol. 81. pp. 149-175.

Burke, E. K., P. Cowling, P. De Causmaecker and G. Vanden Berghe (2001) "A Memetic Approach to the Nurse Rostering Problem," *Applied Intelligence*, vol. 15, no. 3, pp 199-214.

Cattrysse, D. G. and L. N. van Wassenhove (1992) "A Survey of Algorithms for the Generalized Assignment Problem," *European Journal of Operational Research*, vol. 60, pp. 260-272.

Coffman, E. G., M. R. Garey and D. S. Johnson (1996) "Approximation Algorithms for Bin Packing: A Survey," in *Approximation Algorithms for NP-hard Problems*, D. Hochbaum (ed.), pp. 46-93, PWS Publishing, Boston.

Ernst, A. T., H. Jiang, M. Krishnamoorthy and D. Sier (2004) "Staff Scheduling and Rostering: A Review of Applications, Methods and Models," *European Journal of Operational Research*, vol. 153, no. 1, pp 3-27.

Feo, T. and M. G. C. Resende (1995) "Greedy Randomized Adaptive Search Procedures," *Journal of Global Optimization*, vol. 2, pp. 1-27.

Glover, F. and M. Laguna (1997) *Tabu Search*, Kluwer Academic Publishers, Boston.

Klose, A. (1999) An LP-based Heuristic for Two-stage Capacitated Facility Location Problems, *Journal of the Operational Research Society*, vol. 50, pp. 157-166.

Laguna M. and R. Martí (2002) "The OptQuest Callable Library," *Optimization Software Class Libraries*, Stefan Voss and David L. Woodruff (eds.), Kluwer Academic Publishers, Boston, pp. 193-218.

Laguna, M. and R. Martí (2003) *Scatter Search: Methodology and Implementations in C*, Kluwer Academic Publishers, Boston, ISBN 1-4020-7376-3, 312 pp.

Osman, I. H. (1995) "Heuristics for the Generalized Assignment Problem: Simulated Annealing and Tabu Search Approaches," *OR Spektrum*, vol. 17, pp. 211-225.

SOLVING THE TIME DEPENDENT TRAVELING SALESMAN PROBLEM

Feiyue Li,[1] Bruce Golden,[2] and Edward Wasil[3]

[1]*Department of Mathematics*
University of Maryland
College Park, Maryland 20742
lify@math.umd.edu

[2]*R.H. Smith School of Business*
University of Maryland
College Park, Maryland 20742
bgolden@rhsmith.umd.edu

[3]*Kogod School of Business*
American University
Washington, DC 20016
ewasil@american.edu

Abstract In the standard version of the traveling salesman problem (TSP), we are given a set of customers located in and around a city and the distances between each pair of customers, and need to find the shortest tour that visits each customer exactly once. Suppose that some of the customers are located in the center of the city. Within a window of time, center city becomes congested so that the time to travel between customers takes longer. Clearly, we would like to construct a tour that avoids visiting customers when the center of the city is congested. This variant of the TSP is known as the time dependent TSP (TDTSP). We review the literature on the TDTSP, develop two solution algorithms, and report computational experience with our algorithms.

Keywords: Traveling salesman problem; heuristics.

1. Introduction

In the standard version of the traveling salesman problem, we are given a set of customers located in and around a city and the distances between each pair of customers, and need to find the shortest tour that visits each customer exactly once. The TSP has been studied for more than 50 years and a wide variety of heuristics has been developed. Applegate et al. (2004), Johnson and

McGeoch (1997), and Junger et al. (1995) are excellent sources for algorithmic and computational aspects of the TSP.

In this paper, we consider the following variant of the TSP. Suppose that some of the customers are located in the center of the city. Within a window of time, center city becomes congested so that the time to travel between customers takes longer. Clearly, we would like to construct a tour that avoids visiting customers when the center of the city is congested. This variant of the TSP is known as the time dependent traveling salesman problem.

Recently, Bentner et al. (2001) and Schneider (2002) have studied the TD-TSP. They considered the Bier127 problem from TSPLIB (Reinelt, 2001) and defined a region in the city center in which traffic jams occurred in the afternoon. In Figure 1, we show Bier127 with the locations of 127 beer gardens in and around Augsburg, Germany. Congestion occurred in the afternoon for beer gardens in the dashed rectangle (the traffic jam region), so that the time to drive between two locations in the rectangle was multiplied by a jam factor $f > 1$.

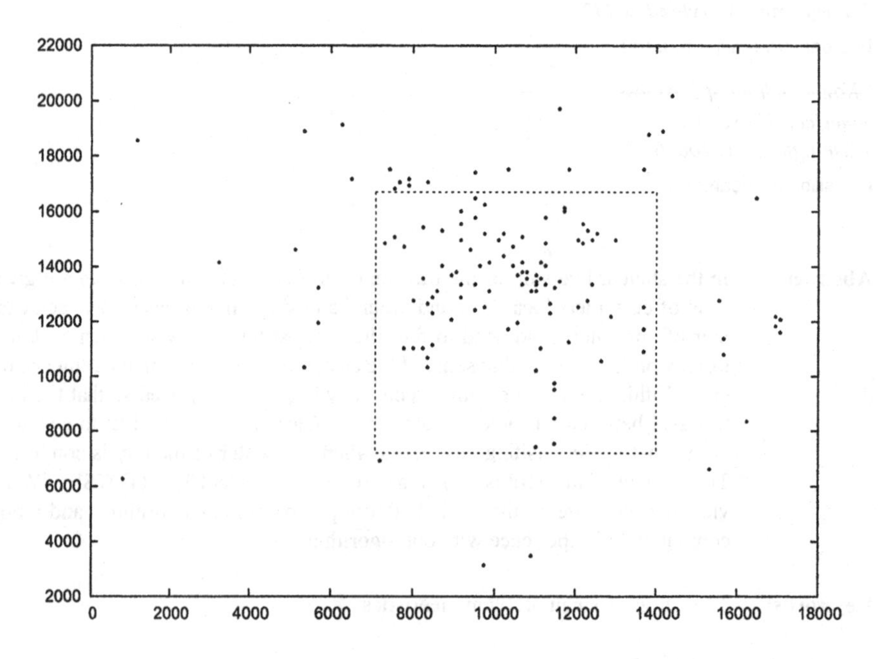

Figure 1. Bier127 with the location of 127 beer gardens in and around Augsburg. Afternoon traffic jams occur in the dashed rectangular region.

Bentner et al. and Schneider varied the value of the jam factor and generated tours for Bier127 using simulated annealing. As the value of the jam factor increased, they found that locations in the jam factor region were typically avoided in the afternoon. In addition, Bentner et al. compared different traffic jam regions and found that, when the region was small, the salesman could detour and avoid the traffic jam without greatly increasing the tour length. However, when the traffic jam region was large, short detours were not always possible.

In Section 2, we develop two algorithms for solving the TDTSP. In Section 3, we conduct computational experiments with both algorithms on Bier127. In Section 4, we consider the time dependent vehicle routing problem (TDVRP) and present limited computational results. In Section 5, we give our conclusions.

2. Algorithms for the TDTSP

In this section, we present two algorithms for solving the TDTSP: one based on record-to-record travel and one based on the chained Lin-Kernighan procedure.

2.1 Record-to-record Travel Algorithm

Our record-to-record travel algorithm (RTR) is based on the procedure that we developed to solve the vehicle routing problem (Li et al., 2004). We describe RTR in Table 1. The initial solution is generated by the modified Clarke and Wright algorithm (Golden et al., 1977). We then use two-opt moves and one-point moves, and allow uphill moves. Finally, we try to improve the current solution by allowing only downhill moves.

2.2 Chained Lin-Kernighan Algorithm

We also developed a variant of the chained Lin-Kernighan algorithm (CLK) to solve the TDTSP (Applegate et al., 1999). Our variant is described in Table 2. In Algorithm 7, the outer loop runs for min{number of nodes/2, 100} iterations. We use a neighbor list with 25 nearest neighbors.

At the end of Algorithm 7, we perturb the current solution. We use the double-bridge kick shown in Figure 2 (see Applegate et al., 1999 for more details). In Figure 2a, we randomly select four pairs of nodes from the current solution. We re-link them as shown in Figure 2b. This changes the structure of the current solution and will hopefully lead to a better local optimum.

In Algorithm 8, we apply an iterative operation on each node that exploits possible two-opt moves. In the traditional two-opt move, each node is examined and only downhill moves are made. In Algorithm 9, we do not finish processing a node immediately if there is no downhill move. Instead, we do the two-opt move and apply Algorithm 9 (recursively) to the new solution. We do this four times. If a downhill move is found, we accept it. Otherwise, we restore the solution that was generated before the recursive call.

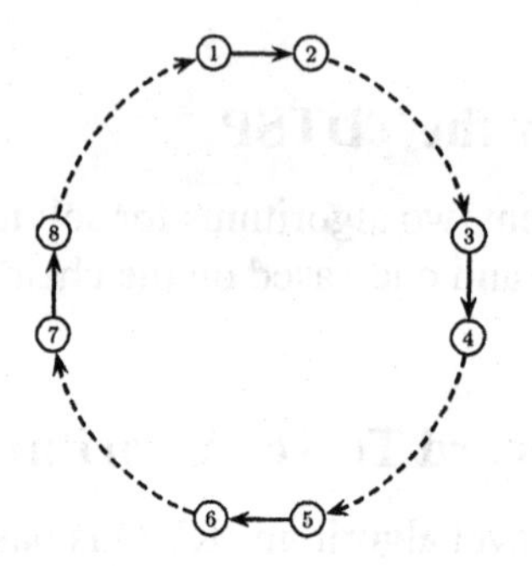

Figure 2a. Four pairs of adjacent nodes.

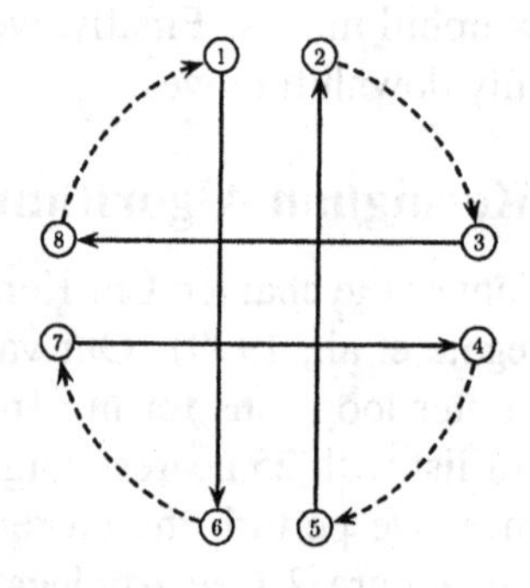

Figure 2b. Re-linked nodes.

Figure 2. Double-bridge kick.

Table 1. Record-to-record travel algorithm for the TDTSP.

Algorithm 1 Main Program with Multiple Trials

```
begin
   Set bestRecord:=null; OptLength:=∞
   for λ := 0.6 to 2.0 step 0.4 do
       generate an initial TSP tour p by the modified Clarke and Wright
       algorithm with parameter λ
       use record-to-record travel on p to improve the solution
       if p.length < OptLength
          then bestRecord:=p; OptLength:=p.length;
end
```

Algorithm 2 Record-to-record Travel for TDTSP

```
input: TSP tour p
output: an improved TSP tour p
begin bestTour:=p; deviation:=0.01*p.length; M=5, I=10
   for counter:= 1 to M do
       for i := 1 to I do (I loop)
           apply two-opt move and one-point move with record-to-record
           travel on p; uphill moves are allowed
           if no move is performed, break I loop
       apply two-opt move and one-point move to the current solution
       only downhill moves are allowed
       if bestTour.length < p.length
          then bestTour:=p;  deviation:=0.01*p.length
end
```

Table 1. (continued)

Algorithm 3 Two-opt Move with Record-to-record Travel

input: record, deviation
begin
  n := number of nodes
  for $i := 1$ **to** n **do** (I loop)
    for $j := i + 1$ **to** n **do** (J loop)
      consider the two-opt move with edge i and j
      if this is a downhill move
        then make the move and continue with the I loop
        else save this move if, after the move,
          tourLength < record + deviation
    make the best move in the J loop
end

Algorithm 4 One-point Move with Record-to-record Travel

input: record, deviation
begin
  n := number of nodes
  for $i := 1$ **to** n **do** (I loop)
    for $j := 1$ **to** $n (j \neq i)$ **do** (J loop)
      insert the ending node of edge i between edge j (this is a one-point move)
      if this is a downhill move
        then make the move and continue with the I loop
        else save this move if, after the move,
          tourLength < record + deviation
    make the best move in the J loop
end

Table 1. (continued)

Algorithm 5 Two-opt Move

```
begin
   n := number of nodes; improved:=true
   while improve do
         improved:= false
         for i := 1 to n do (I loop)
            for j := i + 1 to n do (J loop)
               consider the two-opt move for edge i and j
               if this is a downhill move
                  then improve:=true
                        make the move and continue with the I loop
end
```

Algorithm 6 One-point Move

```
begin
   n:= number of nodes; improved:=true
   while improved do
         improve:= false
         for i := 1 to n do (I loop)
            for j := 1 to n(j ≠ i) do (J loop)
               consider the one-point move for edge i and j
               if this is a downhill move
                  then improve:=true
                        make the move and continue with the I loop
end
```

Table 2. Chained Lin-Kernighan algorithm for the TDTSP.

Algorithm 7 Chained Lin-Kernighan

```
begin
   T = the initial tour. bestObj = length(T)
   while stopping rule is not satisfied do
         Linkern(T)
         if length(T) < bestObj
            then bestObj = length(T)
         perturb T
end
```

Algorithm 8 Linkern

```
input:tour T
begin
   level=0
   put each node of T into a queue q
   while q is not empty do
         t1=q.pop()
         t2=next(t1)
         record = length(T)
         rval=improve(t1, t2, level, record, q)
         if rval > 0
            then push t1, t2 into q
end
```

Table 2. (continued)

Algorithm 9 Improve

```
Input:
 t1, t2 : two consecutive nodes in the tour
 level : current recursive call
 record : best tour length so far
 q : the node queue
Output: 0 if no improvement has been found; positive otherwise
begin
   rval = 0
   if level ≥ 4
      then return 0
      else
            Find a set of nodes S belonging to the neighbor set of t2
             that will yield a promising two-opt move
            for each node t3 ∈ S do
            t4 = prev(t3)
            make two-opt move (t1, t2, t4, t3), that is,
            reverse the nodes between t2 and t4 inclusively
            update the tour length
            if the new length < record
               then update record
                     rval = 1
               rval = rval + improve(t1, t4, level+1, record)
               if rval = 0
                  then undo the two-opt move (t1, t2, t4, t3)
                        restore the old tour length
                  else push t3 and t4 into q
                        return 1
end
```

3. Computational Experiments

In this section, we report the results of two computational experiments. We use the Bier127 problem. The traffic jam region (rectangle) has lower left-corner coordinates (7080, 7200), a width of 6920, and a height of 9490. The starting node (node 1) has coordinates (9860, 14152). A salesman starts at 9 am and finishes at 3 pm. The traffic jam occurs from 12 pm to 3 pm. The travel speed is computed by dividing the total distance of the optimal TSP tour

(118293.524) by the number of hours in the workday (six). The travel speed is held constant for all values of the jam factor. It is not necessary that a tour fills the work day exactly.

3.1 Old Assumption

In Bentner et al. (2001) and Schneider (2002), traffic jams occur on all edges with *both* end points in the rectangle. We refer to this as the *old assumption*.

We apply our record-to-record travel algorithm to Bier127 with the old assumption. The computational results are given in Table 3. We present results for 20 different values of the jam factor. The computation time is in minutes on an Athlon 1 GHz computer. We see that RTR finds the best-known solution for four jam factors (1, 1.20, 1.38, and 1.39) and, on average, is 0.30% above the best-known solution. In Figures 3 and 4, we show the best-known solution for different values of the jam factor f. The salesman starts the tour at the circle. The last edge is not shown in order to indicate the direction of the tour.

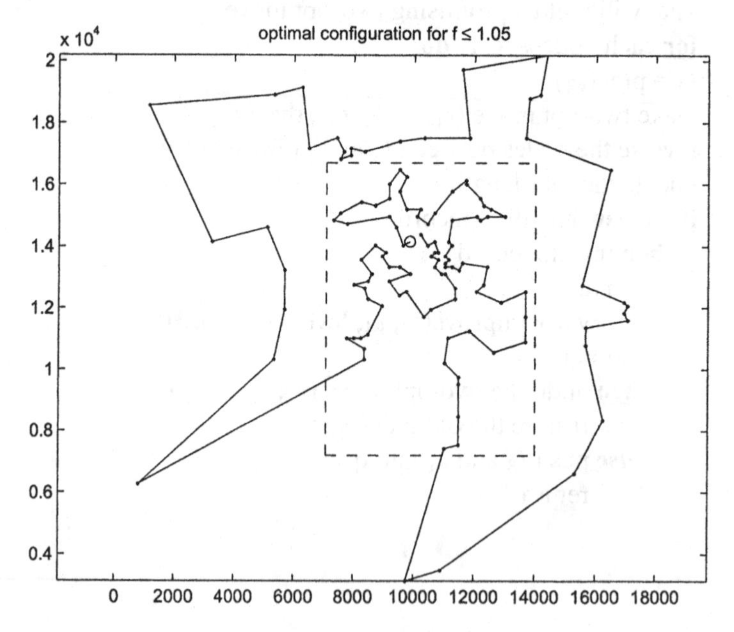

Figure 3. Best-known solution for $f \leq 1.05$.

3.2 New Assumption

With the old assumption, a salesman starts at 9 am and finishes at 3 pm. The traffic jam occurs at noon. An edge ℓ is penalized during the traffic jam only if the following two conditions are satisfied: (1) ℓ is traveled after noon and (2) both end points of ℓ are inside the traffic jam region.

Table 3. Computational results for RTR on Bier127 with the old assumption.

Jam Factor	*Time(min)*	*Tour Length*	*Percent Above Best Known*	*Best-known Solution*
1.00	2.61	118293.524	0.00	118293.524
1.03	4.31	118796.154	0.04	118749.356
1.04	2.74	119971.191	0.90	118901.300
1.05	3.29	119503.279	0.38	119053.244
1.06	3.98	119857.323	0.60	119153.582
1.10	2.88	119957.387	0.54	119313.720
1.20	3.61	119714.065	0.00	119714.065
1.30	3.17	120637.093	0.44	120114.410
1.38	2.73	120434.687	0.00	120434.687
1.39	3.10	120453.554	0.00	120453.554
1.50	4.22	120617.178	0.04	120571.743
1.60	4.55	121108.329	0.36	120679.186
1.70	3.72	120898.269	0.09	120786.630
1.80	3.03	121195.816	0.25	120894.074
1.90	3.49	121148.519	0.12	121001.518
2.02	5.58	121298.538	0.14	121125.195
3.00	4.34	122222.204	0.91	121125.195
10.00	3.67	121167.051	0.03	121125.195
100.00	4.47	122280.886	0.95	121125.195
2000.00	3.84	121417.575	0.24	121125.195
Average	3.66		0.30	

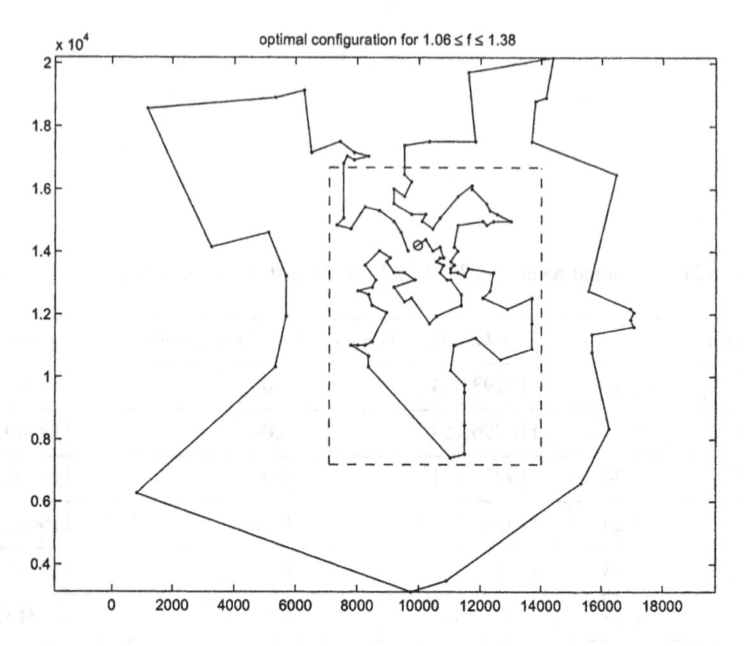

Figure 4. Best-known solution for $1.06 \leq f \leq 1.38$.

The second condition is not realistic in practice. A salesman might travel along an edge with end points i and j after noon, where i is outside the traffic jam region and j is inside the region. Under condition 2 of the old assumption, this edge would not be penalized. We would like to penalize this edge in proportion to the length inside the traffic jam region and use the following revised conditions, called the *new assumption*. An edge ℓ is penalized during the traffic jam only if the following two conditions are satisfied: (1) some part of ℓ is traveled after noon and (2) some part of ℓ is inside the traffic jam region and only that part is penalized. Thus, we penalize that part of ℓ inside the traffic jam region that is traveled after noon.

We apply our record-to-record travel algorithm and our chained Lin-Kernighan algorithm to Bier127 with the new assumption. The computational results are given in Tables 4 and 5. We present results for 13 different values of the jam factor. The computation time is in minutes on an Athlon 1 GHz computer. In Table 5, the results for chained Lin-Kernighan are from 10 runs of the algorithm (randomness is introduced into each run since we use the double-bridge kick). We report the best tour length found in the 10 runs and the total running time for the 10 runs. We see that CLK generates nearly all of the best-known solutions. RTR performs nearly as well — it quickly generates solutions that are, on average, within 0.75% of the best-known solutions.

We now examine the objective function of the TDTSP. Let the value of the objective function be defined by $Obj = L_1 + \alpha f$, where L_1 is the total distance traveled without being penalized, α is the total distance traveled being

Table 4. Computational results for RTR on Bier127 with the new assumption.

Jam Factor	*Time(min)*	*Tour Length*	*Percent Above Best Known*	*Best-known Solution*
1.00	0.83	118838.502	0.46	118293.524
1.05	0.86	119593.645	0.39	119125.478
1.06	1.06	119291.869	0.03	119250.182
1.18	1.15	120449.356	0.44	119923.796
1.19	0.83	120803.542	0.70	119968.895
1.70	0.81	123232.728	1.26	121697.112
1.71	1.16	123067.541	1.10	121728.121
2.42	1.13	125203.619	1.07	123874.262
2.43	1.01	125207.611	1.05	123901.699
3.74	1.24	128904.044	1.11	127491.060
3.75	1.07	129093.330	1.24	127518.019
6.53	1.02	135561.752	0.52	134858.670
6.54	1.07	135390.528	0.38	134883.255
Average	1.02		0.75	

Table 5. Computational results for CLK on Bier127 with the new assumption.

Jam Factor	*Time(min)*	*Tour Length*	*Percent Above Best Known*	*Best-known Solution*
1.00	4.08	118293.524	0.00	118293.524
1.05	4.16	119125.478	0.00	119125.478
1.06	4.53	119250.182	0.00	119250.182
1.18	4.78	119923.796	0.00	119923.796
1.19	4.31	119968.895	0.00	119968.895
1.70	4.20	121697.112	0.00	121697.112
1.71	5.04	121728.121	0.00	121728.121
2.42	4.80	123874.262	0.00	123874.262
2.43	4.96	123901.699	0.00	123901.699
3.74	4.72	127491.060	0.00	127491.060
3.75	4.82	127518.019	0.00	127518.019
6.53	5.19	134916.352	0.04	134858.670
6.54	4.53	134883.255	0.00	134883.255
Average	4.62		0.003	

Table 6. Boundary intervals for the jam factor.

Boundary Intervals for Jam Factor f	L_0	α
[1.00,1.05]	118293.524	16639.086
[1.06,1.18]	118913.374	5613.453
[1.19,1.70]	119325.049	3388.663
[1.71,2.42]	119581.980	3022.734
[2.43,3.74]	119983.542	2739.969
[3.75,6.53]	120256.583	2640.522

penalized, and f is the jam factor. We see that Obj is a linear function of f that can be rewritten as $Obj = (L_1 + \alpha) + \alpha(f - 1)$. The first term represents the objective function of the underlying TSP and the second term represents the time-dependent part. If we denote the first term by L_0, then each configuration in the TDTSP is uniquely determined by the pair (L_0, α).

If we allow the jam factor to change continuously, there are several boundary values for f where the best configuration for the TDTSP changes. In Table 6, we give six boundary intervals for Bier127 with the new assumption. In Figures 5 to 8, we show the best configuration for four different boundary intervals. The bold edges are traveled after noon in the traffic jam region. We see that as the value of the jam factor increases (in moving from Figure 5 to Figure 8) the number of bold edges decreases, that is, the salesman travels fewer edges after noon in the traffic jam region since these edges incur a high penalty. Stated differently, as the value of f increases, L_0 increases slightly and α decreases rapidly.

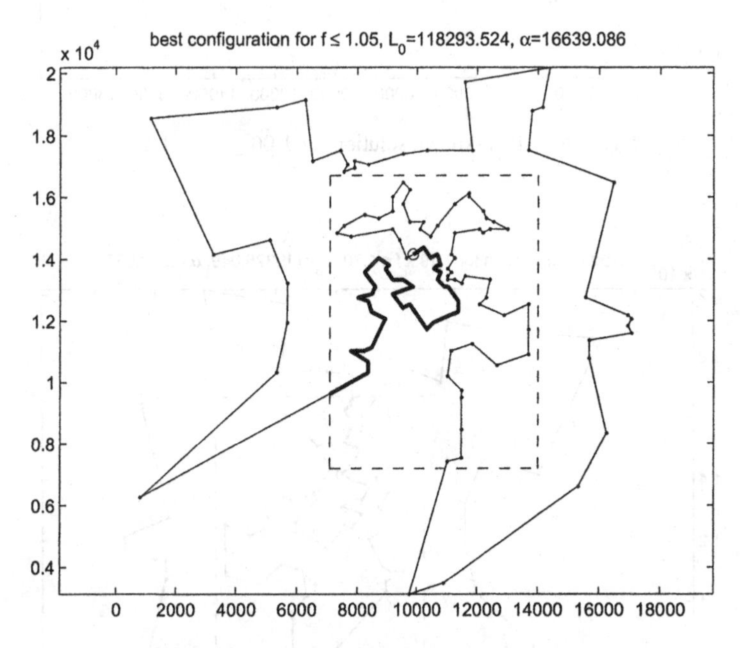

Figure 5. Best-known solution for $f \leq 1.05$.

4. Time Dependent Vehicle Routing Problem

In the traditional vehicle routing problem, we need to generate a sequence of deliveries for fixed-capacity vehicles in a homogeneous fleet based at a single depot so that all customers are serviced and the total distance traveled by the fleet is minimized. In the time dependent vehicle routing problem (TDVRP),

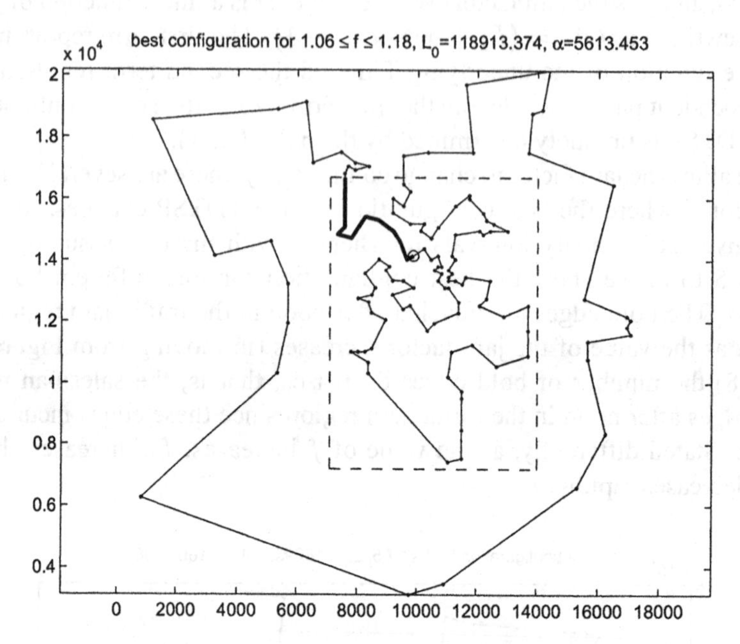

Figure 6. Best-known solution for $1.06 \leq f \leq 1.18$.

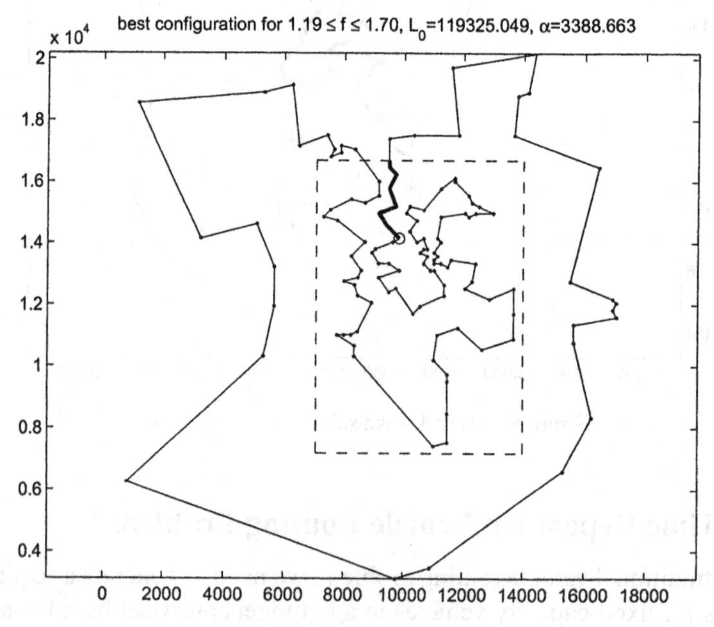

Figure 7. Best-known solution for $1.19 \leq f \leq 1.70$.

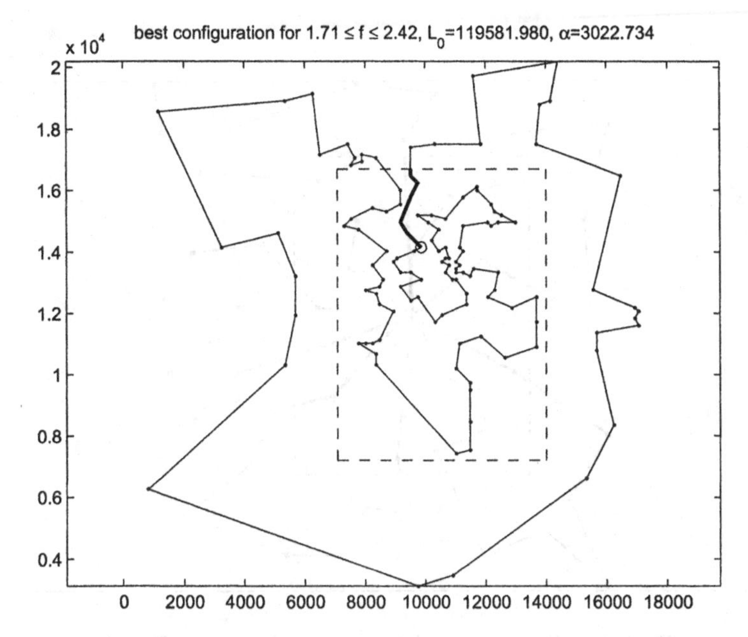

Figure 8. Best-known solution for $1.71 \leq f \leq 2.42$.

we define a traffic jam region, so that, at a specific time, the center of the city becomes congested and travel time between customers in the region takes longer.

To illustrate the TDVRP, we use the 50-node benchmark vehicle routing problem of Christofides et al. (1979). The traffic jam region is a rectangle with lower left-corner coordinates (15, 20), a width of 30, and a height of 40. A truck starts delivery at 8 am and finishes at 5 pm. The traffic jam starts at 12 pm. The travel speed is computed by dividing the distance of the longest route in the optimal solution to the VRP by the number of hours in the work day (nine). The travel speed is held constant for all values of the jam factor. It is not necessary that a route fills the work day exactly.

We applied our record-to-record travel algorithm (Li et al., 2004) to the 50-node problem with the new assumption. The results are illustrated in Figures 9 to 12.

As f increases in value, we see that fewer customers are serviced in the traffic jam region after noon (the bold edges are traveled after noon) and the value of L_0 increases. The average running time for the four different jam factors is about 2.4 minutes on an Athlon 1 GHz computer.

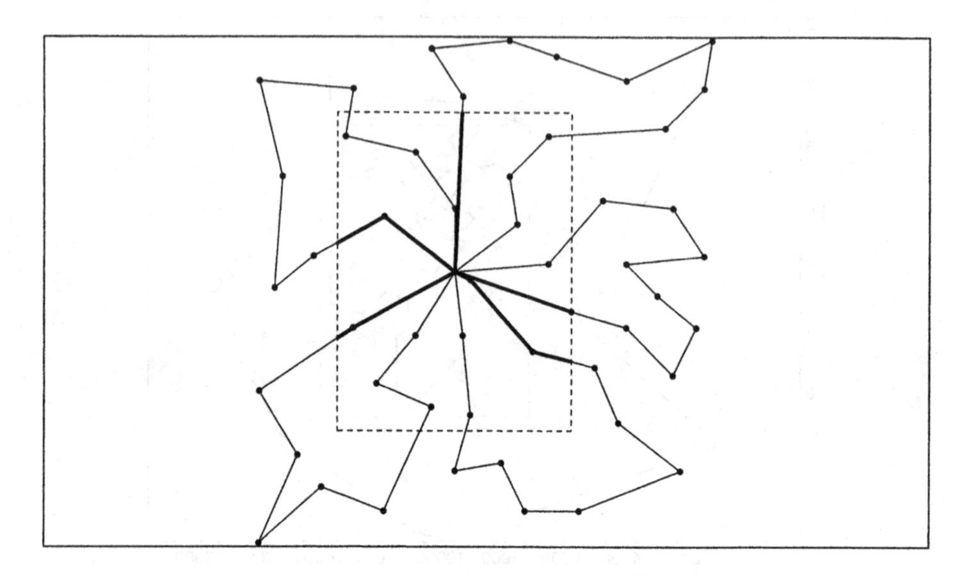

Figure 9. Best-known solution for $f \leq 1.02$, $L_0 = 524.61$.

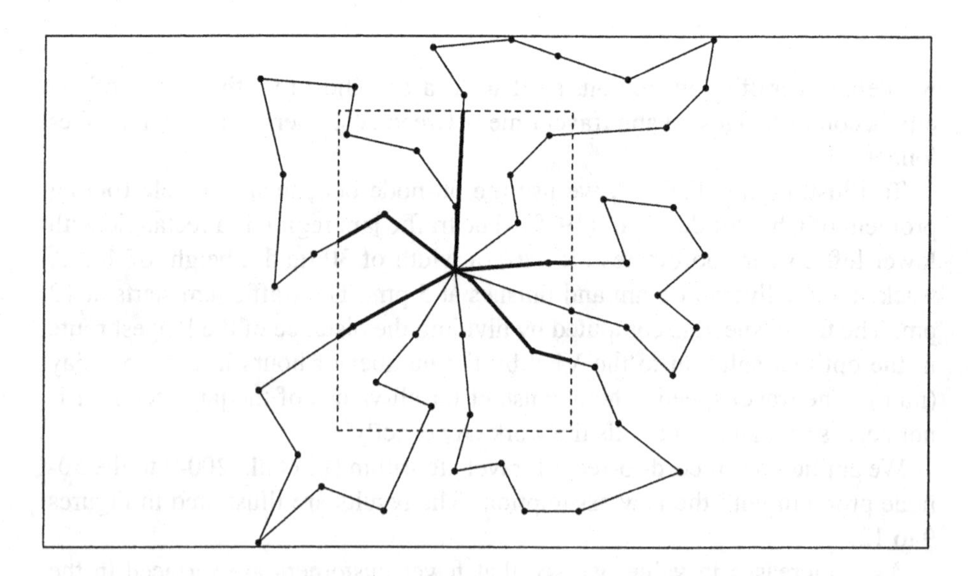

Figure 10. Best-known solution for $1.03 \leq f \leq 1.77$, $L_0 = 524.63$.

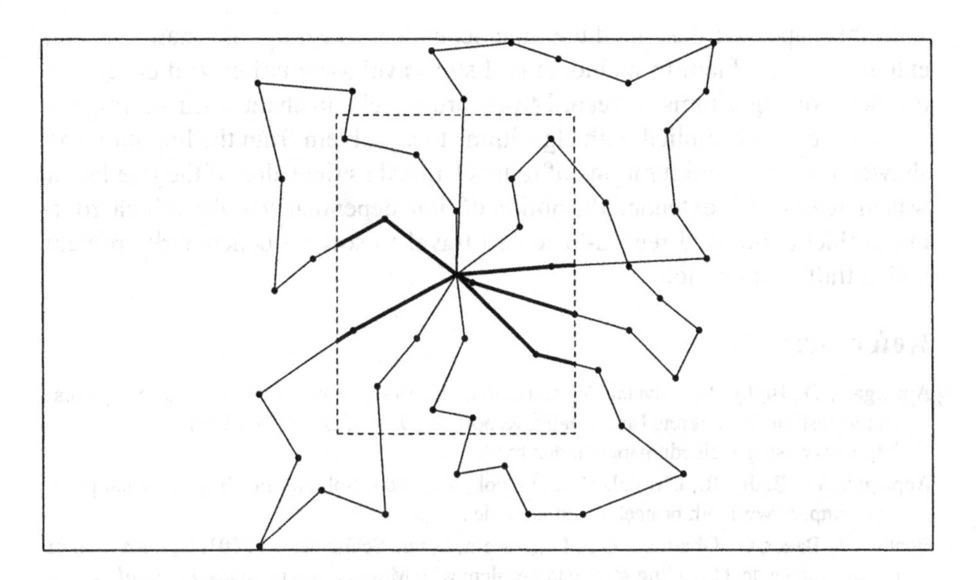

Figure 11. Best-known solution for $1.78 \leq f \leq 2.27, L_0 = 527.98$.

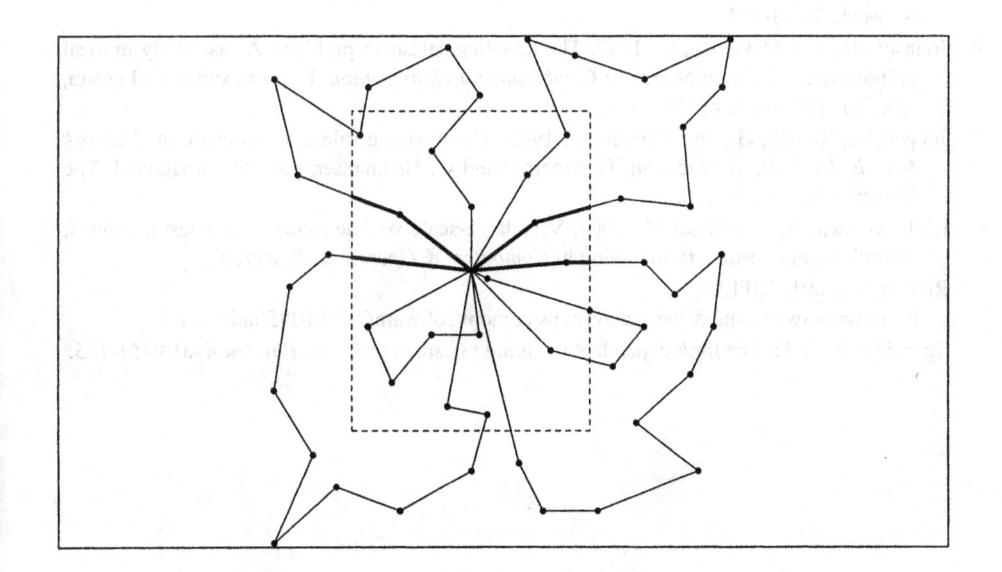

Figure 12. Best-known solution for $2.28 \leq f \leq 3.75, L_0 = 553.88$.

5. Conclusions

In this paper, we described the time dependent traveling salesman problem, extended the problem to include a realistic travel assumption, and developed two solution algorithms — record-to-record travel and chained Lin-Kernighan — to solve it. We applied both algorithms to a problem from the literature and showed how the configurations of tours changed as the value of the jam factor was increased. We extended the notion of time dependency to the vehicle routing problem and used record-to-record travel to solve a benchmark problem with a traffic jam region.

References

Applegate, D., Bixby, R., Chvatal, V., and Cook, W., 1999, Finding tours in the TSP, Forschungsinstitut fur Diskrete Mathematik, Report No. 99885, Universitat Bonn; http://www.tsp.gatech.edu/papers/index.html.

Applegate, D., Bixby, R., Chvatal, V., and Cook, W., 2004, Solving traveling salesman problems; http://www.math.princeton.edu/tsp/index.html.

Bentner, J., Bauer, G., Obermair, G., Morgenstern, I., and Schneider, J., 2001, Optimization of the time-dependent traveling salesman problem with Monte Carlo methods, *Physical Review E*.**64**:036701.

Christofides, N., Mingozzi, A., and Toth, P., 1979, The vehicle routing problem, in: *Combinatorial Optimization*, N. Christofides, A. Mingozzi, P. Toth, and C. Sandi, eds., John Wiley & Sons, Chichester, UK, pp. 315-338.

Golden, B., Magnanti, T., and Nguyen, H., 1977, Implementing vehicle routing algorithms, *Networks* 7:113-148.

Johnson, D., and McGeoch, L., 1997, The traveling salesman problem: A case study in local optimization, in: *Local Search in Combinatorial Optimization*, E. Aarts and J. K. Lenstra, eds., Wiley, London, pp. 215-310.

Junger, M., Reinelt, G., and Rinaldi, G., 1995, The traveling salesman problem, in: *Network Models*, M. Ball, T. Magnanti, C. Monma, and G. Nemhauser, eds., North-Holland, The Netherlands, pp. 225-330.

Li, F., Golden, B., and Wasil, E., 2004, Very large-scale vehicle routing: New test problems, algorithms, and results, forthcoming in *Computers & Operations Research.*

Reinelt, G., 2001, TSPLIB; http://www.iwr.uni-heidelberg.de/groups/comopt/software/TSPLIB95/index.html.

Schneider, J., 2002, The time-dependent traveling salesman problem, *Physica A*. **314**:151-155.

THE MAXIMAL MULTIPLE-REPRESENTATION SPECIES PROBLEM SOLVED USING HEURISTIC CONCENTRATION

Michelle M. Mizumori, Charles S. ReVelle and Justin C. Williams
Johns Hopkins University, 3400 N. Charles St., Baltimore, MD 21210

Abstract: The Maximal Multiple-Representation Species Problem (MMRSP) is formulated here and examined using two different integer program formulations solved by LP plus branch and bound as well as the metaheuristic known as Heuristic Concentration (HC). It is seen that for some instances of this problem, the exact method can be allowed to run for a long time (> 1 day) without termination while HC can find optimal or near-optimal solutions to the same instances in a few seconds. In one such case, LP-IP was allowed to run for 1,000 times longer than HC and still found a worse solution. Furthermore, HC found the optimal solution in 72.3% of cases and had an objective value gap of less than 1% in 94% of cases. Even when HC takes longer than LP-IP,the longest run-time was under 20 minutes. Therefore, HC is a valuable tool for approaching the MMRSP.

Key words: Maximal Multiple-Representation Species Problem, heuristic concentration

1. INTRODUCTION

The original Maximal Covering Species Problem (MCSP) examines nature reserve selection by maximizing the number of species that occur at least once in a set of *p* parcels of land given *n* potential parcels. Practically, this means that if those *p* parcels are chosen as a nature reserve, the greatest number of species will be preserved. This problem was first identified by Underhill (1994), but was not mathematically formulated until Camm et al. (1996) and Church et al. (1996).

Church et al. (1996) recognized this problem as a counterpart to the classic Maximal Covering Location Problem (Church and ReVelle, 1974), and solved it as such. Csuti et al. (1997) also examined this formulation of the MCSP with various heuristic methods as well as LP plus branch and bound. Both found that MCSP was very integer friendly and solved quickly.

The Maximal Multiple-Representation Species Problem (MMRSP) is a variation of the classic MCSP that can require species to be represented multiple times. It is understood by conservation biologists that a single representation of a species in a reserve system does not ensure long-term survival. Multiple representations increase the probability of long-term survival. Thus the MMRSP would require *b* representations for a species to be considered covered. However, some species occur less than *b* times in the entire study area. It is therefore impossible for those species to be represented *b* times in the selected reserve. Such species are considered to be covered when each and every parcel that they occur in is chosen. We define b_i to be the actual goal representation for species *i*, and a_i to be the number of times that species *i* occurs in the original study area. Therefore, if a_i is less than *b*, then we would set the goal representation for species *i*, b_i, to a_i. For the remaining species, the goal representation, b_i, remains equal to *b*.

ReVelle et al. (2002) discussed how many reserve selection models have counterparts in location modeling. The location counterpart to the MMRSP is the Maximum Availability Location Problem (MALP). The MALP was first presented by ReVelle and Hogan (1989) and later examined by Marianov and ReVelle (1996). In this location version, the population covered by at least b_i servers within a time or distance standard is maximized when *p* total servers are selected, in order to maximize the probability of availability of a server. ReVelle and Hogan (1989) and Marianov and ReVelle (1996) both used linear programming plus branch and bound as necessary to solve this problem.

2. FORMULATING THE MMRSP

Two different IP formulations of the MMRSP are explored in this paper. We will call the first formulation the integer-coefficient (IC) model. This model uses the following notation:

i,I = the index and set of species;

j,J = the index and set of available parcels of land;

N_i = the set of parcels, *j*, that contain species *i*;

p = the number of parcels to be selected;

x_j = 1,0; it is 1 if a parcel *j* is selected for a reserve, 0 otherwise; and

y_i = 1,0; it is 1 if species *i* is represented b_i times in the reserve, 0 otherwise.

The IC model can thus be defined mathematically in non-standard form as:

Maximize $Z = \sum_{i \in I} y_i$

Subject to

$$b_i y_i \le \sum_{j \in N_i} x_j \qquad \forall i \in I$$

$$\sum_{j \in J} x_j = p$$

$$y_i, x_j \in \{0,1\} \qquad \forall i \in I, \forall j \in J$$

Marianov and ReVelle (1996) found that for the MALP numerical experience showed that this type of formulation tended to lead to many fractional solutions and thus extensive branching and bounding. They therefore formulated a 'counting variables' version, which has more variables and constraints, but was expected to be more integer-friendly. A similar counting variables (CV) formulation can be used for the MMRSP. The counting variable formulation of the MMRSP would use a new variable:

u_{ik} = 1,0; it is 1 if species i is represented at least k times in the reserve network, 0 otherwise.

The mathematical formulation would then be:

Maximize $Z = \sum_{i \in I} u_{ib_i}$

Subject to:

$$\sum_{k=1}^{b_i} u_{ik} \le \sum_{j \in N_i} x_j \qquad \forall i \in I$$

$$u_{ik} \le u_{i(k-1)} \qquad \forall i \in I, k = 2,3,\ldots b_i$$

$$\sum_{j \in J} x_j = p$$

$$u_{ib_i}, x_j \in \{0,1\} \qquad \forall i \in I, \forall j \in J$$

Marianov and ReVelle (1996) also noticed that only the x_j and the u_{ib_i} need to be defined as 0,1 variables. When u_{ib_i} is equal to 1, all of the other u_{ik} for the same species will also be forced to 1 by the second set of constraints. When u_{ib_i} is equal to 0, the values of the other u_{ik} for the same species are unimportant because they have no impact on the objective value or the values of any other variables.

3. REVIEW OF HEURISTIC CONCENTRATION

Heuristic Concentration (HC) is a two or three stage metaheuristic originally designed to approach large instances of the p-median problem (Rosing and ReVelle, 1997). The basic structure of HC is illustrated in

Figure 1. The first stage consists of q runs of a fairly simple base heuristic with some random element. This randomness allows the multiple runs to find several different good solutions, whereas if the algorithm contained no random element, each run would arrive at the same solution. The multiple good solutions allow for the creation of Concentration Sets, which lie at the heart of HC. These concentration sets are then used to allow a more complex algorithm to be used in stage 2.

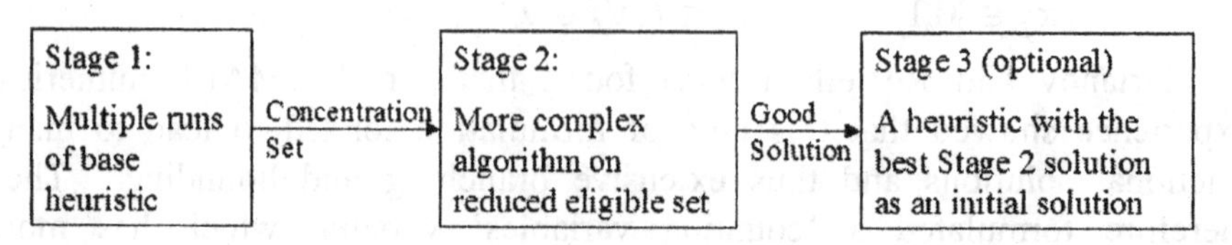

Figure 1. HC Schematic

The main concentration set, CS, is the union of the best m unique solutions from all of the runs of the base heuristic. In other words, it is every item that was ever selected in any of the top m solutions. It is thought that the members of CS possess some good quality to have been selected in the top solutions, and therefore the optimal solution is thought likely to be a subset of CS. Thus, the CS is used as the reduced eligible set of items to be chosen from in the second stage.

The other concentration set, CS_0, is the intersection of the best m unique solutions from all of the runs of the base heuristic. This differs from the CS because it only includes an item if that item was selected in *every single one* of the best solutions. It is thought that if an item was selected in all of these good solutions, it will also be selected in the optimal solution. Therefore, the members of CS_0 are required to be selected in every solution considered in the second stage, thus further reducing the feasible solution set.

The second stage consists of s runs of a more complex algorithm, using the concentration sets to reduce the feasible region as described above. Notice that if this algorithm has no randomness, for example an exact method, the number of repetitions, s, should equal one. However, if there is a random element to the algorithm, then the number of repetitions can be greater than one.

The optional third stage takes the best solution from the second stage and tries to improve upon it by performing another heuristic on the original eligible set. This way, if the second stage algorithm did not find the optimal solution, the third stage allows for further improvement.

For examples and analyses of different versions of HC, see Rosing and ReVelle (1997), Rosing et al. (1998), Rosing et al. (1999), Rosing (2000) and Rosing and Hodgson (2002).

4. APPLYING HC TO THE MMRSP

In this study, we used the version of HC with 1-opt as the first stage, 2-opt as the second stage and 1-opt again as the third stage (Rosing et al., 1999). The 1-opt algorithm begins with a random initial solution of any p selected items. A selected item is then exchanged for each unselected item and, if there was an improvement, the best exchange is kept. The next selected item is then exchanged in the same manner. These exchanges are made iteratively until no one-for-one exchange can be made that improves the objective value. The solution at this point is a local optimum, and the current run of 1-opt is complete. (Teitz and Bart, 1968). Each run of the 1-opt algorithm then uses a new random start.

The second stage of this version of HC consists of a 2-opt algorithm. This algorithm is similar to the 1-opt, except that two-for-two exchanges are used instead of one-for-one exchanges. In this version, the runs of 2-opt began with each of the top s solutions from the first stage. If we used the entire eligible set, the 2-opt algorithm would have had excessive run times, but by using the concentration sets created from the first stage, the amount of computation necessary is reduced to a reasonable level. For the third stage, 1-opt was run again, this time using the best second stage solution as the initial solution and the original study area as the eligible set.

In the first stage, we used q = 30 random starts of 1-opt. We chose to vary the number of unique solutions that enter the concentration sets, m, in order to determine a good value for that parameter. Values of 5, 10 and 15 were used. Finally, we used the s = 5 best stage one solutions as the initial solutions for the 2-opt algorithm in the second stage.

Thus far, the HC literature has only examined the p-median problem for which it was originally designed. This work is the first step in showing that HC can be used to address many combinatorial problems. The versions using 1-opt and 2-opt have the further requirement that the objective function can be defined to be immediately evaluable with the sole constraint that p items are to be selected. Notice that this definition of the objective function need not be linear. The HC formulation of the MMRSP can be written:

Maximize

$$Z = \sum_{i \in I} \min \left\{ 1, \left\lfloor \frac{\sum_{j \in N_i} x_j}{b_i} \right\rfloor \right\}$$

Subject to:

$$\sum_{j \in J} x_j = p$$

where all of the x_j are 0 or 1, and $\lfloor \; \rfloor$ means 'floor', or the largest integer less than the value inside. In the objective function, if, for a given species,

$\sum_{j \in N_i} x_j$ is less than b_i, then the fraction will be less than one, the floor function will yield a zero, the minimum in the brackets will be zero, and that species will not be counted in the summation. On the other hand, if $\sum_{j \in N_i} x_j$ is greater than or equal to b_i, then the floor function will be greater than or equal to one, the minimum in the brackets will be one, and the species will be counted exactly once in the summations. Therefore, the objective function becomes the number of species that occur at least b_i times.

5. DATA AND COMPUTER SPECIFICATIONS

Data on the distribution of 426 terrestrial vertebrate species in the state of Oregon have been developed as part of a cooperative national biodiversity mapping effort known as the Biodiversity Research Consortium (Master et al., 1995). In this data set, 441 regular hexagons with areas of 635 km^2 were defined that completely or partly overlap the political borders of Oregon (White et al., 1992). The likelihood of occurrence of each species in each hexagon was rated in one of the following categories: 1) confident – a sighting of the species has been verified in the past two decades; 2) probable – the parcel contains suitable habitat for the species, there have been verified sightings of the species in nearby sites, and a local expert believes that it is highly probable that the species occurs in the site; 3) possible – there have been no verified sightings of the species at the site, the habitat is of questionable suitability for the species, and a local expert believes it is possible that the species occurs at the site; and 4) not present – the habitat is unsuitable for the species. For the purposes of this study, a species was assumed to be present at the site if it had a ranking of confident or probable, and was assumed to be absent at the site if it had a ranking of possible or not present. This is consistent with the MCSP work done by Csuti et al. (1997).

HC was coded in Visual Fortran version 6 on a 2.66 GHz, Pentium 4 Dell Inspiron with 256 MB of memory. The LP models were solved using XpressMP release 2003f on the same computer.

6. RESULTS

The first comparison to examine is that of the integer-coefficient (IC) model and the counting-variables (CV) model. The run-times and number of branch and bound nodes for b=2,3 can be found in Tables 1 and 2. Notice that b=1 is the classic MCSP, and there is only one version so no comparison is necessary. Some runs did not terminate because either the computer ran out of memory or the model ran for more than a day, and these runs are so noted. It is clear that for most of the runs that did terminate, the integer-coefficient model solved in less time and with fewer branch and bound nodes than the counting variables model. In the cases where the counting variable model terminated more quickly than the integer-coefficient model, the difference in time is generally negligible.

Table 1. Comparison of Exact Models for b=2

p	CV time (s)	CV B&B nodes	IC time (s)	IC B&B nodes
1	32.2	1	4.9	165
2	27.8	371	7.7	51
3	Out of Memory	-	Out of Memory	-
4	69.6	1508	18.8	186
5	Out of Memory	-	108639.6	1226041
6	24.3	357	6.6	38
7	426.7	10529	142	4689
8	13.8	113	5.2	25
9	49.5	1371	26.3	422
10	9.1	62	5.5	27
11	15.6	221	13.9	200
12	12.1	135	16	216
13	37.5	1045	9.5	52
14	42.9	1155	33.6	764
15	51.8	1723	22.6	406
16	26.7	658	17.4	277
17	21.7	800	7.4	80
18	16.2	236	4.4	30
19	6.9	86	5.7	67
20	6.8	116	1.6	11
21	3.6	37	4.3	56
22	3.4	33	2	21
23	2.7	27	3.7	74
24	2.5	44	1.5	15
25	4.7	95	2.3	42
26	2.2	35	2.1	58
27	2.2	25	1.1	25
28	1.9	28	1	31
29	2.4	38	1.5	15
30	1.8	28	0.8	16

Table 2. Comparison of Exact Models for b=3

p	CV time (s)	CV B&B nodes	IC time (s)	IC B&B nodes
1	37.6	5	5.3	201
2	173.2	11131	117.8	9907
3	93.9	1577	41.1	393
4	Out of Time	-	Out of Memory	-
5	Out of Time	-	Out of Memory	-
6	799.7	18251	561.1	18376

p	CV time (s)	CV B&B nodes	IC time (s)	IC B&B nodes
7	Out of Time	-	Out of Memory	-
8	Out of Time	-	Out of Memory	-
9	112.2	1568	67.9	1013
10	131.2	1771	179.5	3263
11	2092.5	34072	1796.1	30716
12	259.4	4085	177.5	4725
13	524.2	8547	323.8	6013
14	856.4	15769	807.9	17341
15	1075.4	21164	2141.4	48739
16	979.4	22549	974.6	22380
17	689.2	12342	188.3	4078
18	239.2	3898	131.2	2390
19	129	1990	67.2	1084
20	49.5	479	46.6	621
21	53.8	807	31.2	423
22	79.7	1347	34.3	570
23	83.8	1318	16.6	101
24	80.9	1605	55.5	928
25	54.6	678	26.1	394
26	33.1	483	48.2	1316
27	37.5	548	30.1	514
28	25.8	260	6.7	33
29	21.7	140	12.6	121
30	15	100	13.6	137

In the HC runs of this problem, various values of *m*, the number of solutions to enter the concentration sets, were used in order to find a good value. The objective values and run times obtained are displayed in Tables 3, 4, and 5. The gray squares indicate that there was an improvement in objective value over the next lower value of *m*. In 14 out of the 83 instances examined, using a value of $m = 10$ yielded a better objective value than using $m = 5$, and yielded the same objective in the other cases. However, using a value of $m = 15$ yielded a better objective value than $m = 10$ in only 3 out of 83 cases. There were instances where using m=10 doubled the run time when compared to m=5 and instances where m=15 further doubled run time when compared to m=10. However, since m=10 yielded so many improvements over m=5 and using m=15 yielded so few improvements over m=10, we decided to use m=10 as a good compromise between run time and objective value.

Table 3. HC performance for various values of m at b=1

p	m=5		m=10		m=15	
	Time (s)	Objective	Time (s)	Objective	Time (s)	Objective
1	0.55	254	0.55	254	0.55	254
2	1.34	318	1.34	318	1.34	318
3	2.59	356	2.59	356	2.59	356
4	3.56	374	3.56	374	3.56	374

p	m=5		m=10		m=15	
	Time (s)	Objective	Time (s)	Objective	Time (s)	Objective
5	4.44	384	4.49	384	4.49	384
6	5.64	390	5.79	390	6	390
7	6.45	395	6.61	395	6.73	395
8	7.65	399	7.98	399	8.47	400
9	8.8	403	9.15	403	9.75	403
10	10.66	406	11.48	406	11.85	406
11	11.4	407	13.64	408	14.4	408
12	12.14	410	12.8	410	13.55	410
13	14.29	412	15.38	412	15.82	412
14	15.48	414	17.28	414	18.39	414
15	15.79	415	20.53	416	21.36	416
16	16.01	417	17.77	417	20.96	417
17	17.66	419	21.02	419	26.77	419
18	23.93	420	39.32	420	40.49	420
19	25.32	421	37.53	421	66.5	421
20	27.49	422	37.86	422	63.88	422
21	25.34	423	43.91	423	64.23	423
22	27.11	423	61.61	424	93.88	424
23	29.62	425	42.85	425	76.79	425

Table 4. HC performance for various values of m at b=2

p	m=5		m=10		m=15	
	Time (s)	Objective	Time (s)	Objective	Time (s)	Objective
1	0.29	1	0.29	1	0.29	1
2	2.08	240	2.1	240	2.19	240
3	3.26	252	3.39	252	3.39	252
4	5.76	308	6.09	308	6.44	308
5	7.7	317	8.1	317	8.57	317
6	10.16	347	10.37	347	10.91	347
7	14.01	353	14.59	353	15.14	353
8	17.7	367	17.94	367	20.91	367
9	18.1	371	18.6	371	20.22	371
10	21.14	378	22.17	378	22.46	378
11	22.97	382	24.7	382	28.32	382
12	25.93	386	29.6	386	32.23	386
13	29.62	388	32.85	389	40.35	389
14	29.15	391	33.4	391	37.42	391
15	32.81	393	38.52	393	49.41	393
16	37.23	396	55.47	397	79.22	397
17	52.6	399	77.76	399	103.1	399
18	60.84	401	80.87	401	102.9	401
19	84.26	403	135	403	193	403
20	80.36	403	140.7	404	246.4	404
21	109.2	407	170.9	407	284.1	407
22	106.8	408	190.8	408	305.8	408
23	77.2	409	157.2	409	242.7	409
24	98.14	410	246.5	410	428.3	410
25	81.27	411	224.3	411	332.6	411

p	m=5		m=10		m=15	
	Time (s)	Objective	Time (s)	Objective	Time (s)	Objective
26	98.42	412	185.4	412	385.9	412
27	143.7	413	376.5	413	787.5	413
28	118.5	414	306.3	414	601.7	414
29	173	414	492.4	414	882.2	414
30	160.1	416	400.7	416	745.6	416

Table 5. HC performance for various values of m at b=3

p	m=5		m=10		m=15	
	Time (s)	Objective	Time (s)	Objective	Time (s)	Objective
1	0.38	1	0.38	1	0.38	1
2	1.02	2	1.02	2	1.02	2
3	4.3	220	4.5	220	4.91	220
4	5.9	245	6.26	245	7.29	245
5	8.98	253	9.37	253	9.78	253
6	13.23	294	16.23	294	18.45	294
7	15.8	305	18.19	305	27.26	308
8	17.69	317	23.21	317	26.72	317
9	28.28	340	30.19	340	35.81	340
10	29.06	340	31.28	340	34.81	340
11	33.5	346	41.71	349	46.91	349
12	43.52	357	56.08	359	63.66	359
13	49.2	362	61.31	362	73.9	362
14	48.24	365	69.59	366	91.18	366
15	60.78	367	70.35	367	78.41	367
16	66.75	374	87.21	374	107.4	374
17	66.34	374	106.5	375	146.2	375
18	93.62	378	136.2	378	178.7	378
19	116.8	380	198.9	381	338.8	381
20	94.31	388	190.7	388	307.6	388
21	97.55	387	210.2	388	335.2	388
22	195.9	391	357.2	392	726.6	392
23	131.3	393	286.8	393	466.5	393
24	207.7	395	478.7	395	892.8	395
25	393.8	396	930.6	396	1665	396
26	186.5	398	472.6	398	841	398
27	171.9	399	524.2	400	1300	400
28	221.5	400	725.7	400	1394	401
29	344.9	403	1008	403	1755	403
30	302.6	404	595.2	404	1255	404

The effect of each stage of HC can also be examined. Table 6 compares the stages of HC in terms of average run time and number of improvements over the previous stage using m=10. As p and b increase, the second and third stages make more improvements and run times increase.

Table 6. Comparison of stages of HC.

b	Stage 1	Stage 2		Stage 3	
	Av. Time	Av. Time	# Improvements	Av. Time	# Improvements
1	11.81	7.026	5	0.078	0
2	38.1	78.24	16	0.442	0
3	55.84	168.5	21	0.611	1

Since it appears that the integer-coefficient model is the better exact model and that using m = 10 is the better HC model, it is now prudent to compare those two methods. Tables 7, 8, and 9 portray run times and objective values obtained by these two solution methods. In the cases where the exact model was cut-off before termination, the value in the Objective column is the objective value of the best integer solution found at the time of cut-off.

Table 7. Comparison of LP-IP and HC for b=1

p	Integer Coefficient LP-IP		HC with m=10		
	Time (s)	Objective	Time (s)	Objective	Gap (%)
1	1.5	254	0.55	254	0
2	1.6	318	1.34	318	0
3	1.3	356	2.59	356	0
4	0.9	374	3.56	374	0
5	0.8	384	4.49	384	0
6	1	390	5.79	390	0
7	2.6	395	6.61	395	0
8	0.9	400	7.98	399	0.25
9	3	403	9.15	403	0
10	2.4	406	11.48	406	0
11	2.3	408	13.64	408	0
12	1.5	410	12.8	410	0
13	1.4	412	15.38	412	0
14	1.1	414	17.28	414	0
15	1.3	416	20.53	416	0
16	0.9	418	17.77	417	0.24
17	1.3	419	21.02	419	0
18	2.2	420	39.32	420	0
19	0.7	422	37.53	421	0.24
20	0.7	423	37.86	422	0.24
21	0.7	424	43.91	423	0.24
22	0.8	425	61.61	424	0.24
23	0.6	426	42.85	425	0.23

In 53 out of 83 cases, HC found the best known solution. In 78 out of 83 cases, the gap (defined as (100*(LP-IP Objective Value – HC Objective Value)/LP-IP Objective Value)) is less than 1%. The biggest overall gap is 5.98%. Furthermore, in one case where LP-IP was cut-off, HC found a *better* solution than LP-IP.

It has been argued that LP-IP plus branch and bound should always be used because even when it does not terminate, it still finds a good integer solution (Onal, 2003; Rodrigues and Gaston, 2002). In order to dispute this argument and show that HC is a good option, we also recorded the best integer solution objective value that LP plus branch and bound found in the time that HC terminated. These objective values are compared with the best overall LP-IP solution known and the HC solution for the cases where HC took less time to terminate than LP-IP in Tables 8 and 9. In 19 cases, HC found a solution with a better objective value that LP-IP in the same amount of time.

Table 8. Comparison of LP-IP and HC for b=2

p	Integer-Coefficient LP-IP		HC with m = 10			LP-IP in
	Time (s)	Objective	Time (s)	Objective	Gap	HC time
1	0.7	1	0.29	1	0	no int sol
2	7.7	241	2.1	240	0.42	no int sol
3	59220.7	252[a]	3.39	252	0	no int sol
4	18.8	308	6.09	308	0	no int sol
5	108629.6	317	8.1	317	0	no int sol
6	6.6	347	10.37	347	0	-
7	142	353	14.59	353	0	326
8	5.2	367	17.94	367	0	-
9	26.3	371	18.6	371	0	369
10	5.5	378	22.17	378	0	-
11	13.9	382	24.7	382	0	-
12	16	386	29.6	386	0	-
13	9.5	389	32.85	389	0	-
14	33.6	392	33.4	391	0.26	-
15	22.6	394	38.52	393	0.25	-
16	17.4	397	55.47	397	0	-
17	7.4	399	77.76	399	0	-
18	4.4	402	80.87	401	0.25	-
19	5.7	403	135	403	0	-
20	1.6	405	140.7	404	0.25	-
21	4.3	407	170.9	407	0	-
22	2	408	190.8	408	0	-
23	3.7	409	157.2	409	0	-
24	1.5	410	246.5	410	0	-
25	2.3	411	224.3	411	0	-
26	2.1	412	185.4	412	0	-
27	1.1	413	376.5	413	0	-
28	1	414	306.3	414	0	-
29	1.5	415	492.4	414	0.24	-
30	0.8	416	400.7	416	0	-

[a] The computer ran out of memory. The bound on the objective at cut-off was 257.969.

Table 9. Comparison of LP-IP and HC for b=3

p	Integer-Coefficient LP-IP		HC with $m = 10$			LP-IP in
	Time (s)	Objective	Time (s)	Objective	Gap	HC time
1	5.3	1	0.38	1	0	no int sol
2	117.8	2	1.02	2	0	no int sol
3	41.1	234	4.5	220	5.98	no int sol
4	73838.6	245[a]	6.26	245	0	no int sol
5	14625.1	252[b]	9.37	253	-0.40	no int sol
6	561.1	300	16.23	294	2.00	298
7	83086.2	309[c]	18.19	305	1.29	288
8	58957.9	317[d]	23.21	317	0	296
9	67.9	340	30.19	340	0	338
10	179.5	347	31.28	340	2.02	342
11	1796.1	350	41.71	349	0.29	345
12	177.5	359	56.08	359	0	357
13	323.8	363	61.31	362	0.28	361
14	807.9	367	69.59	366	0.27	366
15	2141.4	370	70.35	367	0.27	368
16	974.6	374	87.21	374	0	371
17	188.3	378	106.5	375	0.79	377
18	131.2	381	136.2	378	0.79	-
19	67.2	385	198.9	381	1.04	-
20	46.6	388	190.7	388	0	-
21	31.2	390	210.2	388	0.51	-
22	34.3	392	357.2	392	0	-
23	16.6	394	286.8	393	0.25	-
24	55.5	395	478.7	395	0	-
25	26.1	397	930.6	396	0.25	-
26	48.2	399	472.6	398	0.25	-
27	30.1	400	524.2	400	0	-
28	6.7	402	725.7	400	0.50	-
29	12.6	403	1008	403	0	-
30	13.6	405	595.2	404	0.25	-

[a] The computer ran out of memory. The bound on the objective at cut-off was 249.827.

[b] The computer ran out of memory. The bound on the objective at cut-off was 263.765.

[c] The computer ran out of memory. The bound on the objective at cut-off was 312.071.

[d] The computer ran out of memory. The bound on the objective at cut-off was 325.012.

7. DISCUSSION

It is interesting to find that the integer-coefficient IP model seems to perform more quickly and with fewer branch and bound nodes than the counting variables version, because the MALP literature tends to use the counting variable version. Marianov and ReVelle (1996) said that computational experience led them to believe that the counting variable version would be faster, but it would be interesting to perform an extensive comparison on MALP, to see whether this holds true. It is also notable that while the classic MCSP is highly integer friendly, both versions of MMRSP can require a great deal of branching and bounding in some instances.

In the HC runs, as *b*, *p*, and *m* grew, so did the run time. The effects of raising *p* and *m* can be expected because of the combinatorial nature of these parameters. The number of feasible solutions to the problem is ${}_nC_p$, where *n* is the number of parcels, and the notation ${}_aC_b$ is the 'choose' function and equals a!/(b!)(a-b)!. Therefore, the number of combinations ${}_nC_p$ peaks at $p = n/2$ and declines to either side of n/2. Thus, for $p < n/2$, the number of combinations increases as *p* increases. This in turn increases the complexity of the algorithm, and thus the run time.

As *m* is increased, the CS gets larger and CS_0 gets smaller. This is thought to increase the chances that the optimal solution (or at least a better solution) can be found in the second stage, because there are more options to choose from. However, a larger CS and smaller CS_0 also increase the complexity of the algorithm by increasing the number of eligible solutions in the second stage. Notice the number of eligible solutions is ${}_{|CS|-|CS0|}C_{p-|CS0|}$. From the earlier definition of the choose function, it is clear that as ($|CS|-|CS_0|$) grows ($|CS|$ gets larger and/or $|CS_0|$ gets smaller), the numerator grows more quickly than the denominator, so there are more eligible solutions.

Within the HC runs, it seems that a very good value for this problem for the parameter, *m*, the number of solutions that enter the concentration set, is ten. The extra run time compared to *m=5* is generally within the same order of magnitude. This additional run time allows for an improvement in objective value in 16.87% of cases. Therefore, *m*=10 appears to provide a good compromise between objective value and run time. This is the same value that Rosing (2000) chose for the *p*-median problem. It will be interesting to see whether this is a pattern that holds true for other problems.

We compare this version of HC with both the version with no third stage and with a simple 1-opt by examining the improvements and run times of each stage. Notice from Table 6 that Stage 2 makes improvements in 42 out of 83 cases. Since the average run time of Stage 2 is under three minutes and the maximum total run time of HC is under 20 minutes, this high number of improvements seems to justify the additional run time. Stage 3 makes only one improvement, but the average run time is well under 1 second. This time is negligible when compared to the total run time, and therefore seems to validate the use of Stage 3 for the possible improvements.

In comparing HC to the exact method, it useful to compare the *b*=1 case (the classic MCSP) separately from the *b*=2 and *b*=3 cases. In the classic MCSP, the exact method always terminated very quickly, often faster than HC. In the few cases where HC did terminate faster than the LP, the difference in time was negligible. This is likely true because the classic

MCSP problem is very integer friendly, as Church et al. (1996) and Csuti et al. (1997) found. Therefore, for the b=1 case, it would be preferable to use the exact method because it guarantees optimality and solves in a reasonable amount of time.

For the b=2 and b=3 cases, LP-IP tends to have much more branching and bounding than for b=1. Particularly, the cases where b=2 and p=3,5, and where b=3, p=4,5,7,8 had extraordinarily long run times. In the case of b=3, p=5, LP-IP was allowed to run over 1000 times longer than HC, and HC still found a superior solution. This is possible because HC is compared to the best known integer solution obtained in the allowed time, not necessarily the optimal. In the cases of b=3, p=4 up to p=16, HC had substantially lower run times than LP-IP and had a maximum objective value gap of 2.02%. Although in several cases LP-IP is preferable to HC because it guarantees optimality and solves in less time, the above cases also show that there are times when the LP-IP runtime blows up and becomes unreasonable. It is in these cases that HC becomes a valuable tool. Furthermore, Tables 8 and 9 show that in many cases, LP-IP did not find any integer solutions in the time that it took HC to terminate. It is clear from all of the instances that HC finds solutions with near-optimal objective values, and is therefore a good choice for those cases where LP-IP cannot be used. Such instances are difficult to predict, but since HC always found good solutions and had a maximum run time under 20 minutes and in all but 3 cases solved in under 10 minutes, HC can be used in all cases with little lost in time or objective.

8. CONCLUSIONS

We presented the formulation of a new species representation problem, the Maximal Multiple-Representation Species Problem. It is a variation on the classical Maximal Covering Species Problem with the goal of promoting long-term survival by encouraging multiple representations in the reserve system. The mathematical differences also allow MCSP to be very integer friendly, while MMRSP can require substantial branching and bounding.

This problem was solved using both linear programming plus branch and bound and the heuristic known as Heuristic Concentration. In several instance of the MMRSP, the run time of LP plus branch and bound can become quite excessive, while HC can find near-optimal solutions in much less time. All of the cases in this paper indicate that HC finds very good, if not optimal, solutions. Therefore, when time or memory make the branch and bound algorithm infeasible, HC is a very valuable tool.

ACKNOWLEDGEMENTS

This research was supported by a grant from the David and Lucille Packard Foundation, Interdisciplinary Science Program. We gratefully acknowledge their support. Dash Optimization has provided an academic partnership including the use of XpressMP. The research reported here constitutes a portion of the doctoral dissertation of Michelle M. Mizumori.

REFERENCES

Camm, J.D., Polasky, S., Solow, A., and Csuti, B., 1996, A note on optimal algorithms for reserve site selection, *Biological Conservation* **78**:353-355.

Church, R.L., Stoms, D.M., and Davis, F.W., 1996, Reserve selection as a maximal covering location problem, *Biol. Conserv.* **76**:105-112.

Csuti, B., Polasky, S., Williams, P.H., Pressey, R.L., Camm, J.D., Kershaw, M., Kiester, A.R., Downs, B., Hamilton, R., Huso, M., and Sahr, K., 1997, A comparison of reserve selection algorithms using data on terrestrial vertebrates in Oregon, *Biol. Conserv.* **80**:83-97.

Marianov, V., and ReVelle, C., The Queueing Maximal Availability Location Problem: A model for the siting of emergency vehicles, 1996, *Eur. J. of Oper. Res.*, **93**:110-120.

Master, L., Clupper, N., Gaines, E., Bogert, E., Solomon, R., and Ormes, M., 1995, *Biodiversity Research Consortium Species Database Manual*, The Nature Conservancy, Boston.

ReVelle, C., and Hogan, K., Maximum Availability Location Problem, 1989, *Transport. Sci.*, **23**:192-200.

Rosing, K.E., 2000, Heuristic Concentration: a study of stage one, *Envir. and Plan.-B*, **27**:137-150.

Rosing, K.E. and Hodgson, M.J., 2002, Heuristic concentration for the p-median: an example demonstrating how and why it works, *Comp. and Oper. Res.*, **29**:1317-1330.

Rosing, K.E., and ReVelle, C.S., 1997, Heuristic Concentration: Two stage solution construction, *Eur. J. of Oper. Res.* **97**:75-86.

Rosing, K.E., ReVelle, C.S., Rolland, E., Schilling, D.A., and Current, J.R., 1998, Heuristic Concentration and Tabu search: A head to head comparison, *Eur. J. of Oper. Res.* **104**:93-99.

Rosing, K.E., ReVelle, C.S., and Schilling, D.A., 1999, A gamma heuristic for the p-median problem, *Eur. J. of Oper. Res.*, **117**:522-532.

Rosing, K.E., ReVelle, C.S., and Williams, J.C., 2002, Maximizing species representation under limited resources: A new and efficient heuristic, *Envir. Mod. and Assess.* **7**:91-98.

Teitz, M.B. and Bart, P., 1968, Heuristic methods for estimating the generalized vertex median of a weighted graph, *Oper. Res.* **16**:955-961.

Underhill, L., 1994, Optimal and suboptimal reserve selection algorithms, *Biol. Cons.* **35**:85-87.

White, D., Kimerling, A.J., and Overton, W.S., 1992, Cartographic and geometric components of a global sampling design for environmental monitoring, *Cart. and Geog. Info. Sys.*, **19**:5-22.

IV

STOCHASTIC MODELING

FAST AND EFFICIENT MODEL-BASED CLUSTERING WITH THE ASCENT-EM ALGORITHM

Wolfgang Jank*

Department of Decision and Information Technologies, The Robert H. Smith School of Business, University of Maryland

Abstract In this paper we propose an efficient and fast EM algorithm for model-based clustering of large databases. Drawing ideas from its stochastic descendant, the Monte Carlo EM algorithm, the method uses only a sub-sample of the entire database per iteration. Starting with smaller samples in the earlier iterations for computational efficiency, the algorithm increase the sample size intelligently towards the end of the algorithm to assure maximum accuracy of the results. The intelligent sample size updating rule is centered around EM's highly-appraised likelihood-ascent property and only increases the sample when no further improvements are possible based on the current sample. In several simulation studies we show the superiority of Ascent-EM over regular EM implementations. We apply the method to an example of clustering online auctions.

Keywords: Clustering; Monte Carlo EM algorithm; Mixture model.

1. Introduction

The EM (expectation-maximization) algorithm is a very popular tool in many areas of application, in particular for clustering problems. Part of EM's popularity stems from the fact that it can handle situations in which some of the data is unobserved. Clustering problems appeal to this situation by assuming that the data originates from a finite mixture of populations, but that the cluster membership of each data point is not observed.

While the EM algorithm is a very popular method for clustering problems, its practical usefulness is often limited by its computational efficiency. In fact, one of the common criticism of the EM algorithm is, compared to other optimization methods like Newton-Raphson, that it converges only at a linear rate. The convergence can be especially slow if the proportion of unobserved-

*This research was partially funded by the NSF grant DMI-0205489

to-observed information is large (Meng, 1994). In the context of clustering, another drawback of EM is that, in every iteration, it passes through all of the available data. Thus, if the size of the data is very large, even one single iteration of EM can become computationally intense.

One of the reasons for EM's computational intensity is that, in every new iteration, it requires re-evaluation of the conditional expectation of the data. This conditional expectation is calculated by evaluating all points of the database. Related research has shown, however, that an *exact* calculation of this conditional expectation based on all the data may not be necessary. In fact, since the EM algorithm typically takes larger steps in the earlier iterations (especially when the starting values are far from the final values), an approximation may suffice. The Monte Carlo EM algorithm (Wei and Tanner, 1990) is an example of an EM implementation which only uses an approximation to the conditional expectation. In the clustering context, this means that by using only a *sample* of the entire database one can still advance the progress of the algorithm while significantly reducing the computational burden.

While using only a sample of the entire data can lead to enormous efficiency gains, especially in the early iterations, attention has to be paid to the trade-off between computational efficiency and accuracy of the results. Indeed, by using only a small sample of the entire data, one runs the serious risk of obtaining very inaccurate results. In fact, the EM algorithm will not converge to the correct value unless the sample size is increased successively. However, determining the exact amount by which the sample size should be increased is a challenging task (Booth and Hobert, 1999; Levine and Casella, 2001; Levine and Fan, 2003; Caffo et al., 2004). In this work we propose the *Ascent-EM* algorithm that increases the sample size in an intelligent manner. Specifically, motivated by the fundamental likelihood-ascent property of the original EM algorithm, Ascent-EM increases the sample size only if the current sample does not contain any more additional information towards the progress of the algorithm.

This paper is organized as follows. In Section 2 we describe the background for model-based clustering and explain the EM algorithm in that context. Section 3 describes the Ascent-EM algorithm. In Section 4 we provide several simulation studies which show the computational superiority of Ascent-EM over regular EM implementations. Section 5 provides an example of clustering a large database of eBay's online auctions.

2. Model-Based Clustering and EM

We start out be describing the mixture model useful for clustering. We then continue by discussing how this model can be estimated with the help of the EM algorithm.

Finite Mixture Models for Model-Based Clustering

The mixture model assumes that the observed p-dimensional vectors $\mathbf{x} = (\mathbf{x}_1, \ldots, \mathbf{x}_n)$ arise from a mixture of a finite number of g groups in some unknown proportions $\pi_1, \ldots, \pi_g$. Specifically, we assume that the mixture density of the jth data point $\mathbf{x}_j$ $(j = 1, \ldots, n)$ can be written as

$$f(\mathbf{x}_j; \boldsymbol{\theta}) = \sum_{i=1}^{g} \pi_i f_i(\mathbf{x}_j; \boldsymbol{\psi}_i) \tag{1}$$

where the sum of the mixture proportions π_i $(i = 1, \ldots, g)$ is one and the group-conditional densities $f_i(\mathbf{x}_j; \boldsymbol{\psi}_i)$ depend on an unknown parameter vector $\boldsymbol{\psi}_i$. Quite often, the group-conditional densities are assumed to be normal (Ng and McLachlan, 2003). In that case, we can write

$$f_i(\mathbf{x}_j; \boldsymbol{\psi}_i) = \phi(\mathbf{x}_j; \boldsymbol{\mu}_i, \boldsymbol{\Sigma}_i), \tag{2}$$

where $\phi(\mathbf{x}_j; \boldsymbol{\mu}, \boldsymbol{\Sigma})$ denotes the p-dimensional multivariate normal distribution with mean $\boldsymbol{\mu}$ and covariance matrix $\boldsymbol{\Sigma}$. Let $\boldsymbol{\theta} = (\pi_1, \ldots, \pi_{g-1}, \boldsymbol{\psi}_1, \ldots, \boldsymbol{\psi}_g)$ the vector of unknown parameters. Then, the log-likelihood is given by

$$\log L(\boldsymbol{\theta}; \mathbf{x}) = \sum_{j=1}^{n} \log\{\sum_{i=1}^{g} \pi_i f_i(\mathbf{x}_j; \boldsymbol{\psi}_i)\}. \tag{3}$$

The goal of model-based clustering is to determine the parameter vector $\hat{\boldsymbol{\theta}}$ that maximizes the likelihood (3). We will refer to $\hat{\boldsymbol{\theta}}$ as the maximum likelihood estimate (MLE).

One can maximize the log-likelihood in (3) assuming that some of the information is unobserved. Specifically, assume that each of the $\mathbf{x}_j$'s arises from one of the g groups. Let $\mathbf{z}_1, \ldots, \mathbf{z}_n$ denote the corresponding g-dimensional group-indicator vectors. That is, let the ith element of $\mathbf{z}_j$ be one if and only if $\mathbf{x}_j$ originates from the ith group. Notice that the group-indicator vectors $\mathbf{z}_j$ are unobserved. Let us write $\mathbf{x} = (\mathbf{x}_1, \ldots, \mathbf{x}_n)$ for the observed (or incomplete) data and similarly $\mathbf{z} = (\mathbf{z}_1, \ldots, \mathbf{z}_n)$ for the unobserved (or missing) data. Then the complete data is $\mathbf{y} = (\mathbf{x}, \mathbf{z})$ and the complete data log-likelihood of $\boldsymbol{\theta}$ can be written as

$$\log L_c(\boldsymbol{\theta}; \mathbf{y}) = \sum_{i=1}^{g} \sum_{j=1}^{n} z_{ij}\{\log \pi_i + \log f_i(\mathbf{x}_j; \boldsymbol{\psi}_i)\}, \tag{4}$$

where z_{ij} denotes the ith component of $\mathbf{z}_j$.

The EM algorithm

The EM algorithm (Dempster et al., 1977) is an iterative procedure useful to estimate the MLE in incomplete data problems. In each iteration, the EM algorithm performs an expectation and a maximization step. Let $\boldsymbol{\theta}^{(t-1)}$ denote the current parameter value. Then, in the tth iteration of the algorithm, the E-step computes the conditional expectation of the complete data log-likelihood, conditional on the observed data and the current parameter value,

$$Q(\boldsymbol{\theta}|\boldsymbol{\theta}^{(t-1)}) = E\left[\log f(\mathbf{x}, \mathbf{z}; \boldsymbol{\theta})|\, \mathbf{x}; \boldsymbol{\theta}^{(t-1)}\right]. \tag{5}$$

For the mixture model in (1), this conditional expectation simplifies to

$$Q(\boldsymbol{\theta}|\boldsymbol{\theta}^{(t-1)}) = \sum_{i=1}^{g}\sum_{j=1}^{n} \tau_{ij}^{(t-1)}\{\log \pi_i + \log f_i(\mathbf{x}_j; \boldsymbol{\psi}_i)\}, \tag{6}$$

where $\tau_{ij}^{(t-1)} = E(z_{ij}|\mathbf{x}_j; \boldsymbol{\theta}^{(t-1)})$ is the posterior probability that $\mathbf{x}_j$ belongs to the ith component of the mixture. The tth EM update, $\boldsymbol{\theta}^{(t)}$, maximizes (6). That is $\boldsymbol{\theta}^{(t)}$ satisfies

$$Q(\boldsymbol{\theta}^{(t)}|\boldsymbol{\theta}^{(t-1)}) \geq Q(\boldsymbol{\theta}|\boldsymbol{\theta}^{(t-1)}) \tag{7}$$

for all $\boldsymbol{\theta}$ in the parameter space. This is also known as the M-step. Given an initial value $\boldsymbol{\theta}^{(0)}$, the EM algorithm produces a sequence $\{\boldsymbol{\theta}^{(0)}, \boldsymbol{\theta}^{(1)}, \boldsymbol{\theta}^{(2)}, \ldots\}$ that, under regularity conditions (see Boyles, 1983; Wu, 1983), converges to $\hat{\boldsymbol{\theta}}$, the (at least local) maximizer of $L(\boldsymbol{\theta}; \mathbf{x})$ in (3).

In case of the normal mixture model in (2), the E-step and M-step of EM are available in closed form. Specifically, in the E-step we calculate the conditional expectation of the z_{ij}'s via

$$\begin{aligned}\tau_{ij}^{(t-1)} &= E(z_{ij}|\mathbf{x}_j; \boldsymbol{\theta}^{(t-1)}) \\ &= \frac{\pi_i^{(t-1)}\phi(\mathbf{x}_j; \boldsymbol{\mu}_i^{(t-1)}, \boldsymbol{\Sigma}_i^{(t-1)})}{\sum_{l=1}^{g}\pi_l^{(t-1)}\phi(\mathbf{x}_j; \boldsymbol{\mu}_l^{(t-1)}, \boldsymbol{\Sigma}_l^{(t-1)})}\end{aligned} \tag{8}$$

for all $i = 1, \ldots, g$ and $j = 1, \ldots, n$. The normal case allows significant computational advantages by working with the corresponding sufficient statistics,

$$T_{i1}^{(t)} = \sum_{j=1}^{n} \tau_{ij}^{(t-1)} \tag{9}$$

$$\mathbf{T}_{i2}^{(t)} = \sum_{j=1}^{n} \tau_{ij}^{(t-1)}\mathbf{x}_j \tag{10}$$

$$\mathbf{T}_{i3}^{(t)} = \sum_{j=1}^{n} \tau_{ij}^{(t-1)}\mathbf{x}_j\mathbf{x}_j^T. \tag{11}$$

Thus, in the M-step, we update the corresponding parameter estimates using only the sufficient statistics

$$\pi_i^{(t)} = T_{i1}^{(t)}/n \tag{12}$$

$$\mathbf{u}_i^{(t)} = \mathbf{T}_{i2}^{(t)}/T_{i1}^{(t)} \tag{13}$$

$$\mathbf{\Sigma}_i^{(t)} = \{\mathbf{T}_{i3}^{(t)} - T_{i1}^{(t)^{-1}}\mathbf{T}_{i2}^{(t)}\mathbf{T}_{i2}^{(t)^T}\}/T_{i1}^{(t)}. \tag{14}$$

The E-step and M-step are repeated until convergence; that is, generally, they are repeated until the improvements in the parameter estimates and/or the likelihood function are negligibly small. Notice that, in every iteration, the E-step in (8) has to be evaluated for every data point $\mathbf{x}_j$ and for every group i. Thus the computational effort of EM depends linearly on the size n of the data and the number g of groups. For large data sets, EM can therefore be computationally very burdensome, especially when the number of required iterations is large, that is, when the starting values are far form the MLE.

3. Ascent-EM

Since EM makes a pass through all of the available data in every iteration, it can be computationally challenging. In order to improve its computational efficiency, we propose a sampling-based implementation of EM whose motivation is founded in the highly-appraised likelihood-ascent property. Unarguably, one of the most outstanding properties of the EM algorithm is that, unlike many other optimization methods, it guarantees an improvement of the likelihood function in every update of the algorithm. Specifically, it can be shown (Dempster et al., 1977) that any parameter value $\boldsymbol{\theta}^*$ that satisfies

$$Q(\boldsymbol{\theta}^*|\boldsymbol{\theta}^{(t-1)}) > Q(\boldsymbol{\theta}^{(t-1)}|\boldsymbol{\theta}^{(t-1)}) \tag{15}$$

results in an increase of the likelihood function; that is, $L(\boldsymbol{\theta}^*;\mathbf{x}) > L(\boldsymbol{\theta}^{(t-1)};\mathbf{x})$. This *likelihood-ascent property* of EM is a simple consequence of Jensen's inequality applied to (15).

The likelihood-ascent property has many important implications. While (15) implies that the output of EM will *maximize* the likelihood function (which is in contrast to, say, Newton-Raphson, which upon convergence requires additional verification that the output is not a minimizing point), the likelihood-ascent property can be taken advantage of in far more different ways. For instance, in many applications of EM the M-step can be very complicated. In fact, recall that the M-step requires a full maximization of the Q-function in (5), which can be hard or even impossible depending on the complexity of the model. The likelihood-ascent property alleviates this problem. In fact, any parameter value that satisfies (15) will contribute to the overall maximization of the likelihood function. This version of EM is often referred to as a *Generalized* EM (GEM) algorithm (Dempster et al., 1977). (Meng and Rubin,

1993, on the other hand, propose a related version in which the parameter is split into components and in every iteration of EM the conditional likelihood is maximized with respect to the current component, conditional on the other components.) In this work we will take advantage of the likelihood ascent property in a different form. Specifically, we will use the relationship in (15) to estimate the amount of information about the MLE that is contained in only a sub-sample of the entire data.

Let $(\mathbf{x}_{t_1}, \ldots, \mathbf{x}_{t_m}) \subset (\mathbf{x}_1, \ldots, \mathbf{x}_n)$ be a randomly chosen subset of size m $(m < n)$ of the data. We can approximate the Q-function in (6) by

$$\tilde{Q}_m(\boldsymbol{\theta}|\boldsymbol{\theta}^{(t-1)}) = \sum_{j=1}^{m} \sum_{i=1}^{g} \tau_{ij}^{(t-1)} \{\log \pi_i + \log f_i(\mathbf{x}_{t_j}; \boldsymbol{\psi}_i)\}. \qquad (16)$$

In that form, $\tilde{Q}_m$ in (16) resembles the Monte Carlo EM algorithm (Wei and Tanner, 1990). Notice that as $m \to n$, $\tilde{Q}_m \to Q$. Thus, if we use $\tilde{Q}_m$ instead of Q, we sacrifice accuracy (by using only an approximation to the Q-function) for computational efficiency (by using only a small subset $\mathbf{x}_{t_1}, \ldots, \mathbf{x}_{t_m}$ instead of the entire database). In the following, our goal will be to find a good balance between accuracy and computational efficiency for each iteration of EM.

A word of caution is necessary at this point. Notice that for any sample size m, $\tilde{Q}_m$ is an approximation for Q. Thus, one could be tempted to pick a small value of m and simply run EM with Q replaced by $\tilde{Q}_m$, keeping m fixed. While this approach would certainly result in a tremendous computational reduction, that version of EM would never converge to the true MLE because of a persevering sampling error (Booth et al., 2001). For the same reason, EM's likelihood-ascent property also does not carry over. Therefore, the sample size should be increased in an intelligent fashion as the algorithm progresses. Caffo et al., 2004, in the context of MCEM, recently developed a new implementation that increase the sample size in an intelligent manner. (See also Booth and Hobert, 1999, Levine and Casella, 2001, or Levine and Fan, 2003 for related approaches.) In this work, we will modify the approach of Caffo et al., 2004, for a sampling-based implementation of EM.

Let $\tilde{\boldsymbol{\theta}}^{(t-1)}$ denote the previous parameter value and let $\tilde{\boldsymbol{\theta}}^{(t)}$ denote the maximizer of $\tilde{Q}_m(\cdot|\boldsymbol{\theta}^{(t-1)})$ based on the sample $\mathbf{x}_{t_1}, \ldots, \mathbf{x}_{t_m}$. It is well-known that the EM algorithm takes rather large steps in the early iterations, especially when the starting values are far from the MLE (Rubin, 1991). Thus, in the early iterations, a rather rough approximation of Q will suffice to ensure progress of the algorithm and hence we initialize m at a relatively small value. Moreover, for the sake of computational efficiency, we only want to increase m if the current sample does not contain any more additional information about the maximum likelihood estimate $\hat{\boldsymbol{\theta}}$. One way to measure the amount of additional information is via the difference in the likelihood functions

$$\Delta_{t,t-1} := L(\tilde{\boldsymbol{\theta}}^{(t)}; \mathbf{x}) - L(\tilde{\boldsymbol{\theta}}^{(t-1)}; \mathbf{x}). \tag{17}$$

In particular, if $\Delta_{t,t-1} = 0$ then the current parameter value $\tilde{\boldsymbol{\theta}}^{(t)}$ does not provide any more additional information about $\hat{\boldsymbol{\theta}}$ compared to the previous value $\tilde{\boldsymbol{\theta}}^{(t-1)}$. Thus, since $\tilde{\boldsymbol{\theta}}^{(t)}$ is based on only a sub-sample $\mathbf{x}_{t_1}, \ldots, \mathbf{x}_{t_m}$ of the entire data, we conclude that we need a larger sample in order to obtain further information about $\hat{\boldsymbol{\theta}}$ and m is consequently increased.

In practice, while one could compute $\Delta_{t,t-1}$ exactly, it would require the evaluation of the likelihood function in (3) for all of the n points of the database. However, since computational reduction is one of our goals, we would like to *estimate* $\Delta_{t,t-1}$ based on the much smaller sample of size m. We can do so by appealing to the likelihood-ascent property. Specifically, recall that by (15) one can check whether $\Delta_{t,t-1} > 0$ by checking whether

$$Q(\tilde{\boldsymbol{\theta}}^{(t)} | \tilde{\boldsymbol{\theta}}^{(t-1)}) - Q(\tilde{\boldsymbol{\theta}}^{(t-1)} | \tilde{\boldsymbol{\theta}}^{(t-1)}) > 0. \tag{18}$$

Moreover, an estimate of the left hand side of (18) is readily available and given by

$$\tilde{\Delta}_{t,t-1} := \tilde{Q}_m(\tilde{\boldsymbol{\theta}}^{(t)} | \tilde{\boldsymbol{\theta}}^{(t-1)}) - \tilde{Q}_m(\tilde{\boldsymbol{\theta}}^{(t-1)} | \tilde{\boldsymbol{\theta}}^{(t-1)}). \tag{19}$$

Notice that the estimate $\tilde{\Delta}_{t,t-1}$ in (19) is based on the current sample $\mathbf{x}_{t_1}, \ldots, \mathbf{x}_{t_m}$ and is computed automatically throughout the algorithm at no extra computational expense. We also estimate the variability of $\tilde{\Delta}_{t,t-1}$ in order to derive appropriate confidence bounds. Write the right hand side of (19) as

$$\tilde{Q}_m(\tilde{\boldsymbol{\theta}}^{(t)} | \tilde{\boldsymbol{\theta}}^{(t-1)}) - \tilde{Q}_m(\tilde{\boldsymbol{\theta}}^{(t-1)} | \tilde{\boldsymbol{\theta}}^{(t-1)}) = \sum_{j=1}^{m} u_j, \tag{20}$$

where we define $u_j = \sum_{i=1}^{g} \tau_{ij}^{(t-1)} \{\log \pi_i^{(t)} - \log \pi_i^{(t-1)} + \log f_i(\mathbf{x}_{t_j}; \boldsymbol{\psi}_i^{(t)}) - \log f_i(\mathbf{x}_{t_j}; \boldsymbol{\psi}_i^{(t-1)})\}$. Let s_Δ^2 denote the sample variance of the u_j's. Then, an approximate lower confidence bound for $\Delta_{t,t-1}$ is given by

$$\tilde{\Delta}_{t,t-1} - z_\alpha \, s_\Delta, \tag{21}$$

where z_α denotes the α-percentile of the normal distribution.

The Ascent-EM algorithm proceeds as follows. If the lower bound in (21) is positive, then we conclude that there is sufficient evidence that $\tilde{\boldsymbol{\theta}}^{(t)}$ increases the likelihood. Moreover, we conclude that the current sample may contain additional information about the maximum likelihood estimate $\hat{\boldsymbol{\theta}}$; hence, we continue with the next iteration using the same sample as before. Conversely, if the lower bound is negative, then we conclude that current sample does not contain any more additional information about the maximum likelihood estimate. Thus we increase m and continue with the next iteration. We increase the sample size at an exponential rate, setting $m \leftarrow (1 + 1/\kappa)m$, where κ is an integer number, typically 3, 4 or 5. This guarantees that eventually $m = n$. Once $m = n$, we stop the algorithm using standard EM stopping rules, e.g. we stop the method when the relative change in the parameter updates is sufficiently small.

4. Small Simulation Study

We conducted a series of simulation studies to investigate the computational efficiency of Ascent-EM. The general set-up of these simulation studies is as follows. For given values of p, g and $\boldsymbol{\theta}$, we simulate n data points from the mixture density in (1). We first use the regular EM algorithm to estimate the model parameters, running the algorithm from a fixed set of starting values until convergence. Convergence is declared when the relative change in the parameter estimates is less than a threshold value. We then repeatedly apply Ascent-EM to the same set of data using the same starting values as for regular EM. Once Ascent-EM's sample size m reaches n (i.e. once Ascent-EM uses the entire database and hence essentially switches to EM), we apply the same stopping rule as for regular EM. We repeat this experiment for different values of n, g and p.

Table 1. Experiment 1 ($p = 2, g = 2$) and Experiment 2 ($p = 3, g = 3$). The table shows the mean and standard deviation of the run-times of Ascent-EM relative to that of regular EM (in percent) for different data sizes n.

	Experiment 1			Experiment 2		
n	1,000	5,000	10,000	1,000	5,000	10,000
Mean	60.78%	50.28%	43.72%	78.97%	66.26%	54.03%
StDev	9.89%	11.38%	10.37%	7.11%	12.08%	11.18%

We performed two experiments. In the first experiment we simulated data from the mixture distribution with $p = 2$ and $g = 2$ groups, comparing three different data sizes of $n = 1,000$, $5,000$ and $10,000$. We performed a similar second experiment, now changing the parameters to $p = 3$ and $g = 3$ groups.

Table 1 compares the computational effort for the two experiments. We can see that Ascent-EM can reduce the total run-time of EM enormously. In the first experiment, using the smallest data set considered in this study ($n = 1,000$), the average run-time of Ascent-EM is, on average, only about 60% of EM's. This reduction increases even further for larger sets of data. Indeed, while for $n = 5,000$ Ascent-EM converges on average in only about half of

EM's run-time, it is even less than that for $n = 10,000$! Similar results can be found in the second experiment. These findings suggest that a sampling-based approach to EM is particulary effective for large databases. But it is also interesting to note the computational gains that can be achieved already for smaller data sets.

Due to the stochastic nature of Ascent-EM, total run-times vary between two different applications of the method. The standard deviations in Table 1 measure this variability. From a practical point of view, although an *average* superior performance is insightful, it may be even more desirable to investigate the minimum and maximum performance. Figures 1 and 2 show the entire distribution of Ascent-EM's run-times relative to that of EM. We can see that for all three data sizes, Ascent-EM's worst performance is still far better than EM. In fact, for the first experiment, Ascent-EM's worst performance results in a run-time of about 76% of EM's total run-time ($n = 1,000$). For more challenging data problems ($n = 10,000$), Ascent-EM's worst performance is less than 60% of EM's run-time. These finding suggest that Ascent-EM can be a computationally much more efficient alternative than the regular EM algorithm.

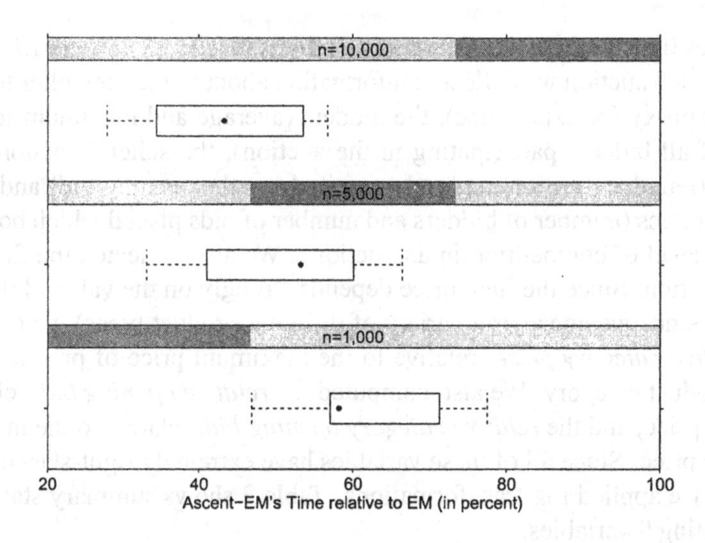

Figure 1. Experiment 1 ($p = 2, g = 2$). The plots show boxplots that display the distribution of the run-times of Ascent-EM relative to that of EM.

5. Application: Clustering Online Auctions

We apply our algorithm to a database of online auctions. This database contains detailed information about 10,078 eBay auctions in a variety of different

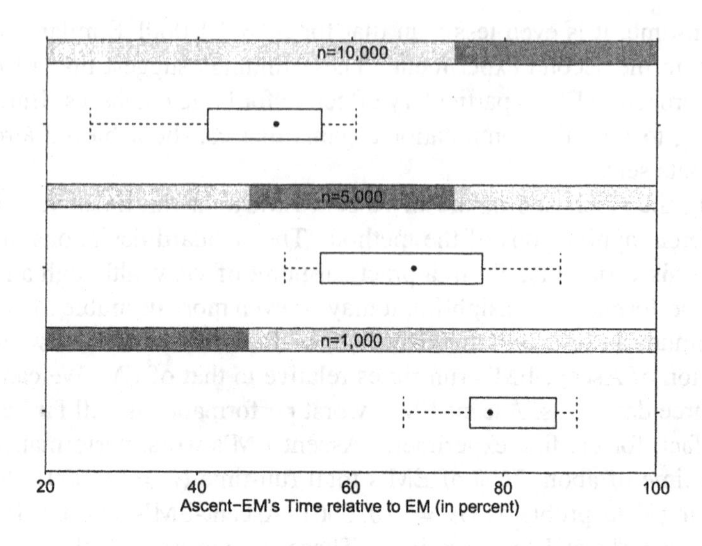

Figure 2. Experiment 2 ($p = 3, g = 3$). The plots show boxplots that display the distribution of the run-times of Ascent-EM relative to that of EM.

categories like *Clothing & Accessories*, *Sports*, *Jewelry*, *Consumer Electronics* etc. For each auction we collected information about the seller (seller feedback rating, a proxy for experience), the bidders (average and maximum feedback rating of all bidders participating in the auction), the seller's auction design choices (length of the auction and magnitude of the opening bid) and market characteristics (number of bidders and number of bids placed which both measure the level of competition in an auction). We also collected the final price of the auction. Since the final price depends strongly on the value of the product (and since we consider a variety of different product types) we computed the *relative category price*, relative to the maximum price of products in the same product category. We also computed the *relative opening bid*, relative to the final price, and the *relative category opening bid*, relative to the maximum category price. Since all of these variables have extremely right-skewed distributions, we applied log-transformations. Table 2 shows summary statistics of the resulting 9 variables.

Auction researchers are interested in what drives bidder-seller interactions and how these interactions affect the final price of an auction. We can explore these relationships by applying cluster analysis to our database of auction characteristics. One of the decisions that has to be made in cluster analysis is the most appropriate number of clusters to consider. Table 3 shows the values of the log-likelihood for different numbers g of clusters. Not surprisingly, as g increases the mixture model provides a better fit to the data and consequently the likelihood increases. Thus we also compute the log-likelihood *gain* (i.e.

Table 2. Summary statistics for 9 auction variables.

Variable	Mean	StDev	Min	Max
log-seller experience	5.10	1.86	0.69	9.91
log-max bidder experience	4.78	1.27	1.09	8.66
log-avg bidder experience	3.72	1.07	1.09	8.26
log-number of bids	2.04	0.72	0.69	4.07
log-number of bidders	1.55	0.56	0.69	3.68
log-length of auction (in days)	1.92	0.30	0.16	2.39
log-(opening bid/price)	0.29	0.22	0.00	0.69
log-(opening bid/ max category price)	0.05	0.08	0.00	0.65
log-(price/ max category price)	0.16	0.14	0.00	0.69

the difference between two adjacent values of g) and the gain relative to the null-model (which is the mixture model with only one component). One can see that 4 clusters improve the likelihood considerably over only 3 clusters, while 5 or 6 cluster do not add much more to the model improvement. Thus, we decided to investigate 4 auction clusters.

Table 3. Log-likelihood values for different number clusters g. The third column shows the difference in adjacent log-likelihood values and the fourth column shows this difference relative to the baseline negative log-likelihood value 28183.

g	LogLike	Gain	Relative Gain
1	-28183	-	-
2	-17042	11141	0.40
3	-11469	5573	0.20
4	-5939	5530	0.20
5	-5228	711	0.03
6	-4696	533	0.02

Table 4 shows the estimated cluster centers as well as the cluster proportions. Cluster 1 is the largest component of the mixture model with about 32% of the data while cluster 4 is the smallest. Interestingly, auctions in cluster 4 achieve the highest price relative to auctions in the same product category. This cluster features bidders with the *smallest* average experience level. Also, the most experienced bidders (as measured by the maximum bidder experience) tend to not participate in auctions of that cluster. It appears that the lack of bidding experience results in higher auction prices. It is also interesting to notice that cluster 2 features the smallest opening bids but the largest number of bidders and bids. Low opening bids attract many bidders which in turn results in a high level of competition for the auction. This can also be seen in the second highest price levels for this cluster.

Table 4. Cluster centers and estimated cluster proportions.

Variable	C1	C2	C3	C4
log-seller experience	5.20	4.96	5.24	4.94
log-max bidder experience	5.09	5.03	4.25	4.61
log-avg bidder experience	4.02	3.66	3.62	3.42
log-number of bids	2.04	2.54	1.29	2.36
log-number of bidders	1.55	1.95	1.01	1.71
log-length of auction (in days)	1.99	1.88	1.88	1.90
log-(opening bid/price)	0.27	0.07	0.58	0.26
log-(opening bid/ max category price)	0.01	0.01	0.12	0.10
log-(price/ max category price)	0.03	0.19	0.14	0.33
Proportion	0.32	0.24	0.24	0.19

References

Booth, James G. and Hobert, James P. (1999). Maximizing generalized linear mixed model likelihoods with an automated Monte Carlo EM algorithm. *Journal of the Royal Statistical Society B*, 61:265–285.

Booth, James G, Hobert, James P, and Jank, Wolfgang (2001). A survey of Monte Carlo algorithms for maximizing the likelihood of a two-stage hierarchical model. *Statistical Modelling*, 1:333–349.

Boyles, Russell A. (1983). On the convergence of the EM algorithm. *Journal of the Royal Statistical Society B*, 45:47–50.

Caffo, Brian S, Jank, Wolfgang S, and Jones, Galin L (2004). Ascent-Based Monte Carlo EM. *Journal of the Royal Statistical Society, Series B, Forthcoming*.

Dempster, A. P., Laird, N. M., and Rubin, D. B. (1977). Maximum likelihood from incomplete data via the EM algorithm. *Journal of the Royal Statistical Society B*, 39:1–22.

Levine, RA and Casella, G (2001). Implementations of the Monte Carlo EM algorithm. *Journal of Computational and Graphical Statistics*, 10:422–439.

Levine, RA and Fan, J (2003). An automated (Markov Chain) Monte Carlo EM algorithm. *Journal of Statistical Computation and Simulation (forthcoming)*.

Meng, Xiao-Li (1994). On the rate of convergence of the ECM algorithm. *The Annals of Statistics*, 22:326–339.

Meng, Xiao-Li and Rubin, Donald B. (1993). Maximum likelihood estimation via the ECM algorithm: A general framework. *Biometrika*, 80:267–278.

Ng, Shu-Kay and McLachlan, Geoffrey J (2003). On some variants of the EM Algorithm for fitting finite mixture models. *Australian Journal of Statistics*, 32:143–161.

Rubin, Donald B. (1991). EM and beyond. *Psychometrika*, 56:241–254.

Wei, Greg C. G. and Tanner, Martin A. (1990). A Monte Carlo implementation of the EM algorithm and the poor man's data augmentation algorithms. *Journal of the American Statistical Association*, 85:699–704.

Wu, C. F. Jeff (1983). On the convergence properties of the EM algorithm. *The Annals of Statistics*, 11:95–103.

STATISTICAL LEARNING THEORY IN EQUITY RETURN FORECASTING

John M. Mulvey[1] & A. J. Thompson[2]
Princeton University
Bendheim Center for Finance &
Department of Operations Research & Financial Engineering
[1]mulvey@princeton.edu, [2]ajt@princeton.edu

Abstract We apply Mangasarian and Bennett's multi-surface method to the problem of allocating financial capital to individual stocks. The strategy constructs market neutral portfolios wherein capital exposure to long positions equals exposure to short positions at the beginning of each weekly period. The optimization model generates excess returns above the S&P 500 , even in the presence of reasonable transaction costs. The trading strategy generates statistical arbitrage for trading costs below 10 basis points per transaction.

Keywords: Statistical Learning Theory, Data Mining, Financial Forecasting, Financial Optimization, Market-Neutral Investing, Hedge Fund Investing.

1. Introduction to Statistical Learning Theory

This paper applies statistical learning theory (SLT) to forecasting financial time series. The general question of the predictability of financial asset prices has been examined from numerous perspectives. As a part of this work researchers have investigated potential implications on market efficiency hypotheses, asset allocation, and factor modeling; these studies have tested a variety of asset classes and time horizons. In some cases, the form of the predictive model is chosen to reflect a particular structural relationship; however, in many others the model selected for convenience. In fact popular parametric models are quite easy to fit, have strong theoretical foundations, and produce results which allow for ready analysis and intuitive interpretation. Nonetheless, they usually prescribe strong structural assumptions on potential relationships in the data. Often, there is no a priori reason to expect the data to adhere to these assumptions. This idea is certainly not unique to economic and financial data as witnessed by the range of non-parametric statistical techniques available for uncovering general statistical relationships. Statistical learning theory falls into this category, thus providing an approach to establish more general predictive models for financial time series.

In this paper, we apply the multi-surface classification method for financial forecasting in equity markets. First we provide a brief background in order to develop the basic framework, describe several common techniques, and illustrate the extension to time series. Next, we discuss our formulation for forecasting equity returns and building portfolios. Last, we present empirical results from historical back-testing, highlighting practical considerations as well as the economic significance of the results.

The general supervised learning problem involves predicting some unknown value, $y \in \mathbb{Y}$, based on a known set of information $\mathbf{x} \in \mathbb{X}$. In the sequel, we refer to $\mathbb{X}$ as the *feature space* and $\mathbf{x} \in \mathbb{X}$ as a *feature vector*. Additionally, $\mathbb{Y}$ is called the *response space*. This problem identifies a function $f : \mathbb{X} \mapsto \mathbb{Y}$, such that $f(\mathbf{x})$ "best" predicts y, where in the general framework we allow both $\mathbf{x}$ and y to be stochastic. Statistical learning theory provides methodologies for finding an estimator of f from a class of functions $(f_\alpha)_{\alpha \in \Lambda}$ that minimizes the risk functional:

$$R(\alpha) = \int_{\mathbb{X} \times \mathbb{Y}} L(y, f_\alpha(\mathbf{x})) \mathrm{d}\mathbb{P}\{\mathbf{x}, y\}, \qquad (1)$$

where L is the *loss* function and $\mathbb{P}$ is the joint probability measure over the product space $\mathbb{X} \times \mathbb{Y}$. These methodologies identify such an f_α using previously observed pairs $(\mathbf{x}, y) \in \mathbb{X} \times \mathbb{Y}$ as *training data*. One obvious approach might be to use the training data to approximate $\mathbb{P}$. As Vapnik [16] discusses, however, this problem of density estimation is in general ill-posed, requiring many training observations to achieve good convergence. Instead, statistical learning theory suggests approaches for recasting the risk minimization problem in (1) to avoid solving the more general density estimation as an intermediate step. Consider the classification case (which we will for the remainder of the paper), where $\mathbb{Y}$ is a discrete space made up of a finite number of *classes*. For a given prediction function f_α, the pre-image $f_\alpha^{-1}(y) \triangleq \{\mathbf{x} \in \mathbb{X} \mid f_\alpha(\mathbf{x}) = y\}$ separates the feature space into different regions corresponding to the classes of $\mathbb{Y}$. The borders of these regions define the decision boundaries of f_α; these boundaries completely specify the prediction function. The SLT approach approximates these boundaries directly, thereby avoiding the need to approximate $\mathbb{P}$.

A popular technique of SLT classification approximate the decision boundaries via affine functions on the feature space. Consider the *two-class* case where, without loss of generality, we assume $\mathbb{Y} = \{1, -1\}$. The resulting decision function is then $f(\mathbf{x}) = \mathrm{sgn}(\mathbf{w} \cdot \mathbf{x} + b)$, [8]. Decision functions of this type are intuitively attractive, computationally tractable, and possess bounds on prediction error. While linear boundaries may seem restrictive, in fact, these models are readily extendable to non-linear decision boundaries. The procedure maps the original feature space into a new, potentially high-dimensional, transformed feature space, and then approximate affine decision boundaries in this transformed space. Depending on the approximation technique, this extension may be simplified further. As alluded, there are several different techniques for finding the "best" affine decision boundaries; we briefly review two of these: the multi-surface method and support vector machines.

Multi-Surface Method

The Multi-Surface Method (MSM) is a separating hyperplane learning model developed by O. Mangasarian and K. Bennett [2]. The methodology finds a unique separating hyperplane if one exists, otherwise it minimizes the average sum of the distance of misclassification. A key advantage is that given n training observations, $(\mathbf{x}_i, y_i)$, it can be formulated as a linear program:

$$\min_{\mathbf{w},b,\xi} \frac{\sum_{i\in\mathcal{A}} \xi_i}{m} + \frac{\sum_{j\in\mathcal{B}} \xi_j}{k} \tag{2}$$

subject to:

$$y_i(\mathbf{w}^{\mathrm{T}}\mathbf{x}_i + b) \geq 1 - \xi_i \quad \forall\, i = 1 \ldots n$$
$$\xi_i \geq 0 \quad \forall\, i = 1 \ldots n$$

where:

$$\mathcal{A} \triangleq \{i = 1 \ldots n \mid y_i = 1\}, \quad \mathcal{B} \triangleq \{j = 1 \ldots n \mid y_j = -1\}$$
$$m \triangleq |\mathcal{A}|, \quad k \triangleq |\mathcal{B}|$$

Once the optimal hyperplane is found, the training data is split by this boundary. The model can be reapplied separately to the training examples falling in each halfspace. This leads to piecewise hyperplanar decision boundaries. The MSM has demonstrated strong predictive ability in a number of cases, including the successful application to breast cancer diagnosis [14].

Support Vector Machines

Support vector machines (SVMs) find the separating hyperplane with the widest margin while controlling the extent of misclassification [16]. Training points closer to the boundary have more influence on the approximated hyperplane. This property has intuitive appeal as it is the prediction of points in these areas that is most sensitive to small changes in the resulting model. More importantly, this sensitivity results in the model's strong generalization ability, allowing SVMs to perform well even in feature spaces with large dimension and over small training samples. Another key advantage involves the properties of the model formulation. In the primal form of the *2-Norm Soft Margin* case shown below, fitting the model is equivalent to solving a convex quadratic programming problem:

$$\min_{\mathbf{w},b,\xi} \|\mathbf{w}\|_2^2 + C\sum_{i=1}^{n} \xi_i \tag{3}$$

subject to:

$$\begin{aligned} y_i(\mathbf{w}^{\mathrm{T}}\mathbf{x}_i + b) \geq 1 - \xi_i \quad \forall\, i = 1 \ldots n \\ \xi_i \geq 0 \quad \forall\, i = 1 \ldots n \end{aligned} \tag{4}$$

where ξ_i represents the distance inside the margin for a point $\mathbf{x}_i$, and C is a predefined "trade-off" constant between the complexity term in the objective function (first term) and the prediction error over the training set (second term).

The dual formulation offers opportunities for computational efficiency [5]:

$$\max_{\boldsymbol{\alpha}\geq \mathbf{0}} \sum_{i=1}^{n} \alpha_i - \frac{1}{2}\sum_{i,j=1}^{N} y_i y_j \alpha_i \alpha_j \langle \mathbf{x}_i \cdot \mathbf{x}_j \rangle - \frac{1}{2C}\langle \boldsymbol{\alpha} \cdot \boldsymbol{\alpha} \rangle \tag{5}$$

subject to:

$$\sum_{i=1}^{n} y_i \alpha_i = 0$$

The corresponding decision function for a new input vector X is then expressed as:

$$f(\mathbf{x}) = \operatorname{sgn}\left(\sum_{i=1}^{n} y_i \alpha_i^* \langle \mathbf{x}_i \cdot \mathbf{x} \rangle + b^* \right) \tag{6}$$

with b^* chosen such that:

$$y_i f(\mathbf{x}_i) = 1 - \frac{\alpha_i^*}{C} \quad \forall\, i \in \{ j \in 1 \ldots n \mid \alpha_j^* \neq 0 \}$$

Three aspects of the dual formulation are especially appealing. First, the constraints are simpler in structure than those of the primal problem. Second, the size of the problem grows only in the number of training examples, not in the dimensionality of the feature space. Third, the feature vectors of the training examples enter the problem only through their inner products with other examples. The last property can be exploited by the non-linear transformation procedure described above to simplify the extension to non-linear decision boundaries.

Also, the dual formulation highlights the fact, mentioned earlier, that support vector methods place more "weight" on points closer to the hyperplane. The dual variables, α_i, correspond to the inequality constraints of equation (4) in the primal problem. From the KKT conditions, for any constraint, i, that is not 'binding at the solution will have a corresponding $\alpha_i^* = 0$ [5]. This non-tightness occurs for points that are outside of the margin of the hyperplane. The fundamental effect of the relationship between the primal constraints and the dual variables is observed through decision function in equation (6). This function depends only on training points, i, with $\alpha_i^* \neq 0$, or those points "close" to the resulting hyperplane. These points, deemed *support vectors*, encode all of the knowledge of the training set via the decision function. Many results on the generalization ability of support vector machines stem from this property; some error bounds depend directly on the number of support vectors.

2. Forecasting Financial Time Series

Next, we discuss the results of a series of empirical experiments. This research focuses on answering the question of whether the separating hyperplane classification models can be successfully trained to predict financial time series innovations. The tests conducted so far examine this question via historical time series of equity returns.

Forecasting equity returns as a classification problem and not a regression is analogous to the strategy of focusing on the decision boundaries of the feature space rather than the conditional probability distribution. In many cases, the primary concern is whether or not a position (long or short) should be taken in a particular security. Thus, formulating the prediction problem as a binary forecast (positive or negative) provides an answer to this question. In addition, it avoids the (potentially) more difficult task of estimating an exact value for the expected return in the coming period.

Historical Testing Scenario

The historical experiments examine one-week forward equity returns. Initial tests employed the AMPL optimization platform (CPLEX optimizer) to fit the separating hyperplane [1][11]. More recently, we migrated to the DASH Optimizer (embedded Xpress-MP solver [6]). The raw data consists of daily price time series from August 1980 to August 2003 for 42 Large-Cap equities[1]. Back-testing results were obtained by making forecasts for R_t, the return received from investing in time t and holding until time $t+1$, at each week historically. To pose the this forecast as a two-class classification problem, we encoded the training examples according to whether the next period return for that example was above or below a pre-specified threshold, ε. Thus for each training example, $y_t \triangleq 1$ if $R_t \geq \varepsilon$ else $y_t \triangleq -1$ if $R_t < \varepsilon$. Initial research efforts focused on the factors in the feature set. Tests were run using all combinations of two to six of the following factors: weekly return last period, R_{t-1}, weekly return for two periods ago, R_{t-2}, cumulative returns over previous 4, 13, and 52 weeks, 10-day and 20-day volatility, and a ratio of two price moving averages with different window lengths.

Similar to other applications, such as in [14], if the feature set consisted of more than two features, there was noticeable degradation in the prediction ability. This may be attributable to problems related to the *curse of dimensionality*, since the density of the training data within the feature space decreases dramatically as the dimension of the feature space increases. Further tests were conducted using the best performing pair of features, R_{t-1} and R_{t-2}, yielding the feature space $\mathbf{x}_t^{\mathrm{T}} \triangleq [R_{t-1}, R_{t-2}]$. In period t, the training data set, $\mathbf{Z}_t$, was drawn from the time series data using the n most recently observed feature–response pairs, yielding $\mathbf{Z}_t^{\mathrm{T}} \triangleq \left[(\mathbf{x}_{t-1}^{\mathrm{T}}, y_{t-1}), \ldots, (\mathbf{x}_{t-n}^{\mathrm{T}}, y_{t-n})\right]$. Figure 1 provides an example of the training data represented in the feature space (the light and dark circles), the optimal hyperplane, and the prediction point for the next week (the light triangle directly left of the origin on the horizontal axis).

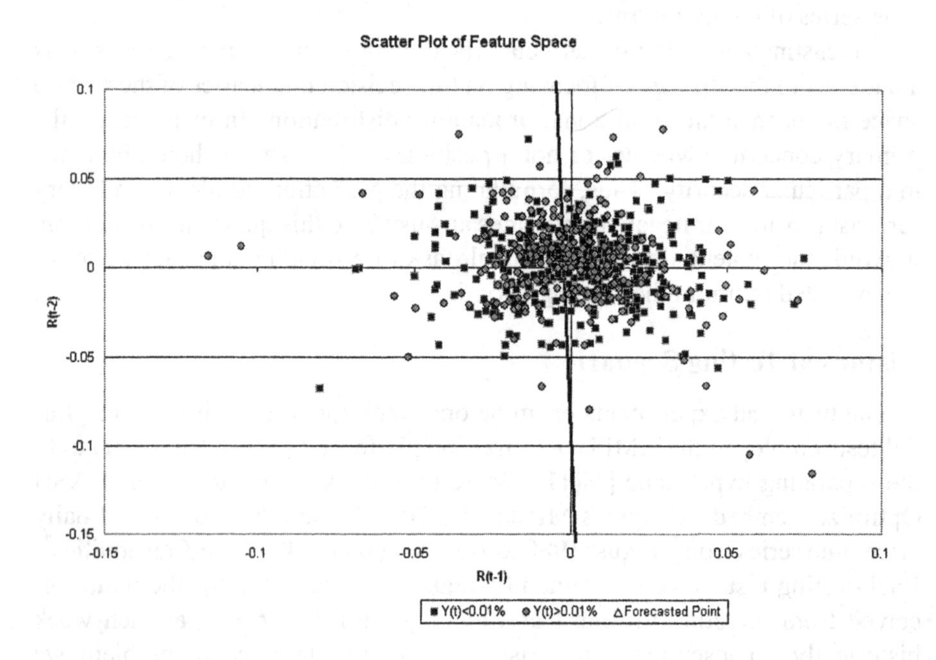

Figure 1. Modeling Scenario

Empirical Results

To apply the model several parameters must be specified. In the empirical results, we show how different specifications can impact the results. To test the predictive performance of these different specifications, we construct a portfolios from the forecasts of the various models. At each backtesting period $t = \{n+1, \ldots, T\}$ the model is re-trained by "sliding" the training data window forward one step. We can then obtain a new forecast based on the re-trained model and present feature vector, $\mathbf{x}_t$, to update the portfolio weights accordingly. Portfolio performance is then calculated using the next-period returns. Because the next-period return data is not used to train the model, all results portfolio results exhibited are *out-of-sample*.

Given historical equity returns in the United States, there is a positive bias in the data. To reduce the impact of this bias, we constructed the portfolios to be *dollar market-neutral* (for every dollar invested into a long position, there is a dollar of exposure in short positions). Whether the portfolio will be long, neutral, or short a particular equity for a given period, is based on the forecasts of two different classification models. The first, termed the "Long Model", where $\varepsilon = \varepsilon^L$, and the second, a "Short Model" where $\varepsilon = \varepsilon^S$. The two models each produce a forecast of 1 or -1 and are denoted by $F^L_{s,t}$ and $F^S_{s,t}$, respectively. To create a single investment signal for each security the two individual forecasts are combined into a single composite forecast, $F_{s,t} \triangleq \frac{1}{2}F^L_{s,t} + F^S_{s,t}$, taking the values 1, 0, or -1 corresponding to a long, neutral, or short position in $s \in \mathcal{S}$. The choice of portfolio weights,$w_{s,t}$, introduces more flexibility into the model specification. To start, we consider an equal weighting strategy where $w'_{s,t} \triangleq 1$ if $F_{s,t} = 1$, else $w'_{s,t} \triangleq -1$ if $F_{s,t} = -1$. Now normalizing the $w'_{s,t}$'s to enforce the market neutrality condition yields the weights:

$$w_{s,t} \triangleq \begin{cases} \frac{w'_{s,t}}{\sum_{s\in\mathcal{S}} w'_{s,t}1_{\{F_{s,t}=1\}}}, & \text{if } F_{s,t} = 1 \\ \frac{w'_{s,t}}{\sum_{s\in\mathcal{S}} w'_{s,t}1_{\{F_{s,t}=-1\}}}, & \text{if } F_{s,t} = -1 \end{cases} \quad \forall s \in \mathcal{S},\ \forall t \in \{n+1, \ldots, T\}. \tag{7}$$

First we compare two different approaches for defining the training set data. In the first method, called the *individual plane* model, we fit the classification scheme described above separately for each security. This straight-forward approach yields a pair of hyperplanes (long and short models). These hyperplanes capture any existing relationships between R_t and (R_{t-1}, R_{t-2}). Since we allow the model to fit a different set of planes for each security, these relationships can vary from security to security. In practice, however, we noticed that the planes lead to frequently similar results. The second approach, termed the *single plane* model, aggregates the training observations cross-sectionally, producing forty-two training observations at each time period. Fitting the model over this training set produces a single pair of hyperplanes at each time period. This pair underpins the forecast *every* security. While this model loses the flexibility to capture differences in the feature-response relationship, the significant increase in the number training points allows us to infer more structural (and potentially persistent) market microstructure effects. Additionally, the increase in the size of the feature space helps suppress the effects of the curse of dimensionality. Finally, since the single plane model fits over a large range of securities, it helps us to deduce more general qualitative properties of equities markets.

Table 1 and figure 2 show a comparison of the individual and single plane models. Also included in this and later figures for bench marking purposes are the S&P 500 and a *Naïve Rebalancing* portfolio, which is a *fixed-mix* portfolio[2] distributed equally across the forty-two stocks being considered. For both the individual and single plane models we set the response encoding parameters to $\varepsilon^L \equiv 1\%$ and $\varepsilon^S \equiv -1\%$ for the long and short sub-models, respectively. We also used $n = 520$ training periods to fit both. As these results indicate, both the individual and single plane models out perform each benchmark. Compared to one another, however, there is not a significant difference in this case. It is promising to see that the single plane performed as well as the individual planes. If the present feature space was extended by adding additional features or applying a non-linear transformation to the feature space, the significantly larger training set in the single plane model would provide it an advantage in face of increased feature space dimensionality.

	Individual	Single	S&P 500	Naïve Rebalancing
Geometric Return	27.39%	27.06%	8.29%	13.83%
Standard Deviation	14.30%	14.43%	15.95%	16.25%
Ratio	1.916	1.875	0.520	0.852

Table 1. Portfolio Statistics – Individual Hyperplanes v/s Single Hyperplane Model

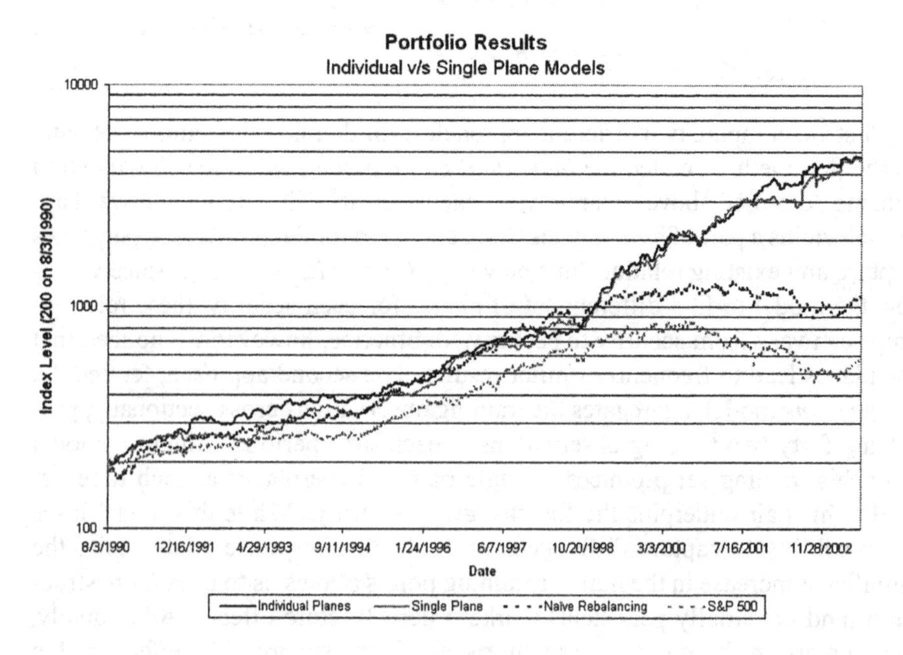

Figure 2. Portfolio Performance – Individual v/s Single Hyperplane Model

It is also interesting to examine the hyperplanes produced by the model to see if they offer any qualitative insight. Figure 3 depicts the hyperplane produced by the single-plane long model for the first trading week of each year in the backtesting period. As expected considering the length of the training period, these planes are reasonable stable through time, even though the training sets of 1991 and 2003 share no data point is common. Also note that the decision boundaries generally have very "steep" slopes and R_{t-1} intercepts near zero. This suggests that the model forecasts are most sensitive to the sign of the prior week return. In particular, the $R_{t-1} \geq 0$ side of the decision boundary corresponds to a *positive* forecast for next weeks return, R_t. As a result the model generally predicts return *reversals*, so the portfolios constructed from these forecasts are highly *contrarian* over a week long horizon. A survey of economics literature would suggest that these findings are not surprising. In [13], Lehmann investigates a reversal trading strategy and finds evidence that such strategies produce significant excess returns. In our case we find similar results using a similar strategy, but our tests were not limited to these reversal strategies, and in fact the model uncovered them on its own. While in this situation we have not discovered a "new" economic relationship, our ability to draw qualitative conclusions from the model suggests that such discoveries may be possible working in more complex or transformed feature spaces. This characteristic stems from the flexibility inherent to SLT techniques and illustrates an important reason they are valuable analytic tools.

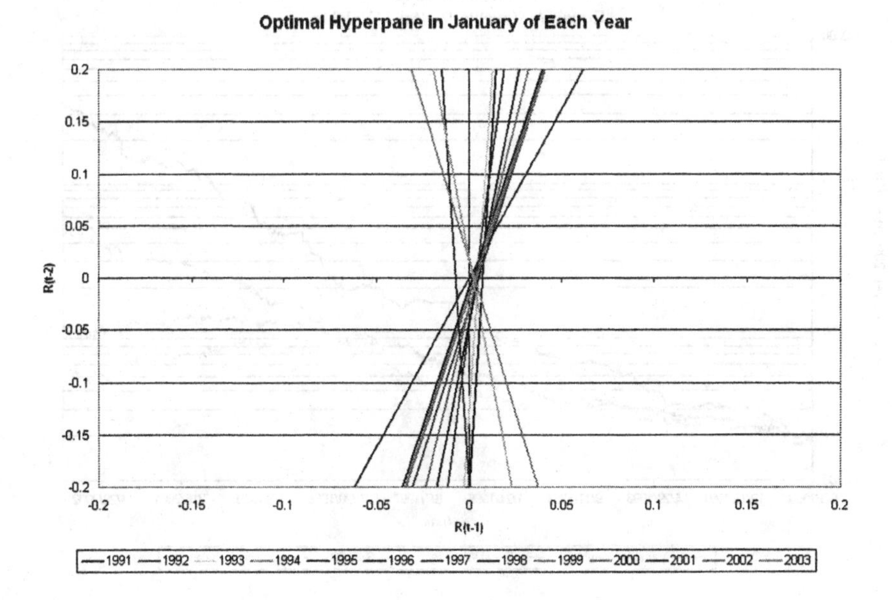

Figure 3. Changes in Optimal Hyperplanes Through Time

Next, we investigate an alternate method for constructing portfolio weights. Recall the general decision function for a linear learning machine $f(\mathbf{x}) = \text{sgn}(\mathbf{w}\cdot\mathbf{x}+b)$. The term within the sign function represents the distance inside the decision boundary. If we interpret this distance as a proxy for the level of forecast confidence we define the alternative portfolio weighting scheme as follows:

$$w'_{s,t} \triangleq \begin{cases} \mathbf{w}_L^{\mathrm{T}}\mathbf{x} + b_L, & \text{if } F_{s,t} = 1 \\ \mathbf{w}_S^{\mathrm{T}}\mathbf{x} + b_S, & \text{if } F_{s,t} = -1 \end{cases}.$$

Again, we normalize according to equation (7) to maintain market neutrality. As shown in table 2 and figure 4, this alternative method for selecting portfolio weights makes a dramatic improvement in the overall performance. These results are particularly striking, because the position direction (long, neutral, or short) for each security is identical at every period, only the size is different. This improvement confirms our interpretation of the distance inside the plane as a measure of confidence in a particular forecast.

	Equal	Variable	S&P 500	Naïve Rebalancing
Geometric Return	27.06%	47.90%	8.29%	13.83%
Standard Deviation	14.43%	19.84%	15.95%	16.25%
Ratio	1.875	2.414	0.520	0.852

Table 2. Portfolio Statistics – Equal v/s Variable Weighted Portfolios

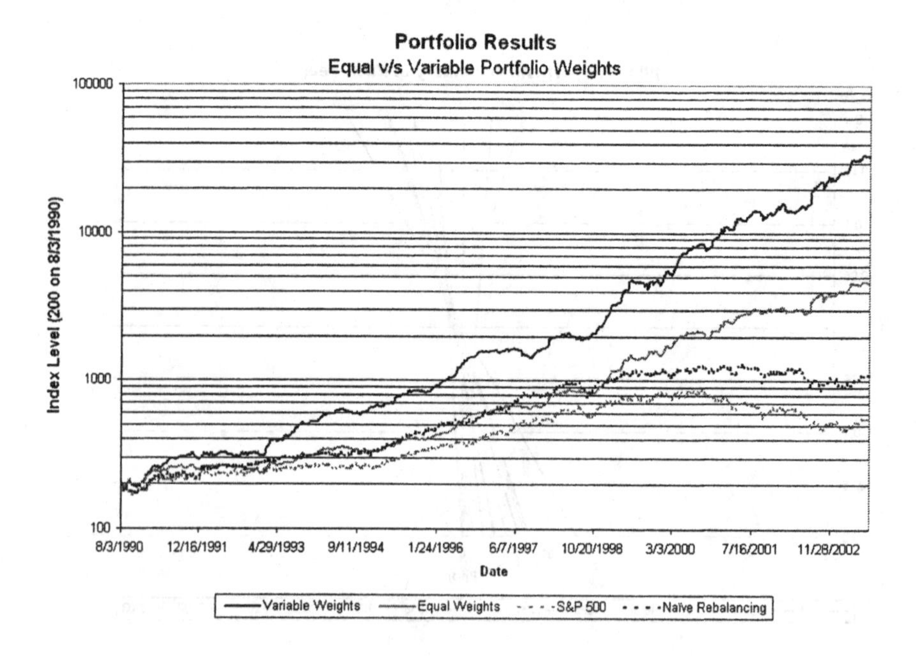

Figure 4. Portfolio Performance – Equal v/s Variable Weighted Portfolios

Table 3 and figure 5 display the model's sensitivity to the encoding parameters ε^L and ε^S. This graph demonstrates the performance for threshold levels of ±0.5%, ±1.0%, and ±2.0% under the equally-weighted portfolio scheme. As evidenced, there is some decrease in performance as the "threshold band" narrows. This is most likely attributable to the fact that the model finds more "investable" securities each period, but these securities are closer to the decision boundary. As discussed above, empirically these positions are not as profitable. This is further substantiated by table 4 and figure 6, where the same sensitivity analysis is shown for the variable weighted portfolios. In this case the portfolio weights are dominated by those points further from the boundaries. Since these weights are less sensitive to small changes in the boundaries, the differences in portfolio composition are small leading to negligible changes in performance.

	±1.0%	**±0.5%**	**±2.0%**	**Naïve Rebalancing**
Geometric Return	27.06%	24.07%	28.22%	13.83%
Standard Deviation	14.43%	14.33%	15.37%	16.25%
Ratio	1.875	1.680	1.836	0.852

Table 3. Long/Short Threshold Analysis – Equally Weighted

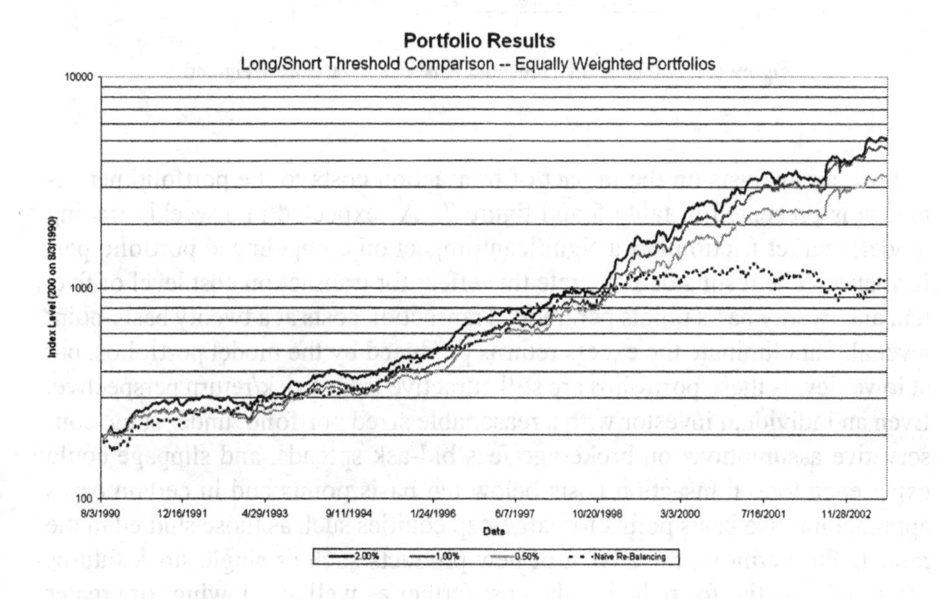

Figure 5. Long/Short Threshold Analysis – Equally Weighted

	±1.0%	±0.5%	±2.0%	Naïve Rebalancing
Geometric Return	47.90%	47.15%	48.96%	13.83%
Standard Deviation	19.84%	19.58%	20.99%	16.25%
Ratio	2.414	2.408	2.332	0.852

Table 4. Long/Short Threshold Analysis – Variable Weighted

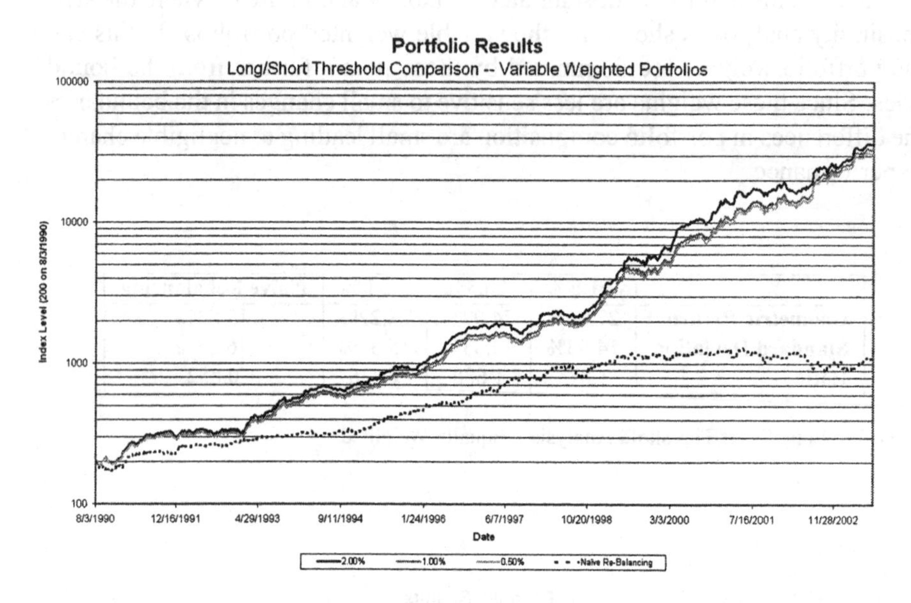

Figure 6. **Long/Short Threshold Analysis – Variable Weighted**

Last, an analysis on the impact of transaction costs to the portfolio performance is presented in table 5 and figure 7. As expected in a weekly trading model, market friction has a significant impact on compounded portfolio performance. The results demonstrate this effect for transaction cost level of five, ten, and twenty basis points per trade. Transaction costs at a twenty basis-point level all but eliminate the excess returns produced by the model portfolios, but at lower levels these portfolios are still attractive from a risk/return perspective. Even an individual investor with a reasonable sized portfolio, under some conservative assumptions on brokerage fees bid-ask spreads, and slippage could experience total transaction costs below ten basis points and in certain cases approaching five basis points for large-cap equities such as those studied in the model. Furthermore, the advent of new products such as single stock futures offer the potential for reducing the cost further as well as allowing for greater leverage.

	No T.C.	5 b.p.	10 b.p.	20 b.p.	Naïve Rebalancing
Geometric Return	47.90%	35.98%	25.00%	5.59%	13.83%
Standard Deviation	19.84%	19.84%	19.85%	19.87%	16.25%
Ratio	2.414	1.813	1.259	0.281%	0.852

Table 5. Portfolio Statistics – Transaction Cost Analysis

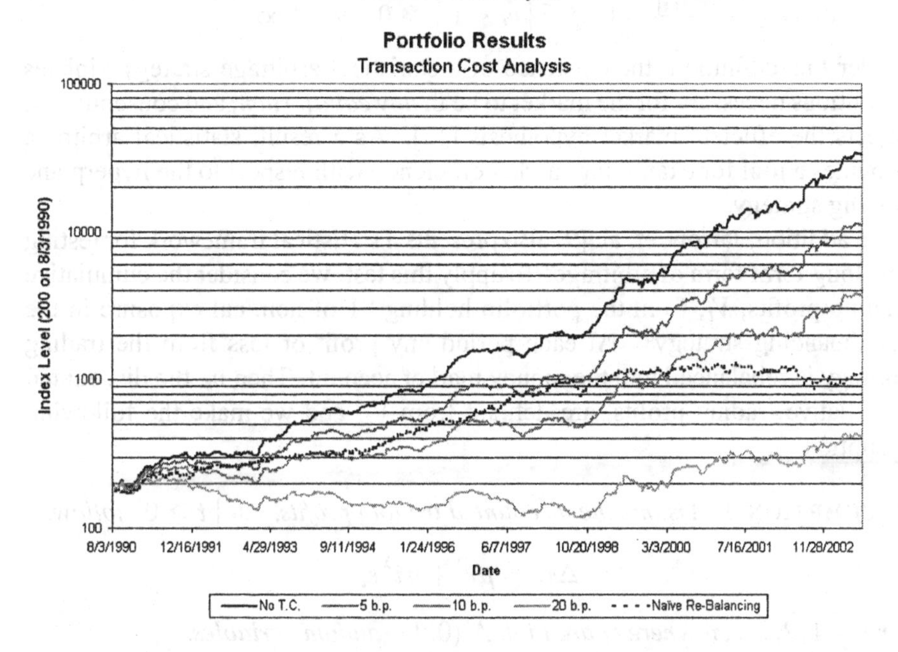

Figure 7. Portfolio Performance – Transaction Cost Analysis

Economic Significance - Testing for Statistical Arbitrage

The aforementioned results offer evidence that the hyperplane portfolio model is capable of generating excess returns at moderate portfolio volatility levels. However, it is not clear that this "strong" performance is economically significant. In particular, the efficient market hypothesis states that it should be theoretically impossible for a model such as ours to provide trading advantages over time. Thus, to draw broader conclusions from the results we must establish their statistical significance. Clearly, $\mathbb{P}\{\text{Portfolio Value}_t < \text{Portfolio Value}_0\} > 0 \quad \forall t > 0$, the hyperplane strategy does not constitute an arbitrage in the traditional sense. In [9], Jarrow, et. al. present the concept of a *statistical arbitrage* for long horizon strategies that generate *riskless* profit over time.

DEFINITION 1 (STATISTICAL ARBITRAGE)
[9] A statistical arbitrage is a zero initial cost, self-financing trading strategy $(x_t \mid t \geq 0)$ *with cumulative discounted value* v_t *such that:*

1 $v_0 = 0$

2 $\lim_{t\to\infty} \mathbb{E}^{\mathbb{P}}\left[v_t\right] > 0$

3 $\lim_{t\to\infty} \mathbb{P}\{v_t < 0\} = 0$, *and*

4 $\lim_{t\to\infty} \frac{Var^{\mathbb{P}}[v_t]}{t} = 0$ *if* $\mathbb{P}\{v_t < 0\} > 0 \quad \forall t < \infty$

Under this definition, the existence of a statistical arbitrage strategy violates conditions necessary for the market to be in *any equilibrium*, and consequently, rejects the efficient market hypothesis [12]. As a result, statistical arbitrage provides a tool for examining market efficiency with respect to the hyperplane trading strategy.

In addition, Jarrow, et. al. [9] also presents a statistical framework for testing this long-term form of arbitrage. To apply this test, we consider the cumulative trading profits, V_t, from the portfolio holding \$1 of nominal exposure in the self-financing strategy[3]. At each period any profit or loss from the trading strategy is rebalanced into the money market account. Then v_t, the discounted cumulative trading profits, are defined from V_t, and we make the following assuption:

ASSUMPTION 1 *Assume the discounted trading profits,* $(v_t \mid t \geq 0)$ *follow:*

$$\Delta v_i = \mu i^{\theta} + \sigma i^{\lambda} \epsilon_i$$

for $i = 1, 2, \ldots, n$ *where* ϵ_i *are i.i.d.* $\mathcal{N}(0,1)$ *random variables.*

Using this assumption, the MLE estimates for μ, σ, θ, and λ, are determined from the log likelihood equation:

$$\log L \; \mu, \sigma, \theta, \lambda \mid \Delta v \; = -\frac{1}{2}\sum_{i=1}^{n} \log \; \sigma^2 i^{2\lambda} \; - \frac{1}{2\sigma^2}\sum_{i=1}^{n} \frac{1}{i^{2\lambda}} \; \Delta v_i - \mu i^{\theta} \; ^2$$

Theorem 1 below establishes values for which the parameter estimates imply the presence of statistical arbitrage, [9].

THEOREM 1 *[9] A trading strategy generates a statistical arbitrage with* $1 - \alpha$ *confidence if the following conditions are satisfied:*

H1: $\hat{\mu} > 0$, H2: $\hat{\lambda} < 0$, H3: $\hat{\theta} > \max\left\{\hat{\lambda} - \frac{1}{2}, -1\right\}$,

with the sum of the p-values forming an upper bound for the test's Type I error. Thus, the sum of the p-values associated with the individual hypotheses must be below α *to conclude that a trading strategy generates statistical arbitrage.*

We calculated the statistical arbitrage parameter estimates for the variable-weighted hyperplane strategy. Table 6 below shows the parameter estimates and total p-value for various levels of transaction costs.

	No T.C.	5 b.p.	10 b.p.	20 b.p.
$\widehat{\mu}$	0.005179	0.002508	0.000516	-0.016340
$\widehat{\sigma^2}$	0.001019	0.001016	0.001016	0.000984
$\widehat{\theta}$	-0.003387	0.081514	0.298900	-1.180119
$\widehat{\lambda}$	-0.113014	-0.112763	-0.112703	-0.109059
p-value	0.0060	0.0205	0.0520	***

Table 6. Statistical Arbitrage Test Parameter Estimates

The parameter and p-values demonstrate support for the hypothesis that the hyperplane strategy represents a statistical arbitrage opportunity even at moderate transaction cost levels. Not surprisingly, however, at higher levels transaction costs present a barrier to exploiting these opportunities. While the existence of statistical arbitrage in the hyperplane trading strategy's historical performance provides evidence against the efficient market hypothesis, these empirical tests were conducted with historical data, and there is no guarantee that such a simple feature space would perform well in a real-world investment system.

3. Conclusion

This paper demonstrates an application of statistical learning theory techniques to the analysis of financial time series. The forecasting model for equity returns is an example of how the models of this class may offer improvements to more traditional parametric models. While the training process for many of the SLT models is more involved, these models impose fewer a priori constraints on the functional form of the predictor and offer techniques capable of *generalizing* structural relationships in the data even over relatively small training sets. Our empirical tests on equity return time series provide an example of the possibilities. There are a number of interesting extensions. The price series of other classes of traded assets could be modeled in a manner similar the equity series in our analysis. There are also questions that arise surrounding the time horizon. While we examine weekly forecasts, there are examples in the neural network literature of modeling intra-day data. Another promising avenue for study involves the cross-sectional relationships between securities or other economic factors. For example, Hong, et. al. [10] finds that significant lead-lag relationships in equity sectors and Sharpe [15] presents a series of risk factors for assessing portfolio management style and measuring performance. SLT techniques are well suited to investigate such dependencies. In particular, the systematic methods for feature selection presented in [3], [4], and [7] can be applied to explore potential nonlinear relationships in economic variables that are non-obvious using traditional measures of dependence such as correlation.

Notes

1. The equities selected were those that have single stock futures (SSF) traded and have historical data running over the period from August 1980 – August 2003.

2. In the fixed mix portfolio weights are rebalanced to constant target proportions at each period

3. At each investment period the hyperplane portfolio model rebalances to a dollar market neurtral position. In this sense the portfolio is zero cost, because the short sales finance the long positions.

References

[1] AMPL Optimization LLC. *AMPL Modeling Language*.

[2] K. Bennett and O. L. Mangasarian. Robust Linear Programming Discrimination of Two Linearly Inseparable Sets. *Optimization Methods and Software*, 1:23–34, 1992.

[3] J. Bi, K. Bennett, M. Embrechts, C. M. Breneman, and M. Song. Dimensionality Reduction via Sparse Support Vector Machines. *Journal of Machine Learning Research*, 3:1229–1243, March 2003.

[4] P. S. Bradley and O. L. Mangasarian. Feature Selection via Concave Minimization and Support Vector Machines. In J. Shavlik, editor, *Machine Learning Proceedings of the Fifteenth International Conference(ICML '98)*, pages 82–90, San Francisco, CA, 1998.

[5] N. Cristianini and J. Shawe-Taylor. *An Introduction to Support Vector Machines*. Cambridge University Press, 2000.

[6] Dash Optimization Inc, Englewood Cliffs, NJ. *Xpress-MP Optimizer*.

[7] G. M. Fung and O. L. Mangasarian. A Feature Selection Newton Method for Support Vector Machine Classification. *Computational Optimization and Applications*, 28:185–202, 2004.

[8] T. Hastie, R. Tibshirani, and J. Friedman. *The Elements of Statistical Learning*. Springer-Verlag, New York, 2001.

[9] S. Hogan, R. Jarrow, M. Teo, and M. Warachka. Testing market efficiency using statistical arbitrage with applications to momentum and value strategies. *Journal of Financial Economics*, forthcoming.

[10] H. Hong, W. Torous, and R. Valkanov. Do Industries Lead Stock Markets? July 2003.

[11] ILOG CPLEX Division, Incline Village, NV. *CPLEX Optimizer*.

[12] R. Jarrow. *Finance Theory*. Pretince-Hall, 1988.

[13] B. Lehmann. Fads, Martingales, and Market Efficiency. *The Quarterly Journal of Economics*, 105(1):1–28, February 1990.

[14] O. L. Mangasarian, W. N. Street, and W. H. Wolberg. Breast Cancer Diagnosis and Prognosis via Linear Programming. *Operations Research*, 43(4):570–577, Jul–Aug 1995.

[15] W. Sharpe. Asset Allocation: Management Style and Performance Measurement. *The Journal of Portfolio Management*, pages 7–19, Winter 1992.

[16] V. Vapnik. *The Nature of Statistical Learning Theory*. Springer-Verlag, New York, 1999.

SAMPLE PATH DERIVATIVES FOR (s, S) INVENTORY SYSTEMS WITH PRICE DETERMINATION

Huiju Zhang[1] and Michael Fu[2]

[1] *University of Maryland*
College Park, MD 20742
huizhang@rhsmith.umd.edu

[2] *University of Maryland*
College Park, MD 20742
mfu@rhsmith.umd.edu

Abstract We consider the problem of simultaneous price determination and inventory management. Demand depends explicitly on the product price p, and the inventory control system operates under a periodic review (s, S) ordering policy. To minimize long-run average loss, we derive sample path derivatives that can be used in a gradient-based algorithm for determining the optimal values of the three parameters (s, S, p) in a simulation-based optimization procedure. Numerical results for several optimization examples are presented, and consistency proofs for the estimators are provided.

Keywords: inventory management, sample path derivatives

1. Introduction

This paper addresses an important problem in the interface between marketing and inventory planning, specifically that of simultaneously finding the optimal price and the optimal inventory control parameters in the face of uncertain price-dependent demands. More particularly, we study a periodic-review, single-product, stationary and infinite-horizon inventory system with the objective of minimizing the average long-term loss rate (maximizing the average long-term profit rate), where the demands faced by the system depend on the constant price p, and the system adopts the (s, S) control policy, in which an order is placed when and only when its inventory level falls below the level s, and the order amount is such that it will bring the inventory level up to S. The (s, S) policy has been proved to be the optimal policy for inventory systems with fixed ordering costs and other inventory costs. Scarf (1960) showed that (s, S) policy was optimal for the finite horizon dynamic inventory system in which the ordering cost was linear plus a fixed reorder cost and holding/penalty costs were convex. Clark and Scarf (1960) extended the results to multi-echelon inventory systems. Iglehart (1963) extended Scarf's study to the infinite horizon case and considered non-zero delivery lead-times, obtaining bounds on s and S, and investigating the limiting behavior of (s, S) pairs. Veinott and Wagner (1963) developed a computational approach for finding an optimal (s, S) inventory policy for the fully backlogged model with fixed set-up cost, linear purchase cost and i.i.d. discrete random demands. A detailed review on the evolution of inventory theory can be found in Scarf's (2002) paper.

Thomas (1974) incorporated pricing decisions into the (s, S) control policy, and the resultant strategy is referred to as an (s, S, p) policy. In this policy, the optimal price p is set to be contingent upon the inventory level and can change from one period to another. Federgruen and Heching (1999) characterized the structure of an optimal combined pricing and inventory strategy for both finite and infinite horizon models with variant price change restrictions. Feng and Xiao (2000) considered a continuous-time model with multiple prices and reversible changes in prices. They found that the optimal price level was based on the length of remaining sales time and on-hand inventory. Recently, Chen and Simchi-Levi (2002) used dynamic programming to determine price and inventory levels simultaneously at the beginning of each period. They showed that an (s, S, p) policy is optimal when the demand model is additive.

Dynamic pricing may not be desirable in some industries or for some companies. Under many circumstances, a more stable pricing policy than the aforementioned inventory-contingent ones is preferred, e.g., Wal-Mart's "Everyday Low Prices". Also for mature products with stable demand that generally incorporate little seasonal effect or advanced technologies, there is relatively little price fluctuation, so the single price model is appropriate. Furthermore, Gallego and Van Ryzin (1994) showed that the optimal fixed-price policy is nearly as good as the optimal inventory-contingent one under rather mild assumptions.

In this paper, we assume there is a fixed price to be selected that influences the future demand levels in a known way, and an (s, S) policy is used to control the inventory. We use a gradient-based search optimization algorithm to find the optimal parameters. Reviews of techniques for simulation-based optimization can be found in Jacobson and Schruben (1989), Safizadeh (1990), and Fu (1994, 2002); see also Spall (1999) for a detailed review on stochastic optimization. To use simulation-based optimization requires sample path derivatives, the main focus of our work.

Fu (1994a) developed sample path derivatives using perturbation analysis (PA) for an inventory system adopting the (s, S) control policy, and Fu and Healy (1997) investigated their use in simulation-based optimization. Systematic and thorough reviews on gradient estimation via perturbation analysis can be found in Glasserman (1991), Ho and Cao (1991) and Fu and Hu (1997). We also use PA to derive our sample path derivatives. Similar to Fu's (1994a) approach, we implement infinitesimal perturbation analysis (IPA) and smoothed perturbation analysis (SPA) to estimate the derivatives. In Fu's model, demand is assumed to be exogenously specified, whereas in our model it depends on product price, which allows demand to be adjusted according to product properties such as price elasticity. The inclusion of price in the model makes the derivation more complicated.

The rest of the paper is organized as follows. Section 2 reviews the (s, S, p) model and the demand structure. The IPA analysis is presented in Section 3, and the SPA analysis is developed in Section 4. Section 5 presents a numerical example where the estimators are used to search for the optimal setting of the parameters (s, S, p). Section 6 concludes the paper with a brief summary. Consistency proofs for the infinite horizon model are contained in the Appendix.

2. Model Formulation

Consider a firm that has to make production and price decisions under stationary independently and identically distributed (i.i.d.) demand that depends on a constant product price. For each period t, $t = 1, 2, \dots, T$, let

$D_t :=$ demand in period t, i.i.d.,with p.d.f. f and c.d.f. F,

$p :=$ selling price,

where the demand function is of the general form

$$D_t = d(p, \varepsilon_t) := \varepsilon_t \gamma(p) + \delta(p) \tag{1}$$

with $\gamma(\cdot)$ and $\delta(\cdot)$ nonincreasing functions and ε_t assumed to be i.i.d. The cases $\gamma(p) = 1$ and $\delta(p) = 0$ are often referred to as the additive and multiplicative models, respectively. We use additive stochastic demand functions in our model with $\delta(p) = b - a * p$, $a, b > 0$, where a is the price elasticity.

In this paper, we assume that the ordering decision is made at the beginning of the period, and the demand for the period is subtracted at the end of the period. We also assume order lead-time is zero, so that the inventory position and inventory level coincide. Let x_t be the inventory level at the beginning of period t before placing an order, and y_t be the inventory level at the beginning of period t after placing an order. Hence, $y_t = x_t$ if no order is made, and $y_t = S > x_t$ if an order is placed at the beginning of period t. The ordering cost includes both a fixed cost and a variable cost proportional to the amount ordered. Demand that cannot be met from inventory on hand is fully backordered. The inventory carrying and stockout costs all depend on the size of the end-of-the-period inventory level and shortfall. The objective is to find $\theta = (s, S, p)$ to minimize long-run average loss L:

$$L_T = \frac{1}{T}\sum_{t=1}^{T} l(y_t, p) \tag{2a}$$

$$= \frac{1}{T}\sum_{t=1}^{T}[I\{x_t < s\}(k + c(S - x_t)) + h(y_t - D_t)^+ + g(y_t - D_t)^- - pD_t(p, \varepsilon_t)]$$

$$L = \lim_{T\to\infty} L_T \tag{2b}$$

where k is the fixed cost of placing an order, c is the variable order cost coefficient, h is the holding cost coefficient, g is the shortage cost coefficient, p is the revenue coefficient, $I\{\cdot\}$ is the indicator function, and x^+ =max$(0, x)$, x^- =max$(0, -x)$.

3. IPA Estimation

Our goal in this section is to develop derivative estimators for L_T with respect to the control parameters θ, where $\theta = s$, S, and p. Without loss of generality, we assume that $y_0 = S$. We define $q = S - s$ for notational convenience in the analysis that follows.

According to the definition of x_t and y_t, the recursive dynamic equation for y_t is given by

$$y_t = \begin{matrix} x_t = y_{t-1} - D_{t-1} & \text{if } x_t \geqslant s, \\ S & \text{if } x_t < s. \end{matrix} \tag{3}$$

That is to say, $y_t = x_t$ if the beginning inventory level is greater than the reorder point s; otherwise, an order is placed so that $y_t = S$. Thus

$$\frac{\partial y_t}{\partial \theta} = \begin{matrix} \frac{\partial y_{t-1}}{\partial \theta} - \frac{\partial D_{t-1}}{\partial \theta} & \text{if } x_t \geqslant s, \\ \frac{\partial S}{\partial \theta} & \text{if } x_t < s. \end{matrix} \tag{4}$$

With the initial condition $y_0 = S = s + q$, we have $\frac{\partial y_0}{\partial s} = \frac{\partial y_0}{\partial q} = 1$, $\frac{\partial y_0}{\partial p} = 0$. According to (4), we have $\frac{\partial y_t}{\partial s} = \frac{\partial y_t}{\partial q} = 1$ for all t, because $D_t = b - a * p + \varepsilon_t$, which is independent of s or q. Thus,

$$\frac{\partial y_t}{\partial p} = \begin{cases} \frac{\partial y_{t-1}}{\partial p} + a & \text{if } x_t \geqslant s, \\ 0 & \text{if } x_t < s. \end{cases} \tag{5}$$

By applying above recursive dynamic equation backwards, for the no-order-decision-made-period, if $x_t \geqslant s, x_{t-1} \geqslant s, \ldots, x_{t'+1} \geqslant s, x_{t'} < s$, we have

$$\frac{\partial y_t}{\partial p} = \frac{\partial y_{t-1}}{\partial p} + a = (\frac{\partial y_{t-2}}{\partial p} + a) + a = \ldots = \frac{\partial y_{t'}}{\partial p} + (t - t')a = (t - t')a,$$

where t' is the most recent period that an order is placed before t. So we can rewrite equation (5) as

$$\frac{\partial y_t}{\partial p} = \begin{cases} (t-t')a & x_t \geqslant s, x_{t-1} \geqslant s, \ldots, x_{t'+1} \geqslant s, x_{t'} < s, \\ 0 & x_t < s. \end{cases} \tag{6}$$

If we place an order in period t, the inventory level will be brought back to S in the same period since ordering lead time is zero. Recall that ordering cost consists of a fixed set-up cost and a variable cost proportional to the ordering amount, so the ordering cost in period t can then be formulated as

$$k\delta(y_t - x_t) + c(y_t - x_t) = \begin{cases} 0 & \text{if } x_t \geqslant s, \\ k + c(S - x_t) & \text{if } x_t < s. \end{cases} \tag{7}$$

The first derivative of ordering cost in period t is straightforward:

$$\frac{\partial(k\delta(y_t - x_t) + c(y_t - x_t))}{\partial\theta} = \begin{cases} 0 & \text{if } x_t \geqslant s, \\ c\frac{\partial(S-x_t)}{\partial\theta} & \text{if } x_t < s. \end{cases} \tag{8}$$

When applying the recursive dynamic relation described in equation (4) and (5) for the case of $x_t < s$, we have

$$c\frac{\partial(S - x_t)}{\partial\theta} = c\frac{\partial(S - (y_{t-1} - D_{t-1}))}{\partial\theta} = c(\frac{\partial S}{\partial\theta} - \frac{\partial y_{t-1}}{\partial\theta} + \frac{\partial D_{t-1}}{\partial\theta}) \tag{9}$$

$$= \begin{cases} c(-\frac{\partial y_{t-1}}{\partial\theta} - a) = -c(t-t')a & \text{if } \theta = p, \\ c(1 - \frac{\partial y_{t-1}}{\partial\theta}) = 0 & \text{if } \theta = s, q. \end{cases}$$

The derivative of holding cost with respect to s and q is the holding cost coefficient itself for all the periods, considering the demand is independent on s or q, and $\frac{\partial y_t}{\partial s} = \frac{\partial y_t}{\partial q} = 1$, specifically, $\frac{h\partial(y_t - D_t)}{\partial\theta} = h\frac{\partial y_t}{\partial\theta} = h$. Similarly, the derivative of shortage cost to s and q is $-g$. Applying equation (6), we obtain the direct differentiation of holding/shortage cost to price p in period t:

$$h\frac{\partial(y_t - D_t)}{\partial p} = h\left(\frac{\partial y_t}{\partial p} - \frac{\partial D_t}{\partial p}\right) = \begin{cases} h(t-t'+1)a & x_t \geqslant s, \\ ha & x_t < s. \end{cases} \tag{10}$$

and

$$g\frac{\partial(D_t - y_t)}{\partial p} = g\left(\frac{\partial D_t}{\partial p} - \frac{\partial y_t}{\partial p}\right) = \begin{cases} -g(t-t'+1)a & x_t \geqslant s, \\ -ga & x_t < s. \end{cases} \tag{11}$$

The sample path derivative of the revenue term $p*D_t$ in period t with respect to s and q is 0, since they rely only on price, not stock levels, whereas the sample path derivatives with respect to p is $-2ap + b + \varepsilon_t$. Combining all the analyses, we obtain the complete IPA estimator for the time-average loss:

$$\left.\frac{\partial L_T}{\partial\theta}\right|_{IPA} = \frac{1}{T}\sum_{l=1}^{T}\frac{\partial l(y_t, p)}{\partial\theta} = \frac{1}{T}\left(\sum_{t:x_{t+1}>0} h - \sum_{t:x_{t+1}<0} g\right) \quad \text{for } \theta = s, q, \tag{12}$$

$$\left.\frac{\partial L_T}{\partial p}\right|_{IPA} = \frac{1}{T}\sum_{l=1}^{T}\frac{\partial l(y_t,p)}{\partial p}$$
$$= \frac{1}{T}[\sum_{t:x_t<s} -c(t-t')a + \sum_{t:x_t\geqslant s,x_{t+1}>0} h(t-t')a + \sum_{t:x_{t+1}>0} ha$$
$$- \sum_{t:x_t\geqslant s,x_{t+1}<0} g(t-t')a - \sum_{t:x_{t+1}<0} ga - \sum_{t=1}^{T}(-2ap+b+\varepsilon_t)]. \quad (13)$$

4. SPA Estimation

In sample path analysis of the (s, S) model, Fu (1994a) concluded that IPA alone is sufficient for estimating the derivative with respect to $\theta = s$, but not for $\theta = q$, where an additional SPA (smoothed perturbation analysis) term must be added. This conclusion also holds for our model, so we need to use SPA.

We consider a positive change in s. Fig.1 shows the perturbation path for a small positive change Δs in the reorder point s. The sample path moves upward by Δs smoothly, i.e., the sample performance is continuous, so IPA alone suffices for s (assuming q is held constant). However, for $\Delta q > 0$, it is possible that an ordering decision changes from order to not order in a period. Fig.2 represents the sample path for change in q, and period t is the order-decision-change period. Since Δq is an infinitesimal amount and demand is finite, the demand during t will lead to an order decision in the next period. The perturbed path for inventory position can be constructed from the nominal path with an appropriate extra period "inserted". The beginning inventory in this period is $y = s - \alpha + \Delta q$. Then, an SPA term based upon conditional expectation is added to smooth the discontinuity:

$$\left.\frac{\partial L_T}{\partial q}\right|_{SPA} = \sum_{t\in M^*(T)} \lim_{\Delta q\to 0}\frac{E_{z_t}[\Delta L_T|\alpha_t \leqslant \Delta q]P_{z_t}(\alpha_t\leqslant \Delta q)}{\Delta q}$$
$$= \sum_{t\in M^*(T)} \lim_{\Delta q\to 0} E_{z_t}[\Delta L_T|\alpha_t\leqslant \Delta q] \cdot \lim_{\Delta q\to 0}\frac{P_{z_t}(\alpha_t\leqslant \Delta q)}{\Delta q}, \quad (14)$$

where $Z_t = y_{t-1} - s$, $\alpha_t = D_{t-1} - Z_t$, and $M^*(T) = \{t \leqslant T : y_t = S\}$ is the set of periods in which orders are placed.

Note that in the rest of the derivation, we will often drop the subscripts for notational convenience. The latter term can be estimated explicitly from the original sample path, given the demand distribution:

$$P_z(\alpha\leqslant x|D>z) = P_z(D-z\leqslant x|D>z) = \frac{P_z(z<D\leqslant z+x)}{P_z(D>z)} = \frac{F(z+x)-F(z)}{1-F(z)}.$$

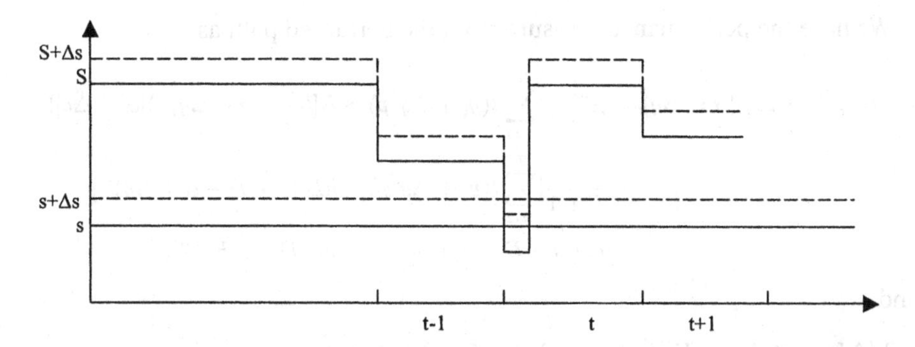

Figure 1. Effect on sample path with p, q fixed and s perturbed

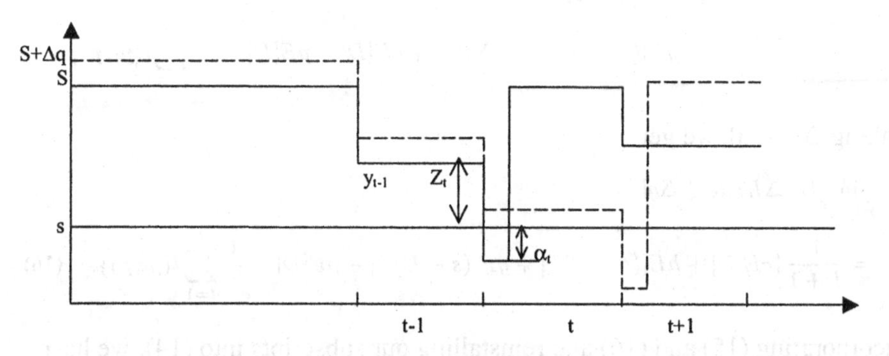

Figure 2. Effect on sample path with s, p fixed and q perturbed

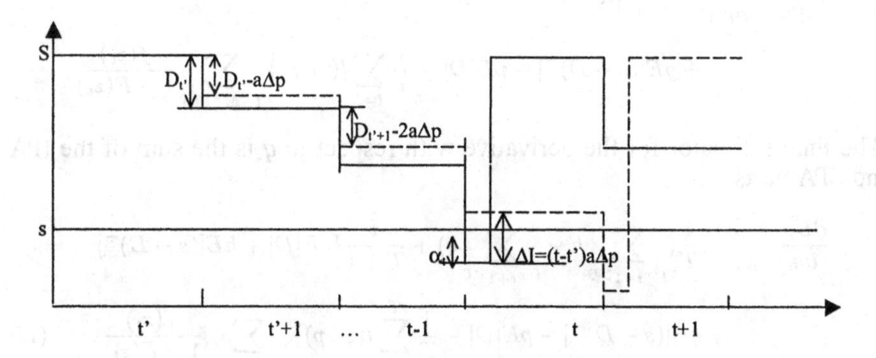

Figure 3. Effect on sample path with s, q fixed and p perturbed

Hence,

$$\lim_{\Delta q \to 0} \frac{P_z(\alpha \leqslant \Delta q)}{\Delta q} = \frac{f(z)}{1 - F(z)}. \tag{15}$$

We have the performance measure from the perturbed path as

$$\begin{aligned} E_z(L_{T+1}(q+\Delta q)|\alpha \leqslant \Delta q) &= \frac{1}{T+1}[\sum_{t=1}^{T} l(y_t + \Delta q, p) + E[l(s - \alpha + \Delta q, p)|\alpha \leqslant \Delta q]] \\ &= \frac{1}{T+1}[\sum_{t=1}^{T} l(y_t + \Delta q, p) + hE[(s - D - \alpha + \Delta q)^+] \\ &+ gE[(s - D - \alpha + \Delta q)^-] + cE[D] - pE[D]], \end{aligned}$$

and

$$\begin{aligned} E_z[\Delta L_T|\alpha \leqslant \Delta q] &= E_z[L_{T+1}(q+\Delta q) - L_T(q)|\alpha \leqslant \Delta q] \\ &= \frac{1}{T+1}\{\sum_{t=1}^{T}[l(y_t + \Delta q, p) - l(y_t, p)] + hE[(s - D - \alpha + \Delta q)^+] \\ &+ gE[(s - D - \alpha + \Delta q)^-] + cE[D] - pE[D] - \frac{1}{T}\sum_{t=1}^{T} l(y_t, p)\}. \end{aligned}$$

Taking $\Delta q \to 0$, we get

$$\begin{aligned} &\lim_{\Delta q \to 0} E_z[\Delta L_T|\alpha \leqslant \Delta q] \\ &= \frac{1}{T+1}\{cE[D] + hE[(s-D)^+] + gE[(s-D)^-] - pE[D] - \frac{1}{T}\sum_{t=1}^{T} l(y_t, p)\}. \end{aligned} \tag{16}$$

Incorporating (15) and (16) and reinstalling our subscripts into (14), we have

$$\begin{aligned} \frac{\partial L_T}{\partial q}\Big|_{SPA} &= \frac{1}{T+1}\{cE[D] + hE[(s-D)^+] \\ &+ gE[(s-D)^-] - pE[D] - \frac{1}{T}\sum_{t=1}^{T} l(y_t, p)\} \sum_{t \in M^*(T)} \frac{f(z_t)}{1 - F(z_t)}. \end{aligned}$$

The final estimator for the derivative with respect to q is the sum of the IPA and SPA parts:

$$\begin{aligned} \frac{\partial L_T}{\partial q}\Big|_{PA} &= \frac{1}{T}(\sum_{t:x_{t+1}>0} h - \sum_{t:x_{t+1}<0} g) + \frac{1}{T+1}\{cE[D] + hE[(s-D)^+] \\ &+ gE[(s-D)^-] - pE[D] - \frac{1}{T}\sum_{t=1}^{T} l(y_t, p)\} \sum_{t \in M^*(T)} \frac{f(z_t)}{1 - F(z_t)}. \end{aligned} \tag{17}$$

As the price changes from p to $p + \Delta p$, demand will decrease in each period by the amount of $a\Delta p$. Figure 3 illustrates the sample path with $\Delta p > 0$.

An additional SPA term is needed for the estimator with respect to p, since the order decision may change. The sample path is similar to that for q, but instead of a change Δq, there is an accumulated change $\Delta I = (t - t')a\Delta p$, where $t \in M^*(T)$. Otherwise, the analysis is the same as for Δq. First we define $\beta_t(p) = D_{t'}(p) + \ldots + D_{t-1}(p) - q = \varepsilon_{t'} + \ldots + \varepsilon_{t-1} + (t - t')(b - ap) - q$. Then we have

$$\begin{aligned}\frac{\partial L_T}{\partial p}_{SPA} &= \sum_{t\in M^*(T)} \lim_{\Delta p\to 0} \frac{E_{z_t}[\Delta L_T|\beta_t(p) \leqslant \Delta I]P_{z_t}(\beta_t(p) \leqslant \Delta I)}{\Delta p} \\ &= \sum_{t\in M^*(T)} \lim_{\Delta p\to 0} E_{z_t}[\Delta L_T|\beta_t(p) \leqslant \Delta I] \lim_{\Delta p\to 0} \frac{P_{z_t}(\beta_t(p) \leqslant \Delta I)}{\Delta p}.\end{aligned} \tag{18}$$

Let $f_\varepsilon(x) = f(x + b - ap)$ and $F_\varepsilon(x) = F(x + b - ap)$ the p.d.f. and c.d.f., respectively, of ε_t. We have

$$\begin{aligned}&P(\beta_t(p) \leqslant x|D_{t'}(p) + \ldots + D_{t-2}(p) = q - z_t, D_{t-1}(p) > z_t) \\ &= P(\varepsilon_{t-1} \leqslant z_t - b + ap + x|\varepsilon_{t'} + \ldots + \varepsilon_{t-2} = q - z_t + (t - t' - 1)(-b + ap), \\ &\quad \varepsilon_{t-1} > z_t - b + ap) \\ &= \frac{F_\varepsilon(z_t - b + ap + x) - F_\varepsilon(z_t - b + ap)}{1 - F_\varepsilon(z_t - b + ap)} = \frac{F(z_t + x) - F(z_t)}{1 - F(z_t)},\end{aligned}$$

where the second equality is due to the independence between ε_t's. Therefore,

$$\lim_{\Delta p\to 0} \frac{P_{z_t}(\beta_t(p) \leqslant \Delta I)}{\Delta p} = \frac{f(z_t)}{1 - F(z_t)} \cdot (t - t')a.$$

We also notice that the inserted period has inventory level $y = s - b + \Delta I$, so

$$\begin{aligned}&E_z(L_{T+1}(p + \Delta p)|\beta \leqslant \Delta I) \\ &= \frac{1}{T+1}[\sum_{t=1}^{T} l(y_t, p + \Delta p) + E[l(s - \beta + \Delta I, p + \Delta p)|\beta \leqslant \Delta I]] \\ &= \frac{1}{T+1}[\sum_{t=1}^{T} l(y_t, p + \Delta p) + cE[D] + hE[(s - D - \beta + \Delta I)^+] \\ &+ gE[(s - D - \beta + \Delta I)^-] - (p + \Delta p)E[D]],\end{aligned}$$

$$\begin{aligned}&E_z[\Delta L_T|\beta \leqslant \Delta I] \\ &= \frac{1}{T+1}\{\sum_{t=1}^{T}[l(y_t, p + \Delta p) - l(y_t, p)] + cE[D] + hE[(s - D - \beta + \Delta I)^+] \\ &+ gE[(s - D - \beta + \Delta I)^-] - (p + \Delta p)E[D] - \frac{1}{T}\sum_{t=1}^{T} l(y_t, p)\},\end{aligned}$$

$$\lim_{\Delta p \to 0} E_z[\Delta L_T | \beta \leqslant \Delta I]$$

$$= \frac{1}{T+1}\{cE[D] + hE[(s-D)^+] + gE[(s-D)^-] - pE[D] - \frac{1}{T}\sum_{t=1}^{T} l(y_t, p)\}. \quad (19)$$

Therefore,

$$\frac{\partial L_T}{\partial p}_{SPA} = \frac{1}{T+1}\{cE[D] + hE[(s-D)^+] + gE[(s-D)^-] - pE[D]$$
$$- \frac{1}{T}\sum_{t=1}^{T} l(y_t, p)\} \sum_{t \in M^*(T)} [\frac{f(z_t)}{1-F(z_t)}(t-t')a]. \quad (20)$$

Summing IPA and SPA terms gives derivative of loss function to p:

$$\frac{\partial L_T}{\partial p}_{PA} = \frac{1}{T}[\sum_{t:x_t<s} -c(t-t')a + \sum_{t:x_t \geqslant s, x_{t+1}>0} h(t-t')a + \sum_{t:x_{t+1}>0} ha$$
$$- \sum_{t:x_t \geqslant s, x_{t+1}<0} g(t-t')a - \sum_{t:x_{t+1}<0} ga - \sum_{t=1}^{T}(-2ap + b + \varepsilon_t)]$$
$$+ \frac{1}{T+1}\{cE[D] + hE[(s-D)^+] + gE[(s-D)^-] - pE[D]$$
$$- \frac{1}{T}\sum_{t=1}^{T} l(y_t, p)\} \sum_{t \in M^*(T)} [\frac{f(z_t)}{1-F(z_t)}(t-t')a]. \quad (21)$$

5. Optimization Example

Our goal is to find θ^* that solves $\min_{\theta \in C} L(\theta)$, where C represents a constraint set defining the allowable values for the parameters θ. For local optimization, a necessary condition when L is continuously differentiable is that θ^* satisfies: $L'(\theta^*) = 0$. Using a Robbins-Monro stochastic approximation algorithm (cf. Kushner and Yin 1997) and the gradient estimator derived in the previous section, we apply the following iterative gradient-based procedure to update the parameter values, in order to reach a local optimum:

$$\begin{bmatrix} s_{k+1} \\ q_{k+1} \\ p_{k+1} \end{bmatrix} = \begin{bmatrix} s_k \\ q_k \\ p_k \end{bmatrix} - e_k \begin{bmatrix} (\frac{\partial L}{\partial s})_{PA} \\ (\frac{\partial L}{\partial q})_{PA} \\ (\frac{\partial L}{\partial p})_{PA} \end{bmatrix}_{s_k, q_k, p_k}, \quad (22)$$

where k is the iteration number, and the gain sequence $\{e_k\}$ is a version of the accelerated harmonic series given by e/E_k, where

$$E_{k+1} = \begin{matrix} E_k + 1 & \text{if } sgn(\frac{\partial L}{\partial \theta})_{PA,k+1} \neq sgn(\frac{\partial L}{\partial \theta})_{PA,k}, \theta = s, q, p, \\ E_k & \text{otherwise.} \end{matrix}$$

with $E_0 = 1$, and sgn is the sign function of a vector of parameters. The step size changes only when all three signs of the vector elements change simultaneously. Since s, q, and p must be positive, we project back to the previous point whenever the algorithm brings s, q or p negative. The values of s, q and p are updated every T periods, with the PA estimator reinitialized at each update. Furthermore, we take the starting point to be $s_0 = q_0 = E[D]/2$, and $p_0 = (p_{min} + p_{max})/2$, where p_{min} and p_{max} are the lower bound and upper bound of price range, respectively. We expect that the parameters e and T greatly affect the initial convergence rate of the algorithm.

Our numerical study is based on the data collected from a specialty retailer of high-end women's apparel (Federgruen and Heching 1999). Table 1 summarizes all parameters for the base scenarios pertaining to the dress. The variables ε_t are normally distributed with zero mean and standard deviation $\delta(p) * cv$, with cv a specified coefficient of variation, and truncated at $-\delta(p)$ to preclude negative demand realizations.

Table 1. Base Parameters for Dress.

Item	b	a	cv	k	c	h	g	price range
Dress	174	3	0.25	0	22	0.22	21.78	25-44

Following the previously described procedure, the initial values of the parameters are set at $(s, q, p) = (35, 35, 34.5)$. We choose the update period $T = 100$ and run 10 replications. Each simulation replication is terminated when the sum of the gradient estimate components for the three parameters is less than 0.001 or the number of iteration is greater than 50,000. Table 2 shows 95% confidence intervals of optimal values for three cases corresponding to different fixed cost k and holding cost h with $cv = 0.25$:

Case 1 – k=0, h=0.22;
Case 2 – k=0, h=5;
Case 3 – k=100, h=0.22.

Table 2. 95% Confidence Interval for $cv = 0.25$.

95% C.I.	$k = 0, h = 0.22$	$k = 0, h = 5$	$k = 100, h = 0.22$
s	65.92 ± 5.49	50.13 ± 0.40	68.55 ± 12.75
S	95.05 ± 19.39	67.80 ± 2.06	181.57 ± 30.72
p	39.17 ± 1.22	39.84 ± 0.22	39.29 ± 2.34
L	-980.14 ± 30.25	-895.43 ± 20.70	-926.01 ± 23.27

The optimal value of q is small for zero fixed ordering cost, substantially smaller than expected demand. In this situation, holding costs dominate, since frequent ordering is not penalized. Comparing case 1 and case 2, we find that s and S decrease as holding cost increases. In case 3, where holding cost is small compared with fixed ordering cost, q is relatively large, decreasing the number of ordering cycles. In all three cases, price doesn't differ much, i.e., price does not appear to be a major determinant for the various inventory-related cost scenarios.

Figures 4, 5, 6 and 7 illustrate the convergence rate of three cases based on one common run for L, p, s and S, respectively. The figures show that the algorithm converges very fast at the beginning. The fluctuations in case 3 are due to a large initial step size.

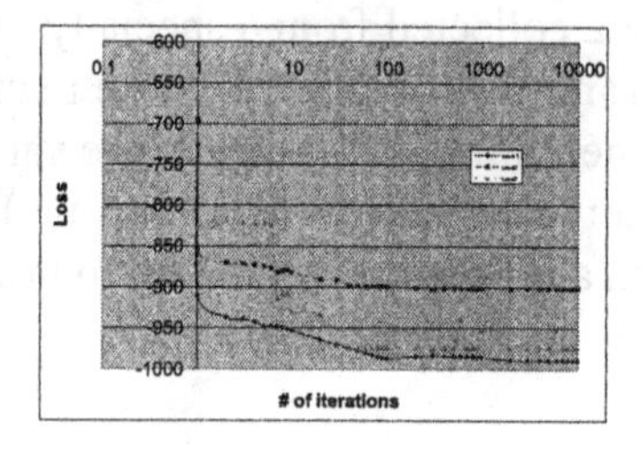

Figure 4. Expected loss as function of iteration number

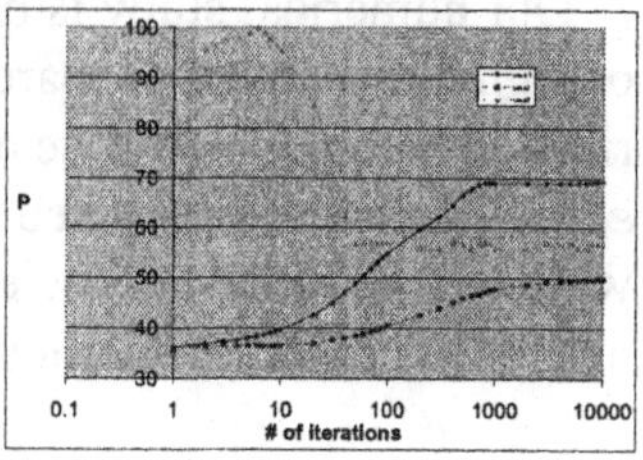

Figure 5. Selling price as function of iteration number

Finally, we investigate the impact of price elasticity by modifying the slope a of the demand function to $a = 1$, $a = 3$, and $a = 5$, with $k = 100$ and $h = 0.22$. Table 3 shows the values of the control parameters and total loss. Price decreases dramatically as a increases. Demand elasticity measures the change of demand to change of price; thus, when it goes up, a manufacturer has to reduce price to attract more consumers so as to increase revenue.

Some general observations from the simulation results:

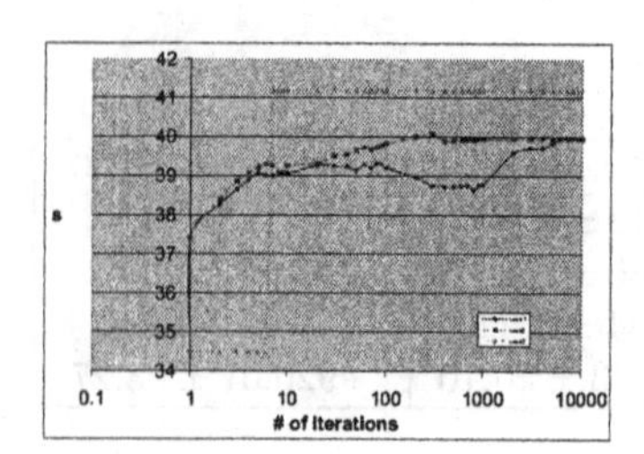

Figure 6. Base stock level as function of iteration number

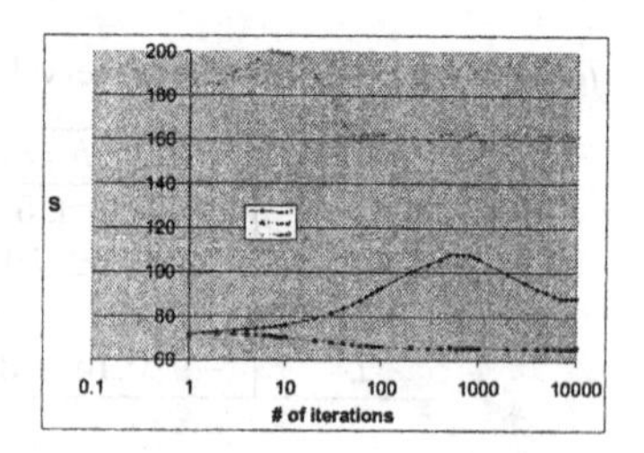

Figure 7. Reorder stock level as function of iteration number

Table 3. Experimental Results for Different Price Elasticity a ($cv = 0.25, k = 100, h = 0.22$)

a	s	S	p	L
1	93.53	166.11	43.91	-2811.44
3	53.70	133.32	40.37	-961.37
5	47.62	121.07	25.04	-123.23

1. $T = 100$ is not sufficient for reaching steady state for large fixed ordering cost, since iterate updates are not carried out at regenerative points, so we have the 'last period effect'.
2. In some simulations, periodic behavior occurs in the iterates, due to the implementation of the gain sequence, which only decreases if all three components in the gradient change signs.

6. Conclusions

This paper presents a period review inventory model with price-dependent uncertain demand. The proposed inventory control policy reflects a common practice in some industries. To minimize the expected loss, management determines both the optimal stock level and price. Using perturbation analysis, we developed sample path derivatives for this (s, S, p) inventory model, which could be incorporated into gradient-based algorithms to select optimal values for the three controllable parameters. Some numerical results for simulation optimization are presented using a Robbins-Monro stochastic approximation algorithm. Consistency proofs are provided for the infinite horizon case. For future research, useful extensions of our model include non-zero lead-time scenarios.

Appendix: Consistency Proof

Our goal is to prove: $(\frac{\partial L_T}{\partial \theta})_{PA} \rightarrow \frac{\partial E[l(y,p)]}{\partial \theta}$, where $l(y_t, p) = (k + c(S - x_t)) \cdot \delta(y_t - x_t) + h(x_{t+1})^+ + g(x_{t+1})^- - pD_t$, loss in the period t.

Let X and Y denote the steady-state random variables for x_t and y_t. Then, we have $\sum_{t=1}^{T} x_t/T \rightarrow E[X]$, $\sum_{t=1}^{T} y_t/T \rightarrow E[Y]$, and $\sum_{t=1}^{T} D_t/T \rightarrow E[D]$. From Fu's derivation (1994a), We already know $P(X = S) = 1/(1 + R(q))$, and $E[Y] = s + \frac{q+\bar{R}(q)}{1+R(q)}$. In our model, we have $X = Y - D$, therefore, $E[X] = s + \frac{q+\bar{R}(q)}{1+R(q)} - E[D]$.

The long run average cost per period for infinite horizon is given by equation (2):

$$L(s, S, q) = \frac{K}{1 + R(q)} + cE[D] + hE[X^+] + gE[X^-] - pE[D].$$

For any stable policy, the average per period production amount is always $E[D]$. According to PA estimator equations (12) and (17), we have w.p.1 that

$$(\frac{\partial \bar{x}_t^+}{\partial s})_{IPA} \rightarrow P(X > 0). \tag{A.1}$$

$$(\frac{\partial \bar{x}_t^-}{\partial s})_{IPA} \to P(X < 0). \tag{A.2}$$

$$(\frac{\partial \bar{x}_t^+}{\partial q})_{PA} \to P(X > 0) + P(Y = S)E[\frac{f(z)}{1 - F(z)}] \cdot [E[(s - D)^+] - E[X^+]]. \tag{A.3}$$

$$(\frac{\partial \bar{x}_t^-}{\partial q})_{PA} \to -P(X < 0) + P(Y = S)E[\frac{f(z)}{1 - F(z)}] \cdot [E[(s - D)^-] - E[X^-]]. \tag{A.4}$$

$$(\frac{\partial(\overline{k * \delta(y_t - x_t)})}{\partial q})_{PA} \to P(Y = S)E[\frac{f(z)}{1 - F(z)}] \cdot [-\frac{k}{1 + R(q)}]. \tag{A.5}$$

$$(\frac{\partial(\overline{(S - x_t) * \delta(y_t - x_t)})}{\partial q})_{PA} \to P(Y = S)E[\frac{f(z)}{1 - F(z)}] \cdot [E[D] - E[D]] = 0. \tag{A.6}$$

$$(\frac{\partial(\overline{-pD_t})}{\partial q})_{PA} \to P(Y = S)E[\frac{f(z)}{1 - F(z)}] \cdot [-pE[D] + pE[D]] = 0. \tag{A.7}$$

According to Fu's paper, we have

$$\frac{\partial E(X^+)}{\partial s} = P(X > 0) = eq.(A.1), \qquad \frac{\partial E(X^-)}{\partial s} = -P(X < 0) = eq.(A.2).$$

$$\frac{\partial E(x^+)}{\partial q} = P(X > 0) + P(Y = S)E[\frac{f(z)}{1 - F(z)}] \cdot [E[(s - D)^+] - E[X^+]] = eq.(A.3).$$

$$\frac{\partial E(x^-)}{\partial q} = -P(X < 0) + P(Y = S)E[\frac{f(z)}{1 - F(z)}] \cdot [E[(s - D)^-] - E[X^-]] = eq.(A.4).$$

We also have $\frac{\partial(k/(1+R(q)))}{\partial s} = 0$.

$$\frac{\partial(k/(1 + R(q)))}{\partial q} = -\frac{r(q) * k}{(1 + R(q))^2} = P(Y = S)E[\frac{f(z)}{1 - F(z)}] \cdot [-\frac{k}{1 + R(q)}] = eq.(A.5).$$

$$\frac{\partial E[D]}{\partial s} = 0, \qquad \frac{\partial E[D]}{\partial q} = 0 = eq.(A.6).$$

$$\frac{\partial(-pE[D])}{\partial s} = 0, \qquad \frac{\partial(-pE[D])}{\partial q} = 0 = eq.(A.7).$$

Therefore, consistency proof is completed for s and q.

Consistency proof for p is more complicated. First we define $N(q)$ to be the counting process for the demand renewal process:

$$N(q) = max\{t | D_1 + D_2 + \ldots + D_t \leq q\},$$

then for $t \in M^*(T)$, we have $t - t' - 1 \sim N(q)$ and $z_t \sim q - (D_1 + D_2 + \ldots + D_{N(q)})$. Hence,

$$P(t - t' - 1 = 0, z_t = q) = 1 - F(q),$$

and for $n = 1, 2, \ldots$, and $z \in [0, q]$,

$$f_{t-t'-1, z_t}(n, z) = f_n(q - z) * (1 - F(z)).$$

Therefore,

$$
\begin{aligned}
&E\ \frac{f(z_t)(t-t')}{1-F(z_t)} \\
&= \frac{(1-F(q))*f(q)}{1-F(q)} + \sum_{n=1}^{\infty}(n+1)*\int_0^q f_n(q-z)*(1-F(z))*\frac{f(z)}{1-F(z)}dz \\
&= f(q) + \sum_{n=1}^{\infty}(n+1)*\int_0^q f_n(q-z)f(z)dz = f(q) + \sum_{n=1}^{\infty}(n+1)*f_{n+1}(q) \\
&= \sum_{n=1}^{\infty} n f_n(q).
\end{aligned}
$$

For PA analysis of price, according to equation (21), we have w.p.1 that:

(i)

$$
\begin{aligned}
&\left.\frac{\partial(\overline{k*\delta(y_t-x_t)})}{\partial p}\right)_{PA} \\
&= \frac{1}{T+1}\sum_{t\in M^*(T)}\left[\frac{f(z_t)}{1-F(z_t)}(t-t')a\right]*\left[-\frac{k}{1+R(q)}\right] \\
&\to \sum_{n=1}^{\infty} n f_n(q)*\left[-\frac{ak}{(1+R(q))^2}\right].
\end{aligned}
$$

(ii)

$$
\begin{aligned}
&\left.\frac{\partial(\overline{(S-x_t)*\delta(y_t-x_t)})}{\partial p}\right)_{PA} \\
&= \frac{1}{T}\sum_{t:x_t<s} -(t-t')a + \frac{1}{T+1}\sum_{t\in M^*(T)}\left[\frac{f(z_t)}{1-F(z_t)}(t-t')a\right]*[E[D]-E[D]] \\
&= \frac{1}{T}\sum_{t:x_t<s} -(t-t')a \to \frac{1}{T}*-T*a = -a.
\end{aligned}
$$

(iii)

$$
\begin{aligned}
\left.\frac{\partial\bar{x}_t^+}{\partial p}\right._{PA} &= \frac{1}{T}\left[\sum_{t:x_t\geq s,x_{t+1}>0}(t-t')a + \sum_{t:x_{t+1}>0} a\right] \\
&+ \frac{1}{T+1}\sum_{t\in M^*(T)}\left[\frac{f(z_t)}{1-F(z_t)}(t-t')a\right]*[E[(s-D)^+]-E[X^+]] \\
&\to \frac{a}{1+R(q)}\sum_{n=1}^{\infty} n\int_0^q f_n(u)F(s+q-u)du + aP(X>0) \\
&+ \frac{a\sum_{n=1}^{\infty} n f_n(q)}{1+R(q)}[E[(s-D)^+]-E[X^+]].
\end{aligned}
$$

The first term on the right hand side is derived as follows: take $t - t' = n$, $n = 1, 2, \ldots$, then we have

$$\frac{1}{T}\sum_{t:x_t \geq s, x_{t+1}>0}(t-t')a \to \frac{a}{1+R(q)}\sum_{n=1}^{\infty} n \cdot P(x_{t'+n} \geq s, x_{t'+n+1} \geq 0)$$

$$= \frac{a}{1+R(q)}\sum_{n=1}^{\infty} n \cdot \int_0^q P(D_{t'} + \ldots + D_{t'+n} < s+q | D_{t'} + \ldots + D_{t'+n-1} = u) f_n(u) du$$

$$= \frac{a}{1+R(q)}\sum_{n=1}^{\infty} n \cdot \int_0^q P(D_{t'+n} < s+q-u) f_n(u) du$$

$$= \frac{a}{1+R(q)}\sum_{n=1}^{\infty} n \cdot \int_0^q f_n(u) F(s+q-u) du.$$

$$(iv) \quad (\frac{\partial \bar{x}_t^-}{\partial p})_{PA} = \frac{1}{T}[\sum_{t:x_t \geq s, x_{t+1}<0} -(t-t')a - \sum_{t:x_{t+1}<0} a]$$
$$+ \frac{1}{T+1}\sum_{t \in M^*(T)} [\frac{f(z_t)}{1-F(z_t)}(t-t')a] * [E[(s-D)^-] - E[X^-]]$$
$$\to -\frac{a}{1+R(q)}\sum_{n=1}^{\infty} n \int_0^q f_n(u)(1-F(s+q-u))du - aP(X<0)$$
$$+ \frac{a\sum_{n=1}^{\infty} n f_n(q)}{1+R(q)} \cdot [E[(s-D)^-] - E[X^-]].$$

The first term on the right hand side is given by the similar derivation as (iii)

$$\frac{1}{T}\sum_{t:x_t \geq s, x_{t+1}<0} -(t-t')a \to -\frac{a}{1+R(q)}\sum_{n=1}^{\infty} n \cdot P(x_{t'+n} \geq s, x_{t'+n+1} < 0)$$
$$= -\frac{a}{1+R(q)}\sum_{n=1}^{\infty} n \cdot \int_0^q f_n(u)(1-F(s+q-u))du.$$

$$(v) \qquad (\frac{\partial \overline{(-pD_t)}}{\partial p})_{PA} = -\frac{1}{T}\sum_{t=1}^{T}(-2ap+b+\varepsilon_t)$$
$$+ \frac{1}{T+1}\sum_{t \in M^*(T)} [\frac{f(z_t)}{1-F(z_t)}(t-t')a] \cdot [-pE[D] + pE[D]]$$
$$\to 2ap - b.$$

For (i), we have

$$F_n(q) = P(D_1 + \ldots + D_n \leq q)$$
$$= P(\varepsilon_1 + \ldots + \varepsilon_n \leq q - n(b-ap)) = F_{\varepsilon,n}(q - n(b-ap)),$$

and $\frac{\partial F_n(q)}{\partial p} = f_{\varepsilon,n}(q - n(b - ap)) * na = a * nf_n(q)$, hence,

$$\begin{aligned}\frac{\partial(k/(1+R(q)))}{\partial p} &= -\frac{k}{(1+R(q))^2}\frac{\partial R(q)}{\partial p} \\ &= -\frac{k}{(1+R(q))^2}\frac{\partial \sum_{n=1}^{\infty} F_n(q)}{\partial p} \\ &= -\frac{ak\sum_{n=1}^{\infty} nf_n(q)}{(1+R(q))^2}.\end{aligned}$$

This is consisted with the limit derived in part (i).
For (ii), we have $\frac{\partial(E[D])}{\partial p} = -a$, which is consisted with the result from (ii).
For (iii), applying the similar approach by Fu (1994a), we have

$$E[X^+] = \frac{1}{1+R(q)}[\bar{F}(s+q) + \bar{F}(s)R(q) + \int_0^q R(u)F(s+q-u)du],$$

and

$$P(X > 0) = \frac{1}{1+R(q)}[F(s+q) + F(s)R(q) + \int_0^q R(u)f(s+q-u)du.$$

where $\bar{F}(x) = \int_0^x F(x)dx$.
Differentiate the expected value with respect to price, we have,

$$\begin{aligned}\frac{\partial E[X^+]}{\partial p} &= \frac{a}{1+R(q)}\sum_{n=1}^{\infty} n\int_0^q f_n(u)F(s+q-u)du + aP(X>0) \\ &+ \frac{a\sum_{n=1}^{\infty} nf_n(q)}{1+R(q)} \cdot [E[(s-D)^+] - E[X^+]].\end{aligned}$$

This is also consisted with the limit derived in part (iii).
For (iv)

$$\begin{aligned}\frac{\partial E[X^-]}{\partial p} &= \frac{\partial E[X^+]}{\partial p} - \frac{\partial E[X]}{\partial p} \\ &= -aP(X<0) + \frac{a}{1+R(q)}\{\sum_{n=1}^{\infty} n\int_0^q f_n(u)F(s+q-u)du - \sum_{n=1}^{\infty} nF_n(q)\} \\ &+ \frac{a\sum_{n=1}^{\infty} nf_n(q)}{1+R(q)}\{E[(s-D)^+] - E[X^+] + E[X] - s + E[D]\} \\ &= -aP(X<0) - \frac{a}{1+R(q)}\sum_{n=1}^{\infty} n\int_0^q f_n(u)(1-F(s+q-u))du \\ &+ \frac{a\sum_{n=1}^{\infty} nf_n(q)}{1+R(q)}\{E[(s-D)^-] - E[X^-]\},\end{aligned}$$

matching the limit in part (iv).
For (v), we have

$$\frac{\partial(-pE[D])}{\partial p} = \frac{\partial(-p(b-ap))}{\partial p} = 2ap - b,$$

which is equal to the result from part (v).

This completes the consistency proof for the PA estimator with respect to price p. □

References

Chen, X. and Simchi-Levi, D., Coordinating Inventory Control and Pricing Strategies with Random Demand and Fixed Ordering Cost: The Finite Horizon Case, working paper, 2002.

Clark, A and Scarf, H., Optimal Policies for a Multi-Echelon Inventory Problem, *Management Science*, 6, pp475-490, 1960.

Federgruen, A. and Heching, A., Combined Pricing and Inventory Control Under Uncertainty, *Operations Research*, 47, pp454-475, 1999.

Feng, Y. and Xiao B., A Continuous-Time Yield Management Model with Multiple Prices and Reversible Price Changes, *Management Science*, 46, pp644-657, 2000.

Fu, M., Sample Path Derivatives for (s, S) Inventory Systems, *Operations Research*, 42, pp351-364, 1994a.

Fu, M., Optimization Using Simulation: A Review, *Annals of Operations Research*, 53, pp199-248, 1994b.

Fu, M. and Healy, K., Techniques for Simulation Optimization: An Experimental Study on an (s, S) Inventory System, *IIE Transactions*, Vol.29, No.3, pp191-199, 1997.

Fu, M., Optimization for Simulation: Theory vs. Practice, *INFORMS Journal on Computing*, Vol.14, No.3, pp192-215, 2002.

Fu, M. and Hu, J., *Conditional Monte Carlo: Gradient Estimation and Optimization Applications*, Kluwer Academic Publishers, 1997.

Gallego, C. and Van Ryzin, G., Optimal Dynamic Pricing of Inventories with Stochastic Demand over Finite Horizons, *Management Science*, 40, pp999-1020, 1994.

Glasserman, P., Gradient Estimation Via Perturbation Analysis, Kluwer Academic Publishers, 1991.

Ho, Y.C. and Cao X. R., *Discrete Event Dynamic Systems and Perturbation Analysis,* Kluwer Academic, 1991.

Iglehart D., Optimality of (s, S) Policies in the Infinite Horizon Dynamic Inventory Problem, *Management Science*, 9, pp259-267, 1963.

Jacobson, S. and Schruben, L. W., A Review of Techniques for Simulation Optimization, *Operations Research Letters*, 8, pp1-9, 1989.

Kushner, H.J., and Yin, G., *Stochastic Approximation Algorithms and Applications*, Springer-Verlag, New York, 1997.

Spall, J. C., Stochastic Optimization, Stochastic Approximation and Simulated Annealing, in Encyclopedia of Electrical and Electronics Engineering (J. G. Webster, ed.), Wiley, New York, 20, pp529-542, 1999.

Safizadeh, M. H., Optimization in Simulation: Current Issues and the Future Outlook. *Naval Research Logistics*, 37, pp807-825, 1990.

Scarf, H., The Optimality of (s, S) Policies in they Dynamic Inventory Problem, Mathematical Methods in the Social Sciences, Stanford University Press, 1960.

Scarf, H. Inventory Theory, *Operations Research*, 50, pp186-191, 2002.

Thomas, L. J., Price and Production Decision with Random Demand, *Operations Research*, 22, pp513-518, 1974.

Veinott, A. and Wagner, H., Computing Optimal (s, S) Inventory Policies, *Management Science*, 11, pp525-552, 1963.

V

SOFTWARE AND MODELING

NETWORK AND GRAPH MARKUP LANGUAGE (NaGML) - DATA FILE FORMATS

Gordon H. Bradley
Operations Research Department, Naval Postgraduate School, Monterey, CA 93943, bradley@nps.edu

Abstract: The Network and Graph Markup Language (NaGML) is a family of Extensible Markup Language (xml) languages for network and graph data files. The topology, node properties, and arc properties are validated against the user's specification for the data values. NaGML is part of a component architecture that reads, validates, processes, displays, and writes network and graph data. Because it implements a family rather than a single xml language, NaGML offers (1) flexibility in choosing property names, data types, and restrictions, (2) strong validation, and (3) a variety of data file formats. This paper demonstrates these points with a sampling of the possible data file formats.

Key words: networks; graphs; Extensible Markup Language; xml; XML Schema; open source; data file format.

1. INTRODUCTION

The Network and Graph Markup Language (NaGML) is the centerpiece of an open source software project, the Network and Graph Project (Bradley, 2004a), that is constructing a suite of tools to read, validate, process, display, and write networks and graphs. NaGML is a family of xml languages to represent the topology (nodes, arcs, node sets, arc sets, and subgraphs), node properties, and arc properties. The author of a data file specifies the name and data type for each property as well as additional restrictions on the data values. The topology and values of the node and arc properties are validated using XML Schema technology. The NaGML family supports a variety of data file formats that correspond to common data formats for networks and graphs.

Figure 1 is a network that is represented in a NaGML xml language. The specific NaGML language is specified in a description that appears in the first few lines of the figure. Figures 2-6 are partial data files of different networks each represented in a different NaGML language. All NaGML data files are processed by a single program that uses the description in each file to automatically construct a XML Schema for the file and then use it to validate the network topology and the property values in the file. Thus NaGML users have the full power of XML Schema validation without mastering the (complex) task of constructing a comprehensive schema for their data. The description of the NaGML language in each file is used by a program that reads the data and constructs data structures to hold the network topology and property values. This paper introduces the NaGML languages and presents a sampling of the possible data file formats.

NaGML is intended to support the widest possible audience of people who use networks and graphs as well as people who construct algorithms for them. The NaGML system has simple entry points that help the casual user construct and manipulate small instances. It is also scales well for the large instances that may be constructed and consumed by production programs. This surprising capability to support a wide range of users with a variety of requirements is based on the flexibility of NaGML. NaGML achieves this by being not a single xml language, but rather a family of xml languages each with its own XML Schema whose construction is hidden from the user. Each user employs the capabilities of a custom-built xml language with a schema that is automatically constructed and applied.

The Network and Graph Project software has a component architecture that includes separate programs to read data files, validate the topology and properties, construct internal data stores, execute algorithms, display static and dynamic views, link to external systems (for example, spreadsheets and databases), and construct data files. The architecture is "loosely coupled" in the sense that it consists of components with well-defined interfaces that allow multiple implementations of each component. This allows users to construct a system from components that best meets their needs and it allows contributors to the Network and Graph Project to construct their own components that work with other components in the system.

Relational databases and SQL are the de facto standards for data storage and data access; xml has become the de facto standard for sharing data among applications. Xml is preferred for data exchange among organizations because providing xml files is preferable to allowing direct access to databases. Also since xml is character based (rather than binary) it is platform and operating system independent and thus ideal for transmission over the Internet. Recently xml has been introduced into document and spreadsheet programs to allow interoperability of content, for example, see

(Goldfarb, Walmsley, 2004). There has been significant interest in developing xml languages for many data domains (Hunter, 2002; Ray, 2001).

NaGML has been designed to support the vigorous and diverse community of people who use networks and graphs for many purposes in a variety of contexts. Applications involve problem instances that range from a handful of nodes to ones with thousands and even millions of nodes and arcs. Network and graph instances are constructed to model, analyze, solve, design, display, and entertain. For example, mathematicians and social scientists analyze structure, operations researchers and computer scientists compute optimal flows and efficient structures, engineers design roads and computer chips, planners develop land use, contractors schedule projects, graphic artists develop static and dynamic displays, scientists map molecules and solar systems, and intelligence analysts try to "connect the dots."

2. NETWORKS AND GRAPHS

A graph is a non-empty set of nodes, together with a set of arcs that are defined as pairs of nodes. The arcs can be directed (one node is the tail, the other the head) or undirected. A network is a graph with one or more properties associated with each node and arc. Each property has a fixed data type; there is a wide range of possible data types, for example, integer, double, boolean, string, etc. Nodes might have properties such as location (x-y or lat-long), cost, and description. Arc properties might be length, name, cost, open-shut, etc.

Common examples are road and street networks, water and power grids, and communications networks. In addition to these obvious physical networks, networks (and sometimes graphs) model a wide variety of other applications such as assigning people to jobs, scheduling, bid evaluation, organization charts, and circuit board layout. Virtually all the applications envisioned for NaGML have node or arc properties (or both) and are thus networks rather than graphs; the "and graphs" was added to distinguish NaGML from markup for other networks that do not have nodes and arcs. In the subsequent discussion we drop the "and graphs."

Network data files are often input for a variety of algorithms that construct shortest paths, determine the flow of goods, schedule activities, design chips, etc. Some applications use networks as a data model to structure data. These applications use networks to store, and perhaps visualize, data. Some applications may involve only nodes and node properties; for example, data about locations (nodes) displayed on a map. NaGML capabilities support these diverse applications.

3. INTRODUCTION TO XML AND NaGML

Xml is a metalanguage for defining xml documents. It is not a language itself; instead it is a set of rules to define and construct xml documents. Markup is text that is added to a document to add meaning and structure that can then be used to automate processing of the document. This modest description hides the full range of capabilities that have been built around and on top of xml and that have lead to the rapid and widespread use of xml and xml-related technologies. See (Hunter, 2002; Ray, 2001) for a discussion of xml, see (Bradley, 2003a; Bradley, 2003b) for a discussion with operations research examples.

Xml is an open source standard that was developed by, and is supported by, the World Wide Web Consortium (W3C) see (World Wide Web Consortium). It is platform independent. There are a number of free, open source, and proprietary tools available for the efficient construction and processing of xml documents.

As shown in Figure 1, xml markup consists of elements and attributes. The elements are enclosed by a start tag: <Node nodeID="1"> that begins with the name of the element and optionally includes name-value pairs called attributes. Each element must be closed with a matching end tag: </Node>. Elements can include other elements and text (also called PCDATA). The text of an element is everything between the start tag and the end tag that is not enclosed in another element. NaGML follows common practice that limits an element to contain other elements or text, but not both. Elements are fully nested in that any start tag must be matched with its end tag inside any enclosing element. An element may also be an "empty" element that combines the start and end tags: <Arc tail ="2" head="1"/>

Each xml document must have a unique "root" element. An xml document is a rooted tree that is almost always written in the indented format of Figure 1. By convention, element names begin with upper case letters. The name of an attribute is always followed by an equal sign and the value of the attribute in double quotes: name="Length" dataType="xs:double". We follow the convention that attribute names begin with lower case letters; in the text of this paper we will italicize attribute names and element names to make them stand out.

```
<?xml version="1.0" encoding="UTF-8"?>
<!-- NetworkAndGraphML for networks and graphs -->
<NaGXML    xmlns:xsi="http://www.w3.org/2001/XMLSchema-instance">
  <Description>
  <NodeProperties>
    <NodeProperty name="Xpixel" dataType="xs:nonNegativeInteger"/>
    <NodeProperty name="Ypixel" dataType="xs:nonNegativeInteger"/>
    <NodeProperty name="Symbol" dataType="xs:string"/>
  </NodeProperties>
  <ArcProperties>
    <ArcProperty name="Length" dataType="xs:double"/>
  </ArcProperties>
  </Description>
  <Data>
    <Node nodeID="1">
      <Xpixel>100</Xpixel>
      <Ypixel>100</Ypixel>
      <Symbol>black circle</Symbol>
    </Node>
    <Node nodeID="2">
      <Symbol>red  circle</Symbol>
    </Node>
    <Node nodeID="15">
       <Xpixel>300</Xpixel>
       <Ypixel>200</Ypixel>
    </Node>
    <Node nodeID="4">
       <Xpixel>350</Xpixel>
       <Ypixel>125</Ypixel>
    </Node>
    <Arc tail ="4" head="15">
      <Length>34.5</Length>
    </Arc>
    <Arc tail ="2" head="4"/>
    <Arc tail ="15" head="4">
      <Length>200.0</Length>
    </Arc>
    <Arc tail ="2" head="1"/>
  </Data>
</NaGXML>
```

Figure 1. Simple data file format that uses defaults for node and arc identification.

Since the network and graph community is international in scope, and since a data file format standard should be constructed to survive many years, it is appropriate that NaGML use xml and fully embrace the internationalization that choice allows. Xml is not "ANSI-centric" nor is it "English-centric." It is character based (as opposed to binary) and contains full support for all character sets and all human languages. In addition to support for English, there are over 35 different encodings. The practical reality is that today (2004) few of the 35+ encodings that have been designed have been implemented into working code. However, by providing for multiple encodings, xml has laid the foundation for full internationalization, and thus made it likely that xml will remain a dominant data store format for decades, if not centuries.

An xml language (also called a tag set, an xml vocabulary, document type, xml dialect) is a specification for a set of xml documents that places restrictions on the names of elements and attributes, the structure of the elements, and the values of the data. Each xml language should have a schema that is a formal specification of these restrictions. An xml document is an instance of a particular xml language if it conforms to the schema that defines the language; this is determined by validating parser (Duckett et al., 2001; Hunter et al., 2001; van der Vlist, 2002).

There are several different kinds of schemas associated with xml (DTD, XML Schema, RELAX NG, etc.) (Duckett et al., 2001; van der Vlist, 2002). The most widely used of the comprehensive schemas (this excludes DTD) is the XML Schema defined by the W3C (World Wide Web Consortium). The schema is distinguished from its competitors by the capitalization of the X, M, L, and S characters. Every XML Schema document is also an xml document. Each kind of schema has its own characteristics that influence the definition of an xml language and determine what errors are found by validation. Here we discuss XML Schema exclusively. XML Schema is good for expressing data types for element and attribute data values. XML Schema provides 44 built-in data types that cover numbers, strings, boolean, time, dates, etc. In the figures the data types all have the prefix "xs:". This indicates the Namespace associated with XML Schema; the prefix will be omitted when discussing the data types in the text. Data values can be further constrained with restrictions to these types, by applying patterns expressed as regular expressions, or by enumerating choices. The figures below give some idea of the range and detail that is possible. XML Schema is a so-called "closed" schema in which the structure of the document and the name of all elements must be included in the schema. A NaGML user selects the names, data types, and restrictions for their data (and thus describes an xml language for their data files) and a NaGSchemaConstruct component automatically constructs an XML Schema to validate the data.

Defining a new xml language is a difficult task that often involves some serious trade-offs. One goal is to make the language comprehensible and flexible, so that as many people as possible adopt it as a standard. Wide adoption supports interoperability of data and can lead to the development of associated software to read, write, process, transform, and visualize data files. Another goal is to make the xml language extensible, so that it can accommodate evolving user requirements. The final goal is that the schema be detailed and comprehensive, so that it enforces strong validation on xml documents.

New and emerging requirements for interoperability of valid data (particularly across the Web) place a new set of demands on network and graph data representations. Fast, high-volume exchange of data between different organizations and tightly coupled applications that do not have humans in the process must be concerned that data file errors (in structure and in data values) do not corrupt downstream processing. Thus, each data producer and consumer must guard against a variety of data errors. Much of this responsibility can be shifted to xml validating parsers and thus greatly reduce the amount of error checking that must be included in application software. The producer of data can validate a data file before sending it to someone or some application for further processing. The validation by the data producer helps identify data errors as the data is produced—this is the time and place where it is most effective to correct errors. Consumers of validated data know that the data file conforms to the schema and thus does not contain certain errors in structure and content.

The schema defines the structure and the content of the data file and identifies as errors only deviations from this. The "strength" of a validation is a measure of the number, scope, and kinds of errors that can be identified. The construction of the schema determines which errors will be caught by validation. For example, a data item that should be "A123" could be validated to be a string (and thus, "this is data" would be valid), or a string with no spaces and beginning with a letter, or as an English upper case letter followed by exactly three digits, etc. The figures below show some of the validation that XML Schema supports. The construction of a schema for an xml language can be a complex engineering task with critical issues such as how much detail to include—more detail identifies more possible errors, but simultaneously reduces the number of documents that conform to the schema. Validation cannot catch all data errors (for example, entering "3187" instead of "3817"), but many errors can be caught and using validating parsers to check the data is preferable to writing application-specific code.

The NaGML design philosophy has been to make schemas as "strong" as possible and thus include as many checks as possible in the validation. This minimizes the error checking that a NaGReader must do. The reference implementation of the NaGSchemaConstruct has achieved this. However, validation for large data files can be demanding on computer time and space resources. It is anticipated that, in particular situations, some of these errors may not be possible given how the data is constructed, thus it is unnecessary to check for them during validation. For some applications (particularly those with larger data files) it may be efficient to move some error checking from the validation to the reader. Also, the effort to validate a data file can be reduced if the user selects only certain data formats.

4. NETWORK AND GRAPH PROJECT LOOSELY COUPLED COMPONENTS ARCHITECTURE

From its inception, NaGML has been much broader than a markup language. The Network and Graph Project includes a design for a comprehensive software system for networks that supports the full spectrum of processing including: construct, read, validate, store, solve, display, link, and write. This includes multiple data structures so that an algorithm can be paired with a data structure that best supports its calculations, visualization of both static structures and dynamic processes so that algorithms can be animated and analyzed, and linking to external systems for display and computation. The system is not "read data, apply algorithm, print solution," instead, the viewpoint is that the network data model is central and read, validate, solve, display, link, and write are just operators that transform the data.

The Network and Graph Project has a component architecture to support the operations mentioned above. The components have well-defined interfaces and are "loosely coupled" in the sense that the interfaces are minimal and abstract, and thus encourage the development of multiple implementations for each component.

The architecture of the Network and Graph Project presumes there will be multiple implementations for each component. This encourages the development of an open source community, where sharing of components allows users to select the combination of components that best supports their application. This also allows individuals who want to develop a new algorithm, innovative data structure, visualization tool, or analysis technique to concentrate on their interest, while using components constructed by others. This component architecture supports innovation by allowing easy access to, and testing of, new components.

One important activity of an open source component architecture project is to construct reference implementations that demonstrate that components that satisfy the interface can be effectively constructed. Currently (summer 2004), there are reference implementations for two NaGReaders, a NaGSchemaConstruct, a hash table based data store, a visualization tool to construct tables, a visualization system to animate time series data over a map (the application mentioned below), and a NaGWriter that constructs NaGML data files and comma-separated output to link to an Excel spreadsheet. NaGReaders access a data file and, guided by information in the file, optionally invoke another program that constructs a schema (XML Schema) for the file; optionally validates the file using a schema; and always processes the file to construct a "dataStore" that is the internal representation of the network. Following xml practice, if the data file fails validation, it should not be processed. The details of the flexible data formats that NaGML allows do not persist into the dataStore, thus an application can select from various data format/NaGReader combinations to construct the same dataStore, which then interacts with the other components.

5. DATA FORMATS

One of the goals of NaGML is to provide flexible data formats that support the full range of common network data formats. The focus of this paper is to demonstrate the variety of data file formats that are supported and the data validation that is provided.

The first thing to notice about the examples is that the user chooses the names for node and arc properties. This seems like a fairly basic requirement, but in fact, the design that permits this is the most innovative part of the NaGML design. XML Schema supports the development of a range of different languages, but in each specific language the names of the elements are fixed. The NaGML mechanism to allow users to select their own names for node and arc properties, and to specify data types and further restrictions for data values, is described in a companion paper (Bradley, 2004b) written for an xml audience. It is not exaggeration to say that the Network and Graph Project would not have been possible without this innovation to common xml practice.

The second thing to notice is the wide variety of data types that are supported. While it might seem sufficient for most operations research applications to have only integer and double properties, the extension to strings, booleans, times, dates, lat-long, URLs, etc. is essential for the full range of network applications. As mentioned below, the first application of NaGML was an operations research analysis where only a few of the properties are integer or double. Finally note the quite different ways the data files can be defined.

The first example is shown in Figure 1. The *Description* element contains the specification of each property. The NaGReader reads the data file and invokes a NaGSchemaConstruct program to construct the appropriate schema and then validates the topology and property data values. It then constructs a dataStore that contains the topology, properties, and data type information.

Each node and each arc has a unique identifier, *nodeID* and *arcID*. In Figure 1, the defaults are used; *nodeID* is an integer and is assigned in each *Node* element. The *nodeID* values need not be in order in the data file or contiguous integers (or even positive). The validation process checks that each *nodeID* is a legal integer and each node has a unique value. The default for *arcID* is integers 1, 2, ... assigned in the order the arc appears in the data file (called "document order"). The validation checks that each head and tail attribute is assigned to one of the nodes declared in the data file. The validation process checks that each *Xpixel* and *Ypixel* data value is a non-negative integer and each *Length* is a legal double value.

This format shows that with only a small amount of training a user can quickly construct a small network. The plans for the Network and Graph Project include having input from spreadsheets, databases, forms, and Web browsers that will allow even simpler access to NaGML capabilities.

This data format allows any number of nodes and arcs. There is no restriction on the ordering of nodes and arcs and no requirement that nodes precede arcs.

In the subsequent figures, the defaults are explicitly included in order to show where and how the parameters of the system are specified. In Figure 2, the *nodeID* is still integer, however, the user has specified the exact number of nodes and their *nodeID* by specifying the attribute *declared* = "10 to 13." The validation enforces that there will be exactly 4 nodes with the specified *nodeIDs*. The *arcID* is specified to be a string and the attribute declared = "Arcs only" indicates that they will be specified in the *Arc* elements. The *arcIDs* can be any string, but the string must be unique for each arc. Duplicate arcs (each with a different *arcID*) are allowed. The validation checks each property value, which now includes string, date, and boolean values. The integer node properties *Xpixel* and *Ypixel* specify pixel locations so that a NaGViewer can display the network. The non-negative integers are further restricted so the nodes will appear inside a window that is 400 by 600 pixels. All the values for node property *DateConstructed* must be correctly formatted XML Schema dates (for example, 2001-04-23). The arc property *RoadOpen* values must be boolean.

Due to page limitations, the following figures are only partial data files. See (Bradley, 2004a) for an extended version of this paper with complete data files, several tables with data from the figures, and links to the files.

```
<NodeID nodeIDDataType="xs:integer" declared="10 to 13"/>
<ArcID arcIDDataType="xs:string" declared="Arcs only"/>

<NodeProperty name="Xpixel" dataType="xs:nonNegativeInteger"
   maxInclusive="400"/>
<NodeProperty name="Ypixel" dataType="xs:nonNegativeInteger"
   maxInclusive="600"/>
<NodeProperty name="Symbol" dataType="xs:string"/>
<NodeProperty name="DateConstructed" dataType="xs:date"/>

<ArcProperty name="Length" dataType="xs:double"/>
<ArcProperty name="RoadOpen" dataType="xs:boolean"/>

<Arc arcID="a-23" tail ="12" head="13">
   <Length>160.0</Length>
</Arc>
 <Node nodeID="10"> <Xpixel>100</Xpixel>
     <DateConstructed>2001-04-23</DateConstructed>
     <Symbol>black circle</Symbol><Ypixel>100</Ypixel>
</Node>
```

Figure 2. Partial data file with *nodeID* 10 to 13 and *arcID* declared in each *Arc* element.

In the previous figures there is a *Node* element for each node. Some networks have no node properties; a common data format for this lists only the arcs with the implication that there should be a node created for the tail and head nodes specified. The *NodeID declared* = "Nodes then Arcs" indicates that a node is created for each *Node* element in the data file. In addition, absent a *Node* element, any node referred to in an *Arc* element is created. This option can be used even if there are node properties; in that case, a *Node* element is required only if the node has a non-empty property.

Figure 3 shows that it is possible to enumerate the possible values that an arc (or node) property can assume. The figure shows this for a string property but this works equally well for other data types.

Many network algorithms use a "forward star" data store where all the arcs with the same tail node are stored contiguously. Sometimes it is convenient to have this reflected in the data file Figure 4 shows data in forward star form. NaGML also supports reverse star, which is not shown. Forward star and reverse star data formats can be combined with any of the formats shown in the previous figures.

```
<NodeID nodeIDDataType="xs:string" declared="Nodes then Arcs"/>
<ArcID arcIDDataType="xs:string" declared="Arcs only"/>

<ArcProperty name="RoadCondition" dataType="xs:string">
  <Enumeration>closed</Enumeration>
  <Enumeration>open</Enumeration>
  <Enumeration>open if dry</Enumeration>
  <Enumeration>unknown</Enumeration>
</ArcProperty>

<Arc arcID="I 40" tail ="New York" head="Chicago">
  <Length>120.4</Length> <RoadCondition>open if dry</RoadCondition>
</Arc>
```

Figure 3. Partial data file with nodes implicitly declared and an arc property enumerated.

```
<NodeID nodeIDDataType="xs:string" declared="list">
    12 34 14 23 -3</NodeID>
<ArcID arcIDDataType="xs:NMTOKEN" declared="list">
    12to34 34to14 14to23 14to23Again
</ArcID>

<Node nodeID="12">
    <StartTime>13:23:59</StartTime>
    <Signal>-345</Signal>
    <ArcTail arcID="12to34" head="34">
        <Cost> 34.56 </Cost>
    </ArcTail>
</Node >
```

Figure 4. Partial data file with forward star format and lists of nodeIDs and arcIDs.

As shown in Figure 4, the user can specify lists of *nodeID*s and *arcID*s that must be used in the *Data* element. The validation checks that these, and only these, are used in the *Data* element. Lists of values in xml must be space-separated (comma-separated lists can be used, but they are treated as a single string and thus are not subject to the validation of the individual values that we demand). This means that values in lists cannot contain leading, trailing, or embedded spaces. Thus, while “San Francisco” can be used as a *nodeID*, *arcID*, or property, it cannot be used in a list. The XML Schema data types include a restricted string type, NMTOKEN, that does not allow leading, trailing, or embedded spaces. If the user intends to forbid spaces in the *nodeID*, *arcID*, or a property, specifying this restricted data type allows the validation to enforce the no-space decision.

The loosely coupled components architecture guarantees that the data file format is completely independent of the dataStore.

One of the advantages of xml is that the data is represented as character data (as opposed to binary formats that may be computer dependent) so it is "human readable." This encourages the descriptive (and often long) names for the properties that help to document the data files. In the resulting data files the ratio of markup to data values can be high. This is appropriate for files that are constructed and read by people. For large data files, and for data files that are constructed and consumed by computer programs without human interaction with the data, it is useful to reduce this markup. (Note: It is not clear that reducing the markup is necessary even for large data files because xml files can be efficiently and effectively compressed.) One mechanism is to shorten the property names as shown in Figure 5. For *NodeProperty* and *ArcProperty* elements, there is an optional attribute *label* that provides a name that can be used in generating reports or when the data is viewed by people.

There are several other ways to reduce markup. As shown for the *Node* element "123," non-empty node data can be included as attributes. This format is also a compact way to keep all the data associated with a node (or arc) together in the data file. Node property *Location* is entered so that the values of a single property are kept together in the data file. The data is in a space-separated list. The order of the data must conform to the order of the nodes in the data file. This ordering is unambiguous for each of the four different ways to assign *nodeID*s (and each of the four ways to assign *arcID*s).

Assigning default values for node or arc properties (see arc property *Value*) can reduce the size of the data file. The *default* value is passed to the dataStore; a NaGReader places the value into the dataStore only if the attribute *defaultInsert* is specified (see arc property *AnotherValue*).

Figure 6 shows the use of *NodeList* elements that offer yet another way to organize the data file. This format is useful if the property values are non-empty for only a recognized subset of the data. In addition to data entry, node sets and arc sets are useful to specify a part of the structure of the network and can be used to specify subgraphs. All node and arc sets and subgraphs are carried on into the dataStore so they can be used in algorithms, displays, and output files.

```
<NodeProperty name="T" dataType="xs:time" label="Time"/>
<NodeProperty name="S" dataType="xs:integer" label="Special Signal"/>
<NodeProperty name="Location" dataType="xs:string"/>
<ArcProperty name="Value" dataType="xs:double" default="100.0"/>
<ArcProperty name="AnotherValue" dataType="xs:integer"
   defaultInsert="0"/>

<Node nodeID="123">
   <NodeValues nodeID="123" S="-345" T="13:23:59"/> </Node>
<NodeListValuesAll>
<Location> A1 A2 B1 B2 B3 C1 C4</Location> </NodeListValuesAll>
<Node nodeID="100">
   <NodeValues nodeID="100" T="00:00:00"/> </Node>
```

Figure 5. Partial data file: short property names, properties as attributes, and property lists.

```
<NodeProperty name="Cost1" dataType="xs:double" label="operations cost"/>
<NodeProperty name="Cost2" dataType="xs:double" minInclusive="4.0"
      maxInclusive="7.0" />
<MultipleNodeProperties name="BothCosts" dataType="xs:double">
      Cost1  Cost2</MultipleNodeProperties>
<MultipleNodeProperties name="RestrictedCost1" dataType="xs:double"
   minInclusive="4.0"
   maxInclusive="5.0">Cost1</MultipleNodeProperties>
<NodeLists>
      <NodeList nodeListName="SomeNodes">2  4  6 </NodeList>
</NodeLists>
<Processing nodePropertyDuplicate="replace"
   arcPropertyDuplicate="replace" errorMessages="prompt"/>

<Node nodeID="1">  <BothCosts>4.3  5.6</BothCosts> </Node>
<Node nodeID="3">  <RestrictedCost1>4.5</RestrictedCost1> </Node>
<Node nodeID="4"> <BothCosts>4.43  5.46</BothCosts> </Node>
```

Figure 6. Partial data file with node list and multiple property elements.

It is sometimes convenient to group some of the node properties into a list. This may be to reduce markup or just to keep related data values together. A *MultipleNodeProperties* element is defined by a list of node properties. The basic philosophy behind NaGML is that every data value should be subject to validation and that validation should be as "strong" as possible. XML Schema validation allows only one data type for items in a list. For this reason, each *MultipleNodeProperties* element must have its own data type (plus any appropriate restrictions). As shown for *MultipleNodeProperties* element *BothCosts*, this usually means a relaxation of the data type to allow values from all the properties. However, it can also be used to further restrict a property value as shown for *MultipleNodeProperties* element *RestrictedCost1*. The property *Cost1* can be entered in a *Cost1* element or a *RestrictedCost1* element; for the former the validation checks that the value is double, for the later additional restrictions are imposed.

The *MultipleNodeProperties* and *MultipleArcProperties* elements are carried on into the dataStore. One use is to restrict the data displays for networks with large numbers of properties or to construct custom displays and reports or to create new data files with a subset of the properties.

The validation of each data file checks the data type and restrictions for each *nodeID*, *arcID*, and property value and checks that *nodeIDs* and *arcIDs* are unique. However, the great variety in the ways that a data value can be included in the data file makes it impractical to check if a data value has been specified in the *Data* element more than once. This can be viewed as a valuable feature in that additional data can be added to the end of a (perhaps large) data file to update a value in the file. The data file author has control over what a NaGReader does whenever a duplicate property value is encountered. The *Notes* element includes an optional element *Processing* that includes directions that are passed on to a NaGReader. The user can specify that a duplicate value replaces the previous value, is ignored, or causes a fatal error. The specification for NaGReaders details the order that the elements must be processed, thus the definition of "previous" is unambiguous.

In addition to the restrictions on the data types that have been shown in the previous figures, it is also possible to specify a regular expression (using the attribute *pattern*) to further restrict the value of *nodeID*, *arcID*, and property values. XML Schema includes a powerful regular expression capability based on UNIX and Perl regular expressions ("based" means there are a few non-obvious exceptions). Regular expressions are a powerful capability to tightly specify the format and values of special data values (for example, lat-long, non-standard times and dates, phone numbers, serial numbers, etc.); however, the processing cost can be significant.

Graphs and networks have been around for hundreds of years and they are used in a wide variety of contexts. It is important for NaGML to support a wide range of data file formats in order to expedite the transition to a standard data file format. The requirement to simultaneously support all the data formats shown in the previous figures is indicated in the attribute *dataFormat* = "general" in the *Description* element. This default *dataFormat* is supported with the reference implementation reader, which uses the

JDOM API (McLaughton, 2001) to access and then process the elements in a fixed sequence. JDOM constructs a tree structure for an xml document in memory. This allows "random" access to all parts of the representation of the data file, thus the *dataFormat* = "general" adds only a small additional computational cost to allow the top-level elements in the *Data* element to appear in any order.

For larger data files the JDOM construction is not practical; the most effective way to read the data file is in a single pass. This can be done using an xml SAX parser or by constructing a program to read the file directly (as done in the second NaGReader reference implementation). The attribute *dataFormat* in the *Description* element specifies the structure that the user has selected for the contents of the *Data* element. As shown in the previous figures, the default choice of "general" offers the greatest choice of structure and order. The choice of "one pass" means that the schema that is constructed for the data file must guarantee that any NaGReader can read the data file in a single pass through the file. One pass imposes several "define before use" restrictions that must be enforced in the validation. In particular, since each arc requires a reference to a tail node and a head node that should be defined before the arc, all node elements must precede all arc elements and *ArcTail* and *ArcHead* elements are not allowed inside *Node* elements. Also the *NodeList* and *ArcList* elements (in the *Data* element) should follow the *Node* and *Arc* elements. The schema that is constructed enforces this ordering.

There are a number of choices that could be made to limit the data file format and thus allow more efficient readers. The only restriction on extending NaGML in this way is that the schema that is constructed must enforce the restrictions on structure and order, and any NaGReader should refuse to read a file that has a format it cannot correctly process.

As discussed earlier, NaGML is intended for an international community and thus has been constructed so that the visible parts of the markup (element and attribute names) can be changed to different languages (xml encodings). In the reference implementation, NaGReaders access the names of elements and attributes through a list of string constants. Thus, any change in this list changes the text of the markup without modifying any of the code. In the same way, NaGML can be modified to construct tools for communities that work with points and lines, but who have a vocabulary that does not include nodes and arcs.

6. RELATED WORK

The first application of NaGML uses many of the components described here (Schneider, 2004). It is an application to track military incidents and then to analyze them for patterns that are changing over time. Each incident (node) has 34 properties, only a few are integers or doubles. Data collection has begun and a software system to collect, visualize, and analyze the data has been deployed.

The author of a NaGML data file specifies the name, data type, and restrictions for each node and arc property. This is the most innovative feature of NaGML. Previous proposals for xml languages for networks and graphs have defined a single xml language and thus have significant limitations on what can be in a data file and/or what validation can be applied (Bradley, 2003a; Brandes, et al.; Fourer, Lopes, Martin, 2004; Holt, Schurr, Sim, Winter; Martin, 2003; Punin, Krishnamoorthy).

There are proposed xml standards for optimization that cover linear, nonlinear, integer, and stochastic programming (Fourer, Lopes, Martin, 2004; Lopes, Fourer; Martin, 2003). Network optimization problems can be modeled as more general optimization problems, however, these system do not have special markup for network problems.

7. CONCLUSIONS

NaGML is a family of xml languages for network and graph data. As shown in the previous figures, NaGML offers flexibility in choosing property names, data types, and restrictions. NaGML has a simple syntax to specify the name of each property and its data type from among the 44 built into XML Schema. The data types can be modified by adding further restrictions and by applying regular expressions. NaGML offers strong validation. Using software from (Bradley, 2004a), a custom XML Schema is automatically constructed and used to validate that each data value confirms to the specified data type and restrictions. NaGML offers a variety of data file formats. As shown in Figures 1, 2, 3, 4, 5, and 6, data values can be specified in a variety of ways. All the possible data file formats are read by a single software component from (Bradley,2004a).

In addition to software components to construct schema, validate data values, and read data files, (Bradley, 2004a) has components to display network data and to write NaGML files. Future plans include components for executing network algorithms and visualizing static and dynamic views of algorithm calculations.

ACKNOWLEDGEMENT

This research has been supported by a grant from the Mathematical Sciences Division, Office of Naval Research.

REFERENCES

Bradley, G., 2003a, "Extensible Markup Language (XML) with Operations Research Examples," tutorial given at the Eighth INFORMS Computing Society Conference, January 2003, Chandler, AZ, diana.or.nps.navy.mil/~ghbradle/xml/PaperMay2003/GBradleyXMLTutorialJan03.zip.

Bradley, G., 2003b, "Introduction to Extensible Markup Language (XML) with Operations Research Examples," *INFORMS Computing Society Newsletter,* Vol. 24, Number 1, Spring 2003, page 1 (14 pages). HTML version with live links: http://faculty.gsm.ucdavis.edu/~dlw/bradleyNewsletter.htm

Bradley, G., 2004a, Network and Graph Project, see http://diana.or.nps.navy.mil/~ghbradle/NetworkAndGraphProject for a description of this open source project and a link to a repository that contains the project code, examples, and documentation.

Bradley, G., 2004b, "Schema Construction for a Family of xml Languages" (in preparation).

Brandes, U. Eiglsperger M., Herman I., Himsolt M., and Marshall, M., "GraphML," http://graphml.graphdrawing.org/.

Common Optimization Interface for Operations Research (COIN-OR), http://www-124.ibm.com/developerworks/opensource/coin/.

Duckett, J., et al., 2001, *Professional XML Schemas*, WROX.

Fourer, R., Lopes L., and Martin K., 2004, "LPFML: A W3C XML Schema for Linear Programming," http://gsbkip.uchicago.edu/fml/fml.html.

Goldfarb, C. F. and Walmsley P., 2004, *XML in Office 203*, Prentice Hall.

Holt, R., Schürr, A, Elliott Sim, S., and Winter A., "Graph Exchange Language," http://www.gupro.de/GXL/.

Hunter, D., et al., 2002, *Beginning XML*, 2nd edition, WROX.

Lopes, L. and Fourer R., "SNOML," http://senna.iems.nwu.edu/xml/.

Martin, K., 2002, "A Modeling System for Mixed Integer Linear Programming Using XML Technologies," December 11, 2002, revised February 27, 2003, 34 pages. http://gsbkip.uchicago.edu/xslt/pdf/xmlmodeling.pdf.

McLaughton, B., 2001, *Java & XML*, 2nd edition, O'Reilly.

Punin, J. and Krishnamoorthy M., "XGMML (eXtensible Graph Markup and Modeling Language)," http://www.cs.rpi.edu/~puninj/XGMML/.

Ray, E.T., 2001, *Learning XML*, O'Reilly.

Schneider, P., 2004, "Multivariate Change Point Detection in Counter-Insurgency Operations," Master thesis in Operations Research, Naval Postgraduate School, Monterey, CA (completion date September 2004).

van der Vlist, E., 2002, *XML Schema*, O'Reilly.

World Wide Web Consortium (W3C), http://www.w3.org.

SOFTWARE QUALITY ASSURANCE FOR MATHEMATICAL MODELING SYSTEMS

Michael R. Bussieck, Steven P. Dirkse, Alexander Meeraus, Armin Pruessner
GAMS Development Corporation, 1217 Potomac Street NW, Washington, DC
{MBussieck, SDirkse, AMeeraus, APruessner}@gams.com

Abstract With increasing importance placed on standard quality assurance methodologies by large companies and government organizations, many software companies have implemented rigorous quality assurance (QA) processes to ensure that these standards are met. The use of standard QA methodologies cuts maintenance costs, increases reliability, and reduces cycle time for new distributions. Modeling systems differ from most software systems in that a model may fail to solve to optimality without the modeling system being defective. This additional level of complexity requires specific QA activities. To make software quality assurance (SQA) more cost-effective, the focus is on reproducible and automated techniques. In this paper we describe some of the main SQA methodologies as applied to modeling systems. In particular, we focus on configuration management, quality control, and testing as they are handled in the GAMS build framework, emphasizing reproducibility, automation, and an open-source public-domain framework.

Keywords: Quality assurance; software; modeling systems; mathematical programming, automation

1. Introduction

Quality Assurance (QA) has become an essential component in most industrial and commercial undertakings and has increasingly become important in most software engineering sectors. See for example, the emergence of organizations focusing specifically on quality ([10] and [2]). Unfortunately, software quality assurance (SQA) has received much less attention in the Mathematical Programming (MP) community. Historically, many innovative solver technologies have emerged from the academic sector, where the emphasis has generally been on *performance* and not on QA as in the commercial sector. On the other hand, the commercial sector has always emphasized *reliability* as the primary goal. Given this focus on performance in the (academic) MP community, it is not surprising that the few papers addressing SQA methodologies in

MP focus mostly on the areas of performance testing and benchmarking. See for example [17], [5], [4], and [11]. Bussieck et. al. [3] have examined the QA steps necessary for reducing the risks of introducing new solver technologies into the community, although many of these procedures and processes focused on full system integration performance-type testing as well.

Traditional, commercial SQA techniques have always emphasized *full life-cycle testing* (as opposed to system integration testing only) and usually rely heavily on *auditing* and *peer-review* inspections. While these techniques are no doubt effective, the latter is quite expensive, and given the small MP industry, economically prohibitive. Note that many commercial MP companies fall into the small business category or into specialized research groups within larger organizations. Therefore, the focus of SQA activities for MP must rely on *reproducible and automated tools and testing.* We will focus on such activities, emphasizing how such tools and procedures improve overall product quality, reduce cycle time for new distributions, and reduce turnaround time for bug fixes. Furthermore, we place many of these tools in the *public domain* for use by customers and researchers alike.

The public availability of such models and testing tools has several advantages. In the absence of formal peer review and auditing activities, the use of public domain models and testing tools is important in that it allows customers and the MP community to become directly involved in the QA validation process. It also speeds up the dissemination of knowledge in the areas of algorithms and solver technologies from academics by allowing them to have direct access to public domain quality assurance tools. Finally, testing activities without the possibility of reproducibility are essentially meaningless since there is no verification ability of specific quality tests by customers at a future date.

While initial formal SQA processes were pushed for by commercial demand, the use of such methods and tools should be of interest to academics, MP software vendors, individual commercial users, as well as commercial companies. Academic users may be interested in performance testing and benchmarking, whereas a commercial client may need to verify that the third party MP software they receive is of quality[1]. The latter may have its own QA department for their products and domain-specific services, but needs assurance of the modeling system to satisfy their customers. Solver vendors want their solver technology to perform well in the market place and want assurance that the modeling system functions appropriately with their respective product. Fi-

[1]There is a certain imbalance between academic publication in MP and commercial jobs. Consider the unscientific quick-and-dirty study of finding jobs at Careerbuilder.com by keywords (May 12, 2004): "Quality Assurance" 3,556 results, and "Mathematical Programming" 8, a ratio of 444 to 1. A Google search by keyword on the same date results in 4,510,000 for "Quality Assurance" and 111,000 for "Mathematical Programming, a ratio of 40 to 1.

nally the individual user is interested in solving models accurately and to be able to focus on modeling instead of the solution process.

Modeling systems (and numerical software in general) differ from most software systems in that a model may fail to solve to optimality without the modeling system being defective. This *algorithm failure* differs from the traditional *software implementation defect* since the modeling system must be able handle the failure mode and provide the user with sufficient return information to determine the return state. Note that many solvers are available to developers of modeling systems only in library form (i.e. no source), essentially limiting interaction between the modeling system and the solver to black box input-output communication. This additional level of complexity requires *MP-specific QA activities* to test for such returns.

This paper is organized as follows: in §2 we describe general SQA principles, focusing on configuration management, testing, and quality control. In §3 we describe QA activities specific to modeling systems, emphasizing how components interact and where errors can occur. In §4 we show how these principles are applied in the GAMS build framework and how the use of client models can further improve quality. In §5 we give examples of client model testing activities and finally in §6, we draw conclusions.

2. Software Quality Assurance Principles

Various software quality assurance principles (or models) have been developed by different organizations to ensure that specific standards are met and to give guidelines on achieving these standards. Although these address the full software lifecycle, we will focus on configuration management and automated testing.

Standard Models

Many *QA standards and models* exist with a large number of choices. According to [12] "there are more than 300 standards developed and maintained by more than 50 different organizations." Popular models are the ISO 9001, which specifies requirements for a quality management system within an organization and the Software Engineering Institute (SEI) Capability Maturity Model (CMM), which provides a framework for continuous software process improvement [16], although many others are used, depending on user goals. The key notion is that they provide guidelines for conducting audits, testing activities, and for process improvement.

The CMM approach classifies the maturity of the software organization and practices into five levels describing an evolutionary process from chaos to discipline [16]:

- ***Level 1: Initial.*** The software process is characterized as ad hoc, and occasionally even chaotic. Few processes are defined, and success depends on individual effort and heroics.

- ***Level 2: Repeatable.*** Basic project management processes are established to track cost, schedule, and functionality. The necessary process discipline is in place to repeat earlier successes on projects with similar applications.

- ***Level 3: Defined.*** The software process for both management and engineering activities is documented, standardized, and integrated into a standard software process for the organization. All projects use an approved, tailored version of the organization's standard software process for developing and maintaining software.

- ***Level 4: Managed.*** Detailed measures of the software process and product quality are collected. Both the software process and products are quantitatively understood and controlled.

- ***Level 5: Optimizing.*** Continuous process improvement is enabled by quantitative feedback from the process and from piloting innovative ideas and technologies.

The challenge for many MP vendors is to move from Level 1, the chaotic, creative and exciting phase to Level 5 without losing creativity and, most importantly, to stay in business.

Configuration Management

Software configuration management (SCM) refers to all activities used to (1) identify change, (2) control change, (3) ensure that change is being properly implemented, and (4) to report changes in the software to others who may need to know of them [14]. This includes all activities related to version control and change control.

Change identification is often in the form of some audit string information, which details the product primary version number, minor version number as well as possible incremental bug fix releases or patches. Change control is necessary to ensure that existing source used for previous releases is not overwritten and only authorized personnel can add new source. *Accurate audit information* becomes increasingly important for software consisting of a high number of modules (GAMS currently has over 45), each of which may depend and build on various other modules. Furthermore, configuration management ensures that authorized changes are actually implemented in the formal build product. Some configuration management elements that are important include

Audit Strings. Audit strings are necessary to determine the exact versions of a particular product, particularly those that rely on several other modules. The audit strings become particularly important when tracking bug reports so that the exact configuration source (of both the actual product and the supporting modules) can be determined and appropriate fixes can be made.

Product versions are consistent. Use of version checking tools to determine if the version used in the previous official distribution, the latest version in the product repository, the version used for the build in the Makefile and the version used in the audit string are identical. The accuracy and unison of versions is necessary for bug tracking as well (See audit strings).

Product versions are frozen automatically for builds. This ensures that developers do not accidentally overwrite existing code and informs them if versions need to be bumped. Because of the large number of products that exist in the GAMS portfolio, each with numerous versions, automation tools (scripts) for tasks such as making current product directories read-only or read-write become increasingly important.

Quality Control and Testing

All testing activities should include processes to uncover defects during the complete life cycle of the software. The focus on *full life cycle testing* (as opposed to system integration testing only) is important, because the cost of defects rises exponentially the later the phase of the cycle [14]. The testing activities include, but are not limited to:

Unit testing. Testing of the individual component (solver and solver link module) using both black-box (input-output only) and white-box (known internal code structure) type tests.

Regression testing. Testing to determine if changes to the software or fixing a defect cause any problems to other components in the system.

System Integration Testing. Testing done to ensure that the entire product (base module plus solver modules) functions as intended and to specifications.

The emphasis of all testing activities is again on automation and reproducibility. Many of the tools are available publicly in the form of model libraries. The models in these libraries can be used by just running them within GAMS. In [3], some of these tools and models that are publicly available are already described.

Metrics

Most engineering processes involve measurements to make accurate assessments of the attributes of the product. The use of metrics is important in that it quantifies attributes of a given process or of the product. In particular, metrics such as number of defects of a particular type (critical, serious, cosmetic, etc.) are used within GAMS to determine if the product is ready to move from beta phase to a shippable product. Since development takes place continuously, metrics should be collected continuously as well to determine quality of the current product.

3. SQA For Modeling Languages and Systems

In this section we describe some of the *special* problems of maintaining software quality in the context of modeling systems which differs from traditional QA principles. Furthermore, we give some background on modeling systems in general to motivate the QA principles.

Basic Technical Principles

In the early days of mathematical programming systems, the existing techniques to construct, manipulate, and solve models required several manual, time-consuming, and error-prone translations into the different, problem-specific representations required by each solution method. Furthermore, the solution methods were usually tied to a specific architecture and platform and portability of models was virtually nonexistent.

Learning from these early techniques, most modeling languages today, including LINDO [15], GAMS [1], and AMPL [7], appearing in that chronological order, adhere to the following basic technical principles: (1) separation of model and solution methods, (2) computing platform independence, and (3) multiple solvers, platforms, and model types. The adherence to these principles has many advantages which are described, for example in [3].

Description of Components

In order to understand where possible defects can occur, an overview of the system architecture of modeling systems is necessary. The *base module* (or execution system) is the core of the system and is designed to translate the human-level algebraic model statements into the scalar formulation passed on to the *solver modules*. The base module includes the language compiler, which performs syntax and other checks to ensure the (grammatical) integrity of the model formulation. If the compilation is successful, the base module expands the model formulation and generates a (sparse) matrix to be passed on to the solver module to solve the problem. Additional user-specified solver options are passed on to the solver modules as well. The solver modules are essentially black box modules which return solve and model status as well as solution information (if it exists). Finally, there exist other *external modules*, which are not solvers, but are linked to the base module in a similar manner, to perform specific tasks. These include, for example, links to Matlab, Excel or data base interfaces, or conversion modules such as CONVERT [3] to translate the model into other modeling language formats or standards.

Chance of Failure

The reliability of a system is sometimes referred to as "mean time to failure." Although in traditional SQA terminology failure implies defect, in the MP world this refers to algorithmic failure to find a solution. While phenomenal progress has been made and we are now able to solve many difficult and large-scale problems, the steady state of mathematical programming software should *conservatively* still be assumed to be the failure state. Unlike other software systems, such as database applications, control systems, or other types of systems, where ever more detailed specifications and detailed testing can continually reduce the chance of defect (hypothetically to zero), no such paradigm for failure-free solves exists for optimization (or other numerical) software. In particular, it is likely that there will always exist models which cannot be solved reliably. Indeed, many of us have likely experienced unpredictable behavior in models, where a change of a single equation or data item suddenly proves to be the culprit in failing to solve. The use of SQA techniques can help minimize this chance of failure and provide graceful return information if no solution is found. It is because of this chance of failure that commercial modeling systems must focus on reliability, in particular by providing informative return information in case of failure. This level of QA complexity is in *addition* to QA activities associated with traditional defect prevention.

Defects (as opposed to failures) can occur at any phase of the process flow and can result in catastrophic malfunction or undesirable return information if a solution is not found. The defects can occur either in the compilation phase, the data manipulation phase, the model generation (matrix generation) phase, the solve phase by the solver modules, in the solve phase by the sub-solver modules (for example, an MINLP solver may make use of an NLP solver), or during the processing of the solution returned by the solver to the base module. A solve may initiate the execution of other modeling system components. At a minimum, any rigorous quality control testing procedures must address the possibility of failure at each of these phases nested several levels deep.

Uniform Return Status

Because of the inherent chance of failure, modeling systems must be adept at dealing with various return states. Unfortunately, solvers are not uniform in the amount or type of information returned to the base module. At a bare minimum this information includes the solution point, if it has been found, but could also include dual information or infeasibilities if the model was found to be infeasible. In order to ensure uniform return information, GAMS requires return codes for both the model and the solver from which one can accurately deduce the overall return state.

The *solver return status* refers to the status of the solver, for example normal, resource (time) interrupt, iteration interrupt, no solution, and error. For simplicity, numerical codes from 1-13 are assigned to each of these states. The *model return status* refers to the state of the entire model or of a particular point: for example, optimal solution, locally optimal, integer solution, infeasible, locally infeasible, unbounded, or error. These are mapped from 1-19. The status of error (13) should ideally never be returned. Rather, any return status should be mapped more specifically to well-defined return status. As described later in this paper, the error return code is to be treated as a serious bug in the software. It should be clear that without uniform return codes obtained from each solver call, regardless of the solver module, it may become increasingly difficult to do consistent error checking.

Status code analysis is handled in GAMS by use of *matrix filtering*. Through careful analysis, acceptable model/solve return status code combinations have been pre-determined and are ranked (a higher value indicates a better return state than a lower value). Return codes can be used in system integration testing by using matrix filtering: flagging all model/solver combinations with model/solver status code combinations having a ranking that is less than the specified threshold. Roughly, the higher the threshold, the stricter the testing pass criteria.

In Table 1, we show all possible acceptable return code combinations with their respective ranking. Solve status codes are in the columns and model status codes on the rows. For example a Model Status / Solve Status code combination of optimal solution / normal completion (1,1) has a high ranking of 9, whereas a combination of Error no Solution / Error Solver Failure (13,13) is never acceptable (denoted by a dot .).

4. Software Quality Assurance in GAMS

GAMS is available on 7 different platforms (Windows, Linux and 5 other UNIX environments) and consists of over 45 different modules and products. Thus the build process is quite complex and requires a single standard repository from which source and libraries are extracted. Software builds for the dif-

Table 1. Matrix Filtering. Ranking of model and solver status return codes

Model Status \ Solver Status	NormalCompletion	IterationInterrupt	ResourceInterrupt	SolverInterrupt	EvalErrorInterrupt	CapabilityProblem	LicensingProblem	UserInterrupt	SetupFailure	SolverFailure	SolverError	SolverSkipped	SystemFailure
1 Optimal	9	.	.	.	.	.	.	.	.	.	.	.	.
2 LocallyOptimal	9	9	9	9	9	.	.	9	.	.	.	.	.
3 Unbounded	9	.	.	.	.	.	.	.	.	.	.	.	.
4 Infeasible	9	.	.	.	.	.	.	.	.	.	.	.	.
5 LocallyInfeasible	9	.	.	.	.	.	.	.	.	.	.	.	.
6 intermediateInfeasible	.	5	5	5	5	.	.	5	.	.	.	.	.
7 IntermediateNonoptimal	.	5	5	5	5	.	.	5	.	.	.	.	.
8 IntegerSolution	9	9	9	9	9	.	.	9	.	.	.	.	.
9 Intermediate_NonInteger	.	5	5	5	5	.	.	5	.	.	.	.	.
10.IntegerInfeasible	9	.	.	.	.	.	.	.	.	.	.	.	.
11 Licensing_NoSolution	.	.	.	.	.	.	5	.	.	.	.	.	.
12 ErrorUnknown	.	.	.	.	.	.	.	.	.	.	3	.	3
13 Error_NoSolution	.	.	.	.	.	.	.	.	3	3	.	3	.
14 NoSolution	4	4	4	4	4	6	.	4	.	.	.	3	.
15 SolvedUnique	9	.	.	.	.	.	.	.	.	.	.	.	.
16 Solved	9	.	.	.	.	.	.	.	.	.	.	.	.
17 SolvedSingular	9	.	.	.	.	.	.	.	.	.	.	.	.
18 Unbounded_NoSolution	9	.	.	.	.	.	.	.	.	.	.	.	.
19 Infeasible_NoSolution	9	.	.	.	.	.	.	.	.	.	.	.	.

ferent platforms take place on different machines, which may exist in-house, or elsewhere.

Because of the complexity of porting to a new machine, which requires multiple compilers and utilities, initialization of a new machine requires installation of various software modules (including specialized GAMS-specific build scripts). This step is unfortunately not automated, although the build environment is relatively stable and new porting machines are only infrequently introduced. The build process can be described as follows for each platform: (1) Copy general build instructions (in the form of Makefiles) from the master repository to local environment, (2) for each product: extract product-specific build instructions and source in the form of Makefiles, build the product, re-deposit built product to master repository, (3) do post-build processes to create installation files.

The description of this process is of course simplified herein, although it should be clear that without the appropriate *build automation tools*, the necessary steps to do a single build would be extremely time consuming and the probability for failure high. Furthermore, because of dependencies of some products on other products, it becomes increasingly difficult to sort through dependencies manually.

The porting environment allows near-automated builds of entire distributions. Indeed, as an added SQA measure, we have fully *automated the build task* for Windows and Linux, so that a full build is completed automatically once a week. Such automated full compilation activities ensures that our porting environment is continually in a *buildable state*. This uncovers potential bugs early in the life cycle and in turn reduces cycle time for full release of new distributions. The automatic build also includes installation and testing activities so a continual analysis of porting source is possible.

Configuration Management

Within the GAMS porting environment, we use configuration management tools to ensure that product versions are consistent and version source integrity through automatic code freezes is maintained. Furthermore, we have a consistent audit string assignment for each module.

All GAMS modules can be easily identified in terms of their version numbers, build date, last source date change, as well as such information for all modules needed to build the product. Sample audit information includes the one line audit string (here for CPLEX)

```
GAMS/Cplex     Jan 19, 2004 WIN.CP.NA 21.3 025.027.000.VIS For Cplex 9.0
```

We should note that many standard source management and version control tools exist, notably for example CVS (`http://www.cvshome.org/`). Although we do not utilize any of these, any SQA activities should include rigorous version control processes to manage source. Our processes mainly focus on the automation process in maintaining the integrity of the porting repository. The argument for use of a simpler system is that our products consist not only of source in a single language (such as C++), but contains products written in various languages (or consists of modules written in various languages), libraries, and other tools. Management of these various products and source is simplified by maintaining a simple directory tree structure with simple maintenance scripts. It should also be noted that some of the libraries (or source codes) are obtained from sources using their own source management system. But an argument for management tools such as CVS can certainly be made.

Bug Tracking

At GAMS we use various bug tracking systems in order to communicate effectively with our external solver developers. In-house, the system consists of an e-mail based system which users can submit bug reports to and change statuses. Reports can also be viewed online internally. Because outside solver developers (or clients, if we are dealing with consulting projects) may have

Figure 1. (Left) Snapshot of weekly bug statistics by module (in percent). (Right) GAMS full test. Solve aggregation by model type.

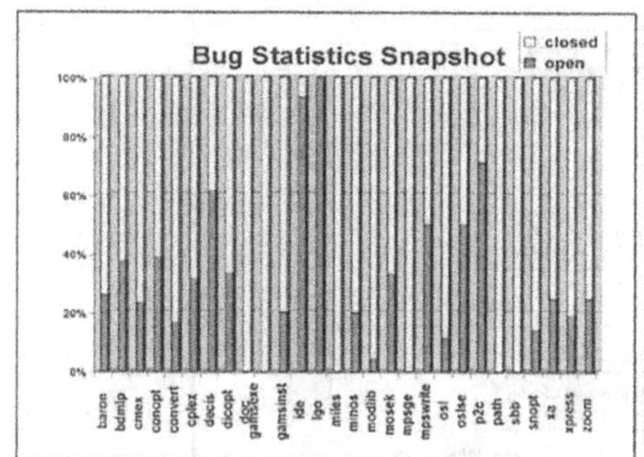

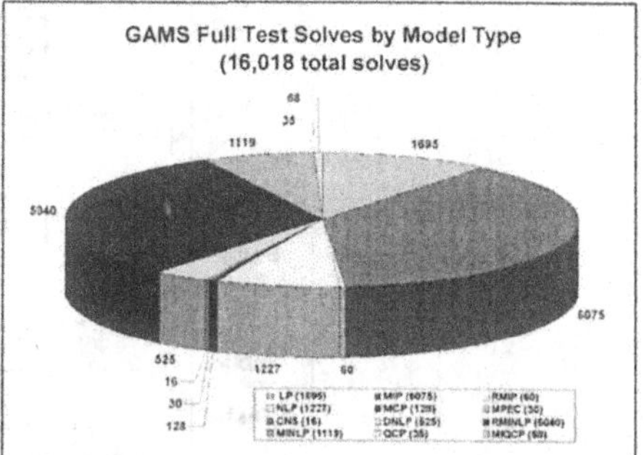

different bug tracking requirements, we also utilize different systems for notification for these parties. Flexibility is the key and we accommodate different systems to coordinate with external contacts.

As an example, our in-house bug tracking system sends *weekly e-mail bug statistics* reports of open, closed, and total number of bug reports for each module, which can be used to determine when a stable distribution is ready. In Figure 1 (left) we show a sample weekly statistics chart of open and closed issues (in percent) for each product.

Testing

Our SQA activities focus mainly on testing, in particular on automated and reproducible tests. In Figure 2 we show a flowchart of our general testing activities. The tests cover the full life cycle of development, build and integration. It should be noted that during each testing phase, potential new relevant test cases are added to the master list of test models, which ensures that (1) previous bugs will not occur again, (2) tests can be run automatically during system integration testing, and (3) tests are publicly available.

This means that, for example, during unit testing, if a particular defect is found, the relevant test to uncover the bug is added to the suite of tests, so that during full integration testing, the bug will be uncovered if it has not been fixed. Such activities are particularly effective in increasing software quality from one build iteration to the next and foster an environment of *continual quality improvement.* Our test cases are available in the following *testing libraries* (or suites), all of which are in the public domain.

Figure 2. GAMS SQA testing activities.

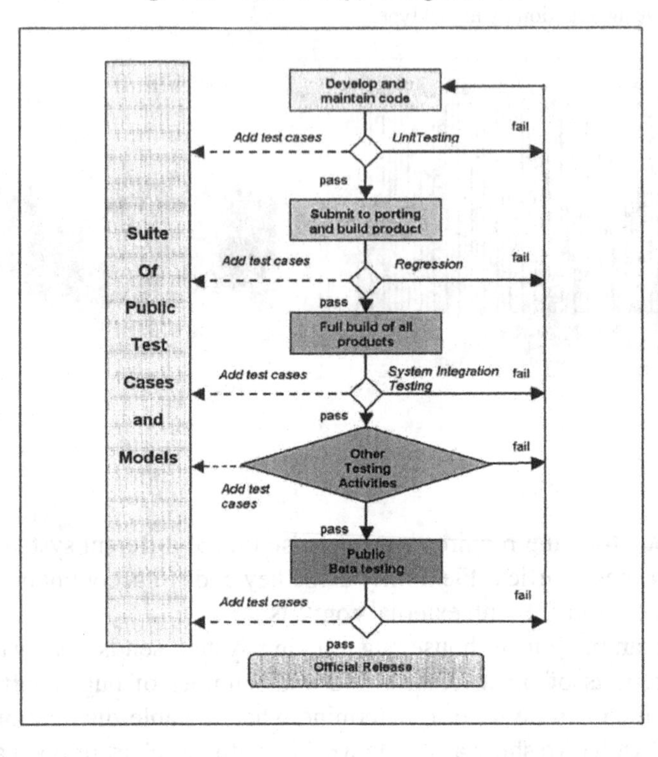

- ***GAMS Model Library.*** A collection of over 300 models, including practical models that can easily be extended to production models. The collection `http://www.gams.com/modlib/modlib.htm.` covers all supported model types.

- ***GAMS Testlib Quality Models.*** A new library of models developed for testing and quality control. These models are designed for use by GAMS staff and solver developers to test solver and base module correctness, special functions, and performance. The quality and correctness tests are integrated with the models and all failed tests are identified in a summary report.

- ***GAMS World Models.*** A collection of real-world models, as well as libraries from the academic literature. See `http://www.gamsworld.org/`.

In order to provide uniform access to these tools and models for users of other modeling systems, models can be converted into other common formats using the CONVERT utility (both as a GAMS and an online utility) for use by non-GAMS users. See [3] for details.

Unit testing is usually performed by the developer on a limited integration system on a local development machine. Regression testing may involve a *GAMS initial test*, which solves a sample model of each model type with all available solvers. It also includes the models from Testlib, which check the GAMS base module for possible defects. If the results of this testing activity are satisfactory, the full system integration test is initiated. Tests are run as pass/fail, where pass/fail is determined by the threshold limit specified for model/solver status return code combinations. See Section 3 (Uniform Return Status).

The full system integration test (referred to as *GAMS full test*) involves solving all demo GAMS models from the model library with all relevant solvers and solver combinations. For details, see the GAMS Library model *slvtest.gms* (`http://www.gams.com/modlib/libhtml/slvtest.htm`). This includes a total of over *16,000 total solves* for each platform. An aggregations of solves per model type is shown in Figure 1 (right).

Furthermore, the full test also includes running the full suite of Testlib quality models. These tests include tests to verify proper behavior for failure modes. For example, we may test for domain input violations of nonlinear functions. The number of quality tests run on each platform differs, since not all solvers are available for all platforms and some modules may not exist for certain platforms. For example, the Excel interface is tested only on the Windows platform. The total number of quality test models is about 140, each containing *numerous* pass/fail tests. New quality tests are added continuously.

The tests are run in fully automated mode, usually after every full build, with results sent via e-mail to the person building the system. The e-mail includes any possible anomalies, as well as a summary of model and solve status combinations. If there are anomalies, the model and solve combination is revealed, so that the bug can easily be reproduced on a local machine. The e-mail reports also serve the function of archiving test results for future analysis and comparison.

A sample excerpt of the full test summary sent via e-mail is shown below. The results are shown in terms of number of solves falling into each model status and solve status return codes. The absence of a model status of 13 (Error No Solution) or solver status code of 13 (Error System Failure) generally indicates a successful system test. In this case, a threshold of 3 has been set, so that any model / solve status return combinations with a ranking less than or equal to 3 are marked as failures (not shown in the Table). See Table 1 for the model/solver return code ranking matrix.

```
Total SOLVE records = 16018  all return codes

solvestatus  15969  RC= 1  1 NORMAL COMPLETION
```

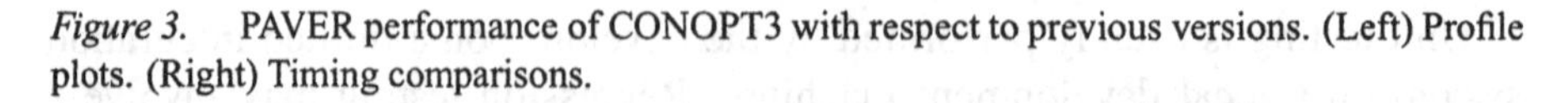

Figure 3. PAVER performance of CONOPT3 with respect to previous versions. (Left) Profile plots. (Right) Timing comparisons.

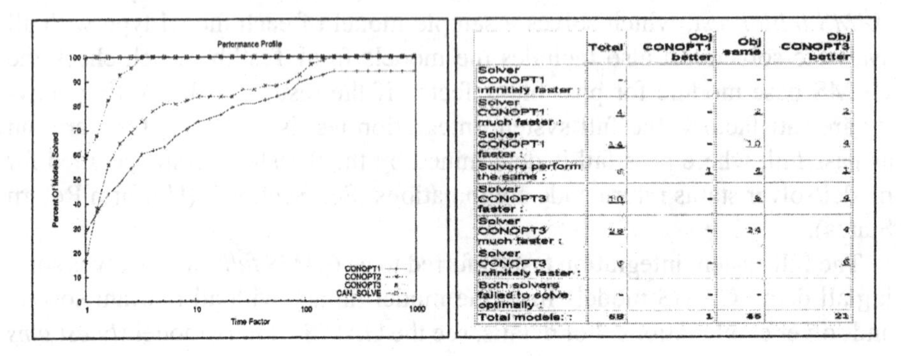

	Total	Obj CONOPT1 better	Obj same	Obj CONOPT3 better
Solver CONOPT1 infinitely faster :	-	-	-	-
Solver CONOPT1 much faster :	4	-	2	2
Solver CONOPT1 faster :	14	-	10	4
Solvers perform the same :	6	1	4	1
Solver CONOPT3 faster :	10	-	6	4
Solver CONOPT3 much faster :	28	-	24	4
Solver CONOPT3 infinitely faster :	6	-	-	6
Both solvers failed to solve optimally :	-	-	-	-
Total models: :	68	1	45	21

```
solvestatus     15  RC= 2   2 ITERATION INTERRUPT
solvestatus     34  RC= 4   4 TERMINATED BY SOLVER

modelstatus   8130  RC= 1   1 OPTIMAL
modelstatus   3210  RC= 2   2 LOCALLY OPTIMAL
modelstatus      1  RC= 3   3 UNBOUNDED
modelstatus   2322  RC= 4   4 INFEASIBLE
modelstatus    912  RC= 5   5 LOCALLY INFEASIBLE
modelstatus   1400  RC= 8   8 INTEGER SOLUTION
modelstatus     26  RC=10  10 INTEGER INFEASIBLE
modelstatus      6  RC=15  15 SOLVED UNIQUE
modelstatus     10  RC=16  16 SOLVED
modelstatus      1  RC=19  19 INFEASIBLE-NO SOLUTION
```

Miscellaneous Tests and QA Activities

The testing activities above uncover many deficiencies, although some other issues may not be addressed by these tests. Quality assurance is further enhanced through the following procedures and using the following tools:

- ***New development using standard I/O libraries.*** New solvers can be attached to GAMS using standard FORTRAN, C or Delphi I/O libraries. The use of standard libraries to communicate with GAMS reduces the number of bugs that are introduced, increases reliability, as well as performance. The libraries have been thoroughly tested for robustness, correctness and performance over many years and provide a reliable way to introduce new solver technologies.

- ***Performance.*** Performance is still an important criterion in SQA, both for solver developers and commercial users. Benchmarks run using large test sets can automatically be examined using automation tools, such as the public-domain PAVER Server [13] to verify solver performance and compare to other solvers. For example, in [6], performance of a new version of CONOPT (CONOPT3) is compared to existing versions (CONOPT1 and CONOPT2). The resulting performance profile plots [4] (created automatically using the PAVER Server) verified that CONOPT3 indeed does have tremendous performance increases from previous versions. See the results in Figure 4. The left side shows the performance profiles and the right figure the resource timings of CONOPT1 and CONOPT3. The use of such tools enables users to quickly identify trends in large data sets.

- ***Client models.*** Since GAMS has a heavy commercial client base, which demands reliable software, the use of model library models may not be sufficient. The client models may be interfaced with complex databases and make use of varying modules that interact in specific ways. Thus, GAMS Development has started an initiative of running client models in their full interfaced capacity. This ensures that models will solve on new distributions as they did with previous releases.
- ***Independent Solution Verification.*** The EXAMINER tool [8] can be used for examining points and making an unbiased, independent assessment of their merit. It is also useful when comparing the solutions returned by two different solvers. Finally, a tool like the EXAMINER allows one to examine solutions using different optimality tolerances and optimality criteria in a way that is not possible when working with the solvers directly.

5. Client Model Testing

The use of client models for SQA testing is important because it generally involves much more complex interfacing between different modules and gives us a better idea of real, large-scale commercial applications. It also provides a unique motivation: it gives clients assurance that their optimization applications are compatible with new GAMS system releases and gives the expected results. By expected results, we mean not only the ability to solve, but also to find the same solution[2] and have similar performance in doing so. The latter is important because even minor changes in solvers (for example changes in default settings), may unexpectedly cause tremendous increases in solve time, which may be unacceptable for the particular application time frame. In this section we describe a client application, which we use for our in-house QA client testing.

[2]To define what we mean by having the 'same solution' is often an unexpectedly difficult problem and a source of frustration for the innocent user. In general, it is impossible to exactly reproduce solutions to optimization models between different software releases, computing platforms, or real-time events. The term 'same solution' needs to be defined for each application.

MARKAL [9] is one of the most widely used energy/environment/economy planning models, playing a central role in Climate Change analysis for numerous countries and communities around the world. Development is coordinated by an international group, the Energy Systems Analysis Program (ETSAP). For current information, see `http://www.etsap.org/markal/main.html`. The model and its data are managed by several different application environments and the model can be operated in different modes offering an almost infinite number of possible model instances. One particular application environment, ANSWER, a graphical interface is shown in Figure 5. Because of the interfacing capabilities and the large number of components in the MARKAL suite, additional failures are possible, which may not be covered by the simpler models in the testing suites described previously. The simpler integration testing also does not do any performance analysis, which is often vital in practical commercial modeling applications. A typical model run may involve dozens of optimization steps and intricate data manipulation, all on very large data sets. To make automatic testing possible, substantial restructuring of the internal MARKAL model management is required. Isolation of the core components and communication with the application environment is done via the GAMS Data Exchange (GDX) format. It is now possible to automatically generate complex job streams that can be operated outside the MARKAL application environment. Those job streams are then used in routine application support and may end up in the customer test suite.

The client model test set consists of a number specific of job streams with additional pass/fail tests designed and implemented jointly with the client. The job streams and all required data and subsystems are archived and become part of a QA support contract. The execution of such test suites is fully automated and QA certificates are generated and archived to allow audits on the QA process. Exceptions are tracked, and a GAMS system will not go into a beta release unless all exceptions have been resolved which, in some cases, may require a redefinition of tests or even making changes in the client models. In a more traditional manufacturing environment, one would call this *preventive* maintenance.

Proprietary and confidential aspects of data and solution processes add some additional complications. In some cases it may be necessary to encrypt certain parts of the models and/or data before a client is ready to share a job stream. GAMS provides a number of tools to extract and transform or hide model and data components to meet the client's need for not disclosing vital information.

The client models such as the MARKAL test suite become the manifestation of the commitment to the continued process of quality assurance for both the client and the software developer. The test suites and the automatic testing procedures are shared by the client and the developer and thus define precisely what quality means for a specific application. Client model testing has become

Figure 4. MARKAL: (Left) Integrated model overview. (Right) ANSWER user interface, data spreadsheet and graph.

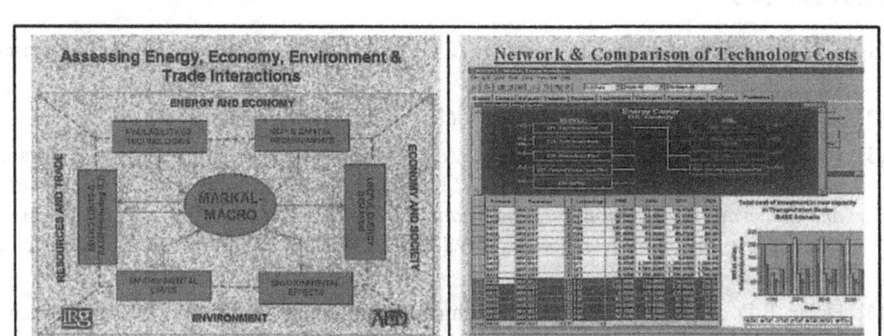

a win for both client and developer and has been made possible by automating the testing process and by sharing the test instances.

6. Conclusions

We have addressed some of the procedures necessary for implementing software quality assurance in mathematical modeling systems, showing how SQA for MP software differs from SQA activities for more traditional software systems. The key steps are automated testing and automated configuration management tools, which foster continual quality improvement. In particular, the focus of our testing activities is on reproducible full lifecycle testing. Our testing framework, which includes model collections, data collection tools, and data analysis tools, allows seamless full lifecycle testing inside the GAMS system, with many components available for outside use by non-GAMS customers and researchers in general. We hope that these tools will illustrate the importance of QA activities for academics, commercial users, solver vendors, and modeling system developers and hope they can be of benefit to the mathematical community in improving their software quality assurance methodologies.

While some of the procedures described may seem disconnected, the focus is on reaching the highest level of maturity defined by the CMM. Most MP vendors, like GAMS, operate in a very small niche market and do not have the resources to follow conventional QA processes used in large enterprises that have revenues thousands of times larger than the entire MP niche market. How can we evolve from Level 1, the exciting but chaotic phase, to Level 5, the mature phase, and become a credible partner for other industries?

The main thrust should be to: (1) Automate the QA process and certification, (2) build the tools into the software and share the QA process (3) make the QA process transparent and reproducible, and (4) involve the clients (academic and commercial) and make them part of the QA process.

References

[1] Brooke, A. Kendrick, D., and Meeraus,A. (1988). *GAMS: A User's Guide*, San Francisco, CA: The Scientific Press.

[2] American Society for Quality. (2004). Online at http://www.asq.org/.

[3] Bussieck, M.R., Drud, A.S., Meeraus, A. and Pruessner, A. (2002). Quality Assurance and Global Optimization. C. Bliek, C. Jermann, A. Neumaier, eds. *Global Optimization and Constraint Satisfaction, First International Workshop on Global Constraint Optimization and Constraint Satisfaction, COCOS 2002*, LNCS2861. Springer Verlag, Heidelberg Berlin, (223-238).

[4] Dolan, E.D. and More, J.J. (2002). Benchmarking optimization software with performance profiles, *Math. Programming*, 91 (2) (201-213).

[5] Dolan, E.D. and More, J.J. (2000). Benchmarking optimization software with COPS, *Technical Report ANL/MCS-TM-246*, Argonne National Laboratory, Argonne, Illinois.

[6] Drud, A.S. (2002). Testing and Tuning a New Solver Version Using Performance Tests, INFORMS San Jose, *Session on "Benchmarking & Performance Testing of Optimization Software."*. See http://www.gams.com/presentations/present_performance.pdf.

[7] Fourer, R. and Gay, D.M. (1993). *AMPL: A Modeling Language for Mathematical Programming*, Redwood City: The Scientific Press.

[8] GAMS Development Corporation. (2004). *GAMS - The Solver Manuals*. GAMS Development Corporation, Washington, DC: http://www.gams.com/solvers/allsolvers.pdf.

[9] Hamilton, L.D., Goldstein, G.A. et al. (1992). MARCAL-MACRO: An Overview, Biomedical and Environmental Assessment Group, *Technical Report BNL-48377*. Analytical Sciences Division, Department of Applied Science, Brookhaven National Laboratory, Associated Universities.

[10] International Organization for Standardization. (2004). Online at http://www.iso.org.

[11] Mittelmann, H.D. (2003). An Independent Benchmarking of SDP and SOCP solvers. *Mathematical Programming*. 95, (407-430).

[12] Moore, J.W. (1998). *Software Engineering Standards: A User's Road Map*. IEEE Computer Society, Los Alamitos, CA.

[13] PAVER Server (2004). Online at `http://www.gamsworld.org/performance/paver`.

[14] Pressman, R.S. (1997). *Software Engineering: A Practitioner's Approach*, 4th Edition, Boston, MA: McGraw-Hill.

[15] Schrage, L.S. (1991). *Lindo - An Optimization Modeling System*, Scientific Press series, Fourth ed., Danvers, MA: Boyd and Fraser.

[16] Software Engineering Institute. (1994). *The Capability Maturity Model: Guidelines for Improving the Software Process*. Reading, MA: Addison-Wesley.

[17] Shcherbina, O. Neumaier, A. Sam-Haroud, D. Vu, X.-H. and Nguyen, T.V. (2003). Benchmarking Global Optimization and Constraint Satisfaction Codes. C. Bliek, C. Jermann, A. Neumaier, eds. *Global Optimization and Constraint Satisfaction, First International Workshop on Global Constraint Optimization and Constraint Satisfaction, COCOS 2002, LNCS2861*. Heidelberg Berlin: Springer Verlag. (223-238).

MODEL DEVELOPMENT AND OPTIMIZATION WITH *Mathematica*™

János D. Pintér[1] and Frank J. Kampas[2]
[1] *Pintér Consulting Services, Inc., Halifax, Nova Scotia, Canada*
jdpinter@hfx.eastlink.ca *http://www.pinterconsulting.com* *http://www.dal.ca/~jdpinter*
[2] *WAM Systems, Inc., Plymouth Meeting, PA, USA*
fkampas@wamsystems.com *http://www.wamsystems.com*

Abstract: *Mathematica* is an integrated scientific and technical computing system, with impressive numerical calculation, programming, symbolic manipulation, visualization and documentation capabilities. In recent years *Mathematica*'s optimization related features have been significantly expanded, both by in-house development and by application packages. Such developments make it an increasingly useful tool also in Operations Research studies. We review and illustrate these features, placing added emphasis on nonlinear (global and convex) optimization, and – within this context – discussing the application packages *MathOptimizer* and *MathOptimizer Professional*.

Key words: *Mathematica*; built-in optimization functions; modeling and optimization packages; *MathOptimizer*; *MathOptimizer Professional*; illustrative examples and applications.

1. INTRODUCTION

Mathematica – an integrated scientific and technical computing environment by Wolfram Research (2004) – is arguably one of the most sophisticated software products available today. Its capabilities and range of applications are documented in the massive *Mathematica* tome (Wolfram, 2003) and in the supplementary documentation. Further information is found in nearly 400 topical books, and in thousands of articles and presentations. According to Wolfram Research, the software is used by well over a million people worldwide.

Mathematica can also increasingly meet the needs of Operations Research professionals, including business analysts, model, algorithm and software developers, researchers, professors, and students. O.R. related features include data analysis and management, model prototyping, concise programming (in several paradigms), advanced computing, visualization, and documentation – all in the same 'live' notebook document, if preferred. Such notebooks can also be directly converted to tex, html, xml, ps, and pdf file formats. *Mathematica* also supports direct links to external application packages, to other software products, and to the Internet. A significant further advantage is portability across a broad range of hardware platforms and operating systems, due to the standardized notebook document format.

For further general information, visit the websites of Wolfram Research, specifically including the *Mathematica* Information Center (2004) that provides extensive details and links. We also refer to a recent review of *Mathematica* in *ORMS Today* (Sodhi, 2003), as well as to an illustrative list of *Mathematica* books with a modeling and/or optimization related content (Bahder, 1995; Schwalbe and Wagon, 1996; Gass, 1998; Bhatti, 2000; Maeder, 2000; Jacob, 2001; Hollis, 2003; Pemmaraju and Skiena, 2003; Kampas and Pintér, 2004). Let us note here that *MathReader*, a freely available viewer, can be used to display and print *Mathematica* notebooks, animate graphics, play sounds, and copy information from notebooks to other documents; *MathReader* can also be used in most web browsers.

In this work we review and illustrate *Mathematica*'s O.R. modeling and optimization related features. Within the broad category of optimization models, we see particularly strong application potentials for *Mathematica* in the analysis of (possibly complex) nonlinear systems when the corresponding decision model can not be brought to simple standard forms. In such cases, problem-specific modeling and code development are essential, and using *Mathematica* as the development platform can be a good choice. For this reason, here we shall place added emphasis on nonlinear (global and convex) optimization, where – in addition to built-in functionality – our packages *MathOptimizer* and *MathOptimizer Professional* can be put to good use.

2. NUMERICAL OPTIMIZATION IN *Mathematica*

We start with a concise summary of built-in optimization functionality. Most of the related *Mathematica* functions can be invoked in several variations, and have a number of optional settings. Here we shall use their basic forms with default settings; for further details, consult (Wolfram, 2003) and the *Mathematica* help system. We shall also refer to several closely related articles and presentations. For simplicity, only minimization problems are considered: several functions also have a maximization equivalent, with identical solver functionality.

In the illustrative statements we shall use **`bold Courier`** fonts for displaying *Mathematica* input and `regular Courier` fonts for *Mathematica* output; however, in the explanatory text we retain the standard (Times New Roman) fonts used in this article. All input/output statements and calculations presented in this work are directly imported from a corresponding *Mathematica* notebook.

2.1 LinearProgramming

The function LinearProgramming[c, A, b] finds a vector x that solves the LP problem stated as min $c^T x$ subject to $Ax \geq b$ and $x \geq 0$. Here c and x are (real) n-vectors; b is an m-vector and A is an m-row, n-column matrix. We will not discuss the ConstrainedMin (and ConstrainedMax) functions since these are also LP solvers, and both became obsolete since the release of *Mathematica* version 5.

A simple example of using LinearProgramming is shown below. Let us remark that in *Mathematica* vectors are denoted by lists: each component of a list is followed by a comma, and the entire list is enclosed by curly braces {}. The next three lines describe the model data (semicolon is used to suppress *Mathematica* output that in this case would simply echo the input lines shown):

```
c={1,2,1,1,3};
A={{2,-3,3,5,4},{-1,2,1,-4,-2},{2,2,2,1,1}};
b={3,8,12};
```

The solution is then simply obtained by entering the statement

```
xopt=LinearProgramming[c,A,b]
{0,2,4,0,0}
```

The result (i.e., the listed components of *xopt*) is shown in the row immediately following the *Mathematica* input statement. The solution is verified and the optimum value obtained by the following statements (the symbol . denotes the matrix-vector and vector-vector (dot) products):

```
A.xopt
{6,8,12}

c.xopt
8
```

The solution time for this 'mini-problem' is less than 0.001 seconds. *Mathematica* timings are usually displayed in one-thousandth of a second precision. All illustrative timing information in this article is measured using a Pentium 4 1.6 GHz processor based desktop machine that runs under Windows XP Professional; we are using *Mathematica* version 5.0.

Let us note here that recent LP related development includes the *Mathematica* implementation of the LAPACK package that has been used worldwide to solve the most common tasks in numerical linear algebra

(Leyk, 2003). Another notable development is discussed by Hu (2003): a new interior point algorithm option has been added to LinearProgramming that is now capable of solving large-scale linear optimization problems with hundreds of thousands of variables and equations.

2.2 FindMinimum

The function FindMinimum locally solves unconstrained nonlinear optimization problems, optionally using various methods that include conjugate gradient and BFGS quasi-Newton search strategies. As a simple illustration, we shall demonstrate its application in the form FindMinimum[f, {x, x0},{y, y0}] that uses the initial solution estimate {x0, y0} in solving the two-variable problem min $f(x,y)$. The multiplication symbol * is used below for clarity: it could be replaced by a space between the multiplier constant and the variable.

```
FindMinimum[Sin[x^2-x]+3*y^2, {{x,3},{y,1}}
{-1., {x → 2.72764, y → -2.91001 × 10^-11}}
```

In the result received, -1 is the objective function value, and → denotes a symbol-to-value assignment. FindMinimum is a local search method: hence, this could be – in fact, is – only one of the local or global solutions (most likely, the one closest to the starting point). This point is illustrated by

```
FindMinimum[Sin[x^2-x]+3*y^2, {{x,13},{y,11}}
{-1., {x → 11.6509, y → -1.49268 × 10^-9}}
```

2.3 NMinimize

The *Mathematica* function NMinimize[{f, cons},{x, y,...}] attempts to find the global minimum of f, subject to the listed constraints *cons*. The following simple example illustrates its application; notice the double equality sign == that denotes a strict equality constraint:

```
NMinimize[{(x1^2-x2)^2,
x1-x1*x2==0,  x1≥-10,  x1≤20,  x2≥-15,  x2≤10},
{x1,x2}]
{2.46519 × 10^-30, {x1 → 1., x2 → 1.}}
```

We will use NMinimize later on in some illustrative comparisons.

3. MODELING AND OPTIMIZATION PACKAGES

There is a range of application packages offered by Wolfram Research and by independent developers with apparent O.R. relevance. A brief review of these is provided below, for simplicity in alphabetical order. We will not mention or display the (quite possibly changing) version numbers, when discussing the packages: for further details see the related references and visit the website of Wolfram Research.

All packages discussed can be seamlessly integrated into *Mathematica*, when properly installed: in particular, their documentation can be directly invoked from *Mathematica*'s help system. Since all packages present detailed application examples, these can be directly used and customized to create new model development and optimization projects.

Needless to say, we do not intend to specifically endorse any of these applications, and – in lack of access to all listed packages – we rely partly on the product descriptions provided by Wolfram Research and the developers. Packages will be referred to using *italics* fonts.

Advanced Numerical Methods expands the functionality of the *Control System Professional* package with an extensive collection of numerical algorithms. These algorithms solve a wide class of control and linear algebra problems.

Combinatorica extends *Mathematica*'s capabilities by over 450 new functions: these serve to construct graphs and other combinatorial objects, and to display them. The detailed guide to *Combinatorica* is Pemmaraju and Skiena (2003) that can also be used as a course textbook.

Control System Professional Suite is an extensible framework of integrated *Mathematica* application packages for handling common, interdisciplinary control problems that arise in engineering, as well as in chemistry, biology, economics and financial studies.

Database Access Kit brings *Mathematica*'s data analysis and management tools to large data sets. These capabilities can be interfaced with relational databases (including Oracle, Microsoft Access, SQLServer, and DB2) and to a number of flat-file databases (like Excel or dBase files).

DiffEqs is a collection of individual packages that accompanies the textbook by Hollis (2003): the book presents an introduction to *Mathematica*, and to differential equations.

Experimental Data Analyst integrates a set of programs that help to analyze experimental data, from error analysis and data fitting capabilities to data visualization and transformation. A collection of examples based on real experimental data is included.

Fuzzy Logic provides a set of tools for creating, modifying, and visualizing fuzzy sets and fuzzy logic-based systems. It also includes practical examples that introduce the basic concepts and demonstrate the numerical solution of various system design problems.

Global Optimization offers a collection of functions for constrained and unconstrained nonlinear optimization, as well as several tools of interest for statistical studies.

Industrial Optimization is designed to solve a range of O.R. models, by providing algorithms for linear, pure and mixed integer linear, and convex optimization, as well as some heuristic techniques such as genetic programming.

Mathematica Link for Excel provides Excel users with a seamless connection to *Mathematica*: one can directly activate a range of advanced *Mathematica* calculations and functions from the calling spreadsheet.

MathOptimizer and *MathOptimizer Professional* are our own application packages (Pintér, 2002b; Pintér and Kampas, 2003): these will be discussed in more details later on.

ModelMaker serves to build and analyze finite element (FE) models. The package permits building parametric models, where the FE database contains both numeric data and symbolic *Mathematica* expressions which can be used to morph the model geometry.

Neural Networks provides tools to define, train, visualize, and validate neural network models. It supports a set of network structures; it also implements training (unconstrained local optimization) algorithms.

Operations Research offers tools for solving linear optimization, quadratic programming, shortest path, and combinatorial optimization problems, including both exact and heuristic approaches.

Optimization Toolbox contains programs that accompany Bhatti's well-written textbook (2000), targeted primarily to an undergraduate and graduate student (and instructor) readership. Optimization theory is presented in an informal style; pedagogical *Mathematica* algorithms are presented and illustrated by examples.

Parallel Computing Toolkit brings parallel computation tools to a computer network, or to multiprocessor machines. It implements parallel programming primitives and includes high-level commands for the parallel execution of operations such as animation, plotting, and matrix manipulation.

VisualDSolve has been developed along with the textbook by Schwalbe and Wagon (1996) that serves as its reference manual. The book covers many of the topics in a first course in ordinary differential equations, and provides a wide variety of tools for visualizing solutions.

4. *MathOptimizer*

4.1 Introduction and Usage

MathOptimizer (Pintér, 2002b) is a native *Mathematica* software package that serves to solve general – global or local – nonlinear optimization models stated in the form

$$
\begin{array}{lll}
(1) & \min f(x) & f\colon D_0 \to \mathbf{R}^1 \\
 & g(x)=0 & g\colon D_0 \to \mathbf{R}^{m1} \\
 & h(x)\leq 0 & h\colon D_0 \to \mathbf{R}^{m2} \\
 & D_0 := \{x\colon xl \leq x \leq xu\} & x, xl, xu \in \mathbf{R}^n
\end{array}
$$

It is assumed that all functions f, g, h are at least continuous, and that xl, xu are finite (known) real n-vectors. All bound, equality and inequality constraints are interpreted component-wise. Notice that the equality and inequality constraints are treated separately: their number is denoted by $m1$ and $m2$, respectively.

In addition to *MathOptimizer*'s built-in local solver methodology, a special emphasis is placed on finding the global solution of models that may have a number of local solutions. Fairly comprehensive reviews of global optimization are presented e.g., in the *Handbooks* edited by Horst and Pardalos (1995), Pardalos and Romeijn (2002); see also the topical website of Neumaier (2004).

MathOptimizer consists of two core solver packages and a solver integrator package. One of these solver components is called MS, abbreviating MultiStart (global search). MS serves for the – as a rule, approximate – global optimization of an exact penalty function that aggregates f, g, and h in the given n-dimensional interval range. MS uses an adaptive stochastic search method, combined with a statistical bounding procedure. The second component package – called CNLP, abbreviating Constrained NonLinear Programming (for local search) – implements a Lagrangian approach that is aimed at finding a (global or local) solution that satisfies the Karush-Kuhn-Tucker optimality conditions. (Note that, theoretically, this component requires smooth problem structure.) CNLP is used for 'precise' local optimization, based on a given initial solution: the latter is either produced by the global search phase, or it can be directly provided by the user. The solver integrator package, called Optimize, supports the individual or combined use of the two solver packages. It is planned to add further solver components to *MathOptimizer*: the presence of the integrator package directly supports this objective.

The *MathOptimizer* User Guide is a *Mathematica* notebook (currently consisting of over 70 printed pages) that can be directly invoked through *Mathematica*'s online help system. The manual presents installation and technical notes, provides concise mathematical background information and modeling tips, and discusses a number of test problems as well as several more advanced applications.

MathOptimizer is invoked by the following *Mathematica* statement. Observe the notation used to identify the entire package and the integrator package component: the latter then indirectly activates both MS and CNLP.

```
Needs["MathOptimizer`Optimize`"];
```

The following *Mathematica* code illustrates the definition of a small non-convex optimization model that is made up by decision variables (denoted below as vars); lower/upper bounds and nominal (initial) values of the variables (varlb, varub, and varnom); objective function to minimize (objf); and the separate lists of equality constraints (eqs) and inequality constraints (ineqs, by assumption, are stated in ≤0 form).

```
vars={x1,x2};
varlb={-10,-15};
varub={20,10};
varnom={8,-14};
objf=10*(x1^2-x2)^2+(x1+3*x2-4)^2;
eqs={x1^4-x1*x2^3};
ineqs={3*x1+4*x2^2-8};
```

The next statement calls *MathOptimizer* to solve the model:

```
Optimize[objf, eqs, ineqs, vars, varnom, varlb, varub]
```

$\{\{1., 1.\}, 6.23774\times10^{-21}, \{-3.94744\times10^{-11}\}, \{-1.\}, \{3.94744\times10^{-11}, 2.73735\times10^{-9}, 0.\}\}$

The result shows the composite list of the following elements: the list of global solution components (x1=x2=1), the optimum value (a close numerical approximation to the theoretical value 0), as well as the lists of constraint function values at the solution, and finally the list of violation levels with respect to feasibility, the Kuhn-Tucker equation (defined by the gradient of the Lagrangian), and the complementary slackness condition at the solution found. The *MathOptimizer* runtime is less than 0.5 seconds.

Note that it is very easy to make changes to the model, and then to immediately repeat the solution procedure. For example, we can replace the constraints by defining (over-writing)

```
eqs={x1^4-Sin[1-x1*x2^3]-1};
ineqs={3*x1-4*x2^2+1};
```

After evaluating these statements – on MS Windows machines, by using the Shift-Enter key combination while pointing anywhere in the

Mathematica cell that includes the above input (so that they can be evaluated in a single move) – we can run *MathOptimizer* again. Observe passing by that the optimum value should be the same, except numerical rounding errors, since the previously found global solution {1, 1} meets also the new constraints.

```
Optimize[objf, eqs, ineqs, vars, varnom, varlb,
varub]
{{1., 1.}, 5.68314×10^-19, {8.07199×10^-11},
 {-1.07488×10^-9}, {8.07199×10^-11, 9.53077×10^-9, 0.}}
```

The numerical solution received is essentially the same as the one found above. For comparison, now we attempt to solve this model by using the built-in function NMinimize (in default mode, similarly to *MathOptimizer*). The NMinimize formulation for the model is slightly different:

```
NMinimize[objf,
x1^4-Sin[1-x1*x2^3]-1==0, 3*x1-4*x2^2+1≤0,
x1≥-10, x1≤20, x2≥-15, x2≤10}, {x1,x2}]
{2532.29,{x1→1.1892, x2→-10.354}}
```

The solution found by NMinimize is obviously sub-optimal. Of course, this finding is not sufficient per se to draw far-reaching conclusions. However, it certainly shows that the solution of nonlinear models can be tricky, even in (very) low dimensions.

4.2 Applications

In addition to a number of relatively simple numerical test examples, the *MathOptimizer* User Guide discusses illustrative applications from the following areas: chemical equilibrium modeling, industrial design, acoustic engineering design, and two numerical mathematics challenges (Problems 4 and 9 from Trefethen (2002)). In solving some of these – specifically, the sonar transducer model formulated by Purcell and the numerical integration problem of Trefethen – it is essential that *MathOptimizer* can handle arbitrary computable (preferably also continuous) *Mathematica* functions. This feature makes it suitable to handle 'black box' models defined by functions that are evaluated by complex, numerically intensive procedures. Pintér and Purcell (2003) discuss the sonar transducer design problem: its solution requires a combination of the *ModelMaker* (Purcell, Dai, and Xue, 2001) and *MathOptimizer* packages.

To mention other areas of application, Kampas and Pintér (2002) solve configuration analysis and design models using *MathOptimizer*: such problems arise e.g. in applied mathematics, statistics, physics, chemistry, and robotics. Pintér (2003c) discusses nonlinear model calibration: the illustrative numerical results demonstrate that *MathOptimizer* produces superior results to local search based model fitting. The article then reviews several case studies in which global optimization has been applied to model calibration problems related to water quality, environmental engineering, time series analysis and photoelectron spectroscopy applications.

5. *MathOptimizer Professional*

5.1 Introduction and Usage

MathOptimizer Professional (Pintér and Kampas, 2003) is another *Mathematica* model development and nonlinear optimization package: however, it is based on an entirely different approach from the native *Mathematica* solver systems reviewed and discussed above. *MathOptimizer Professional* solves globally or locally nonlinear optimization models stated in the following general form (notice that the m-vector function g below now includes both equality and equality constraints):

$$
\begin{array}{lll}
(2) & \min f(x) & f\colon D_0 \to \boldsymbol{R}^1 \\
 & g(x) \le 0 & g\colon D_0 \to \boldsymbol{R}^m \\
 & D_0 := \{x\colon xl \le x \le xu\} & x, xl, xu \in \boldsymbol{R}^n
\end{array}
$$

The core of the package is the LGO external solver system that is activated and then used via *MathLink*, a general-purpose interface that supports communication between *Mathematica* and external programs. LGO – originally abbreviating the Lipschitz (continuous) Global Optimizer – can handle general (continuous) nonlinear optimization models, using a suite of global and local search algorithms. The currently implemented LGO algorithm options include branch-and-bound (BB), global adaptive random search (single-start, GARS) and multi-start (MS) based search strategies, as well as a (local) generalized reduced gradient (GRG) method. Note that in the global search phase the model functions are aggregated applying an exact penalty function; in the local search phase – that either automatically follows one of the global search modes or is used as a 'local search only' option – all constraint functions are treated individually.

The global search methods are, in theory, globally convergent (deterministically, or with probability 1, at least for box-constrained global optimization models). The actual code implementations are numerical approximations of the underlying theory. Due to the usage of an aggregated merit function, the automatic 'switching point' from global to local search, and other parameter settings, there are heuristic elements in LGO (similarly to most – if not all – numerical optimization methods). The optional choice of global methods often helps in solving difficult models, since BB, GARS, and MS apply different search strategies. The parameterization of these component algorithms (e.g., intensified global search) can also help to solve difficult models, although the internally set default search effort typically produces a close numerical approximation of the global solution. The latter statement has been verified by solving some difficult global optimization problems in which the solution is reproducible and publicly available: some examples will be mentioned later on.

Note also that all LGO search algorithms are derivative-free: specifically, in the local search phase central differences are used to approximate gradients. This choice reflects our objective to handle models with merely computable, continuous functions, including 'black box' systems.

LGO has been developed and maintained for well over a decade (as of 2004), and the software is discussed in details elsewhere: consult, e.g., Pintér (1996, 2001, 2002a, 2004a), or the peer review by Benson and Sun (2000). LGO is currently available for essentially arbitrary C and Fortran compiler platforms, with seamless links to Excel, GAMS, Maple, *Mathematica*, and TOMLAB (the latter provides a solver interface and a collection of solvers for optimization using MATLAB). The details of these implementation versions are described in the corresponding documentation: see Frontline Systems and Pintér Consulting Services (2001); Pintér (2003a); Pintér (2004b); Pintér, Holmström, Göran and Edvall (2004).

The computational study (Pintér, 2003b) reviews the performance of LGO in comparison to several state-of-art local nonlinear solvers linked to the GAMS platform. This evaluation has been done in a fully automated and reproducible manner using publicly available GAMS model libraries: hence, it can be considered as reasonably objective, even if the collection of models and other circumstances (solver options and parameters) always carry elements of arbitrariness and subjectivity. The numerical experiments described in this study show that global optimization tools are needed to solve nearly half of the GAMS models from the chosen library, even when – possibly quite useful – initial solution points are provided to the local solvers. (We conjecture that providing random starting points from a search box that contains the feasible region would demonstrate even more pronounced need for global scope search.)

MathOptimizer Professional combines the model development power of *Mathematica* with the LGO solver suite: this leads to enhanced nonlinear solver capabilities, and a performance (solution speed) that – especially on larger models – is comparable to compiler-based solver implementations.

The functionality of *MathOptimizer Professional* is summarized by the following steps (all steps are fully automatic, except the first one):

- model formulation in *Mathematica*
- translation of the *Mathematica* optimization model into C or Fortran code (LGO model function file)
- generation of LGO input parameter file

- compilation of the C or Fortran model code into object code or dynamic link library (dll): this step needs a suitable compiler
- call to the LGO solver engine: the latter is typically provided as object code or an executable program that is linked together with the model object or dll file
- model solution and report generation by LGO
- report of LGO results back to the calling *Mathematica* notebook.

Obviously, the approach outlined supports 'only' the solution of models defined by *Mathematica* functions that can be directly converted into C or Fortran program code. This, however, still allows the handling of a broad range of continuous nonlinear optimization models. A 'side-benefit' of using *MathOptimizer Professional* is that *Mathematica* models are automatically translated into C or Fortran format: this can be useful e.g., in generating new test models.

Following installation, the *MathOptimizer Professional* User Guide (Pintér and Kampas, 2003) can be directly invoked as part of *Mathematica*'s help system. The package is activated by the following statement

```
Needs["MathOptimizerPro`callLGO`"];
```

Upon executing this statement, on MS Windows machines a command window opens that serves to monitor the *MathLink* connection that support external system calls to/from LGO. In our case, this window will display the background compiler and linker operations.

The numerical solution of an optimization model now can be launched by a single *Mathematica* statement of the form callLGO[f, g, {x, xl, xn, xu}]. Here we use the notation corresponding to (2); in addition, *xn* is the nominal setting of *x* (used in the first model function evaluation and/or as a starting point of the 'local search only' LGO solver mode). The following call illustrates the basic *MathOptimizer Professional* functionality:

```
callLGO[x^2+3y^2, {x+Sin[y] ≥ 1}, {{x, -2, 0, 2}, {y, -2, 0, 2}}]
{0.753796,{x→0.757485,y→0.244957},0}
```

The result shows (again, in *Mathematica* list format) the optimum value found, the list of corresponding variable settings, and the maximal model function infeasibility at the solution: all values are numerical approximations, of course. Note that the function callLGO currently has 15 optionally set parameters: these are all documented and illustrated in the User Guide, but their discussion is outside of the scope of this paper. For further details, consult the manual or Pintér and Kampas (2004).

5.2 Applications

For over a decade, LGO has been applied in a variety of professional, as well as academic research and educational contexts. In recent years, LGO has been used to solve models in up to a few thousand variables and constraints. Some recent applications and case studies – including e.g., model fitting in econometrics and laboratory analysis, potential energy models in computational chemistry, laser design, cancer therapy planning, and non-uniform sphere packings – are discussed by Pintér (2001a, b, 2002a), Isenor, Pintér, and Cada (2003), Tervo et al. (2003), Kampas and Pintér (2004a), Pintér and Kampas (2004). Note additionally that some of the LGO software users in the financial industry, process industries, biotechnology, etc. develop other advanced (but confidential) applications. We expect essentially similar performance from the recently released *MathOptimizer Professional* that enables the solution of sizeable, sophisticated nonlinear models formulated in *Mathematica*. The role of communication overhead between *Mathematica* and the external solver suite becomes relatively less significant in solving larger models, in which the external LGO solver time dominates.

The *MathOptimizer Professional* User Guide (an approximately 150-page document when printed) describes several tens of test problems starting with simple LP problems, through convex and non-convex nonlinear models, to a number of fairly challenging optimization models originating from mathematics, physics, chemistry, engineering and economics. For illustration, we shall consider here a pair of transcendental equations:

```
eq1 = (x - Sin[2x + 3y] - Cos[3x - 5y])^2;
eq2 = (y - Sin[x - 2y] + Cos[x + 3y])^2;
```

We wish to find a solution in the region $-2 \le x \le 3$, $-2.5 \le y \le 1.5$, or to numerically verify that there is no solution in the region specified. The surface and contour plots of eq1+eq2 (this corresponds to the squared l_2-norm based error function) reveal the rather complex multi-extremality of the induced optimization model: see Figures 1 and 2.

Let us apply *MathOptimizer Professional* to solve this problem. First, we define the equations (eqs), the constraints (cons: note here that the relations eq1=eq2=0 can be expressed by using the *Mathematica* function Thread), and the variables with bounds and nominal values (varswithbounds). Then we call LGO.

```
eqs={eq1,eq2};
cons=Thread[eqs==0];
varswithbounds={{x,-2,1,3},{y,-2.5,1,1.5}};
callLGO[0,cons,varswithbounds]
{0, {x→ -0.173363, y→ -0.256098}, 1.44819×10^-9}
```

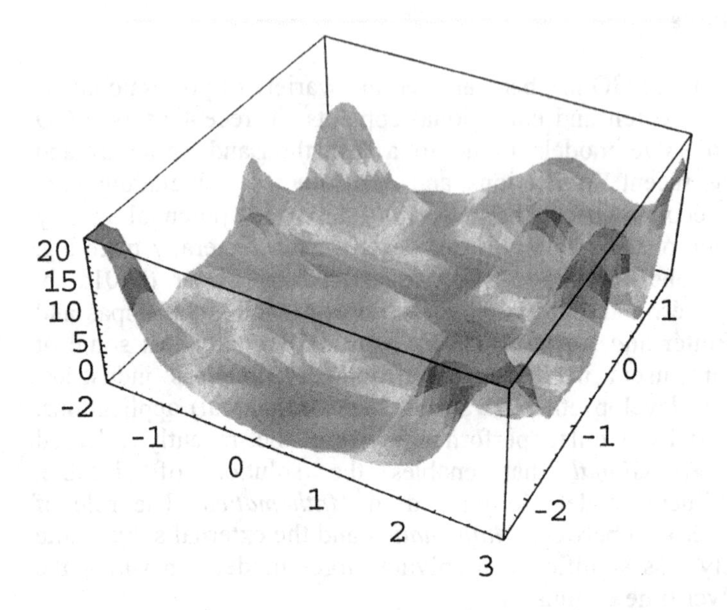

Figure 1. Surface plot of error function in solving a system of equations.

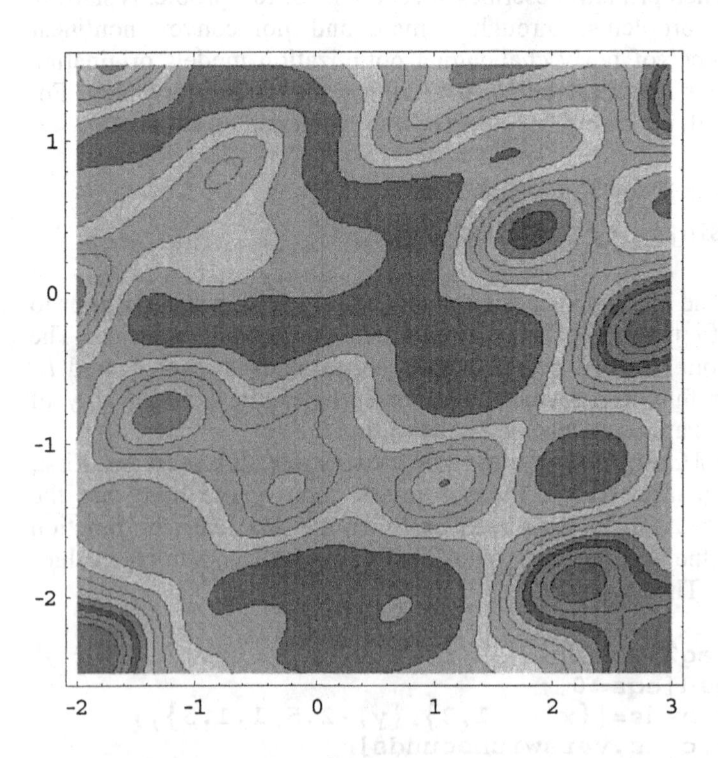

Figure 2. Contour plot of error function in solving a system of equations.

As the result shows, *MathOptimizer Professional* finds a numerical solution that is precise to about $1.45 \cdot 10^{-9}$, when substituted into the equations. The external LGO runtime is 0.03 seconds. (In total, 7843 search steps – model function evaluations, including gradient estimates in the local search phase – are done in using the default MS+LS search mode with default parameterization; all results are exactly reproducible.) Note also that the User Guide addresses the issue of finding (possible) multiple solutions to systems of equations and inequalities.

As for another illustrative application, in (Kampas and Pintér, 2004a) we state and solve a challenging new model type: our objective is to find the 'best' non-overlapping arrangement of a set of given non-uniform size circles in an embedding circle. The best packing is defined here by a combination of two criteria: the size (radius) of the circumscribed circle, and the average pair-wise distance between the centers of the embedded circles. The relative weight of the two objective function components can be selected as a model-instance parameter.

Detailed numerical results are reported in (Kampas and Pintér, 2004a) for circles defined by the radii $r_i=i^{-0.5}$, $i=3,\ldots,N$, up to 40-circle configurations. For illustration, the configuration found for the case N=20 circles using *MathOptimizer Professional* is displayed below. In this example, equal consideration (weight) is given to minimizing the radius of the circumscribed circle and the average distance between the circle centers.

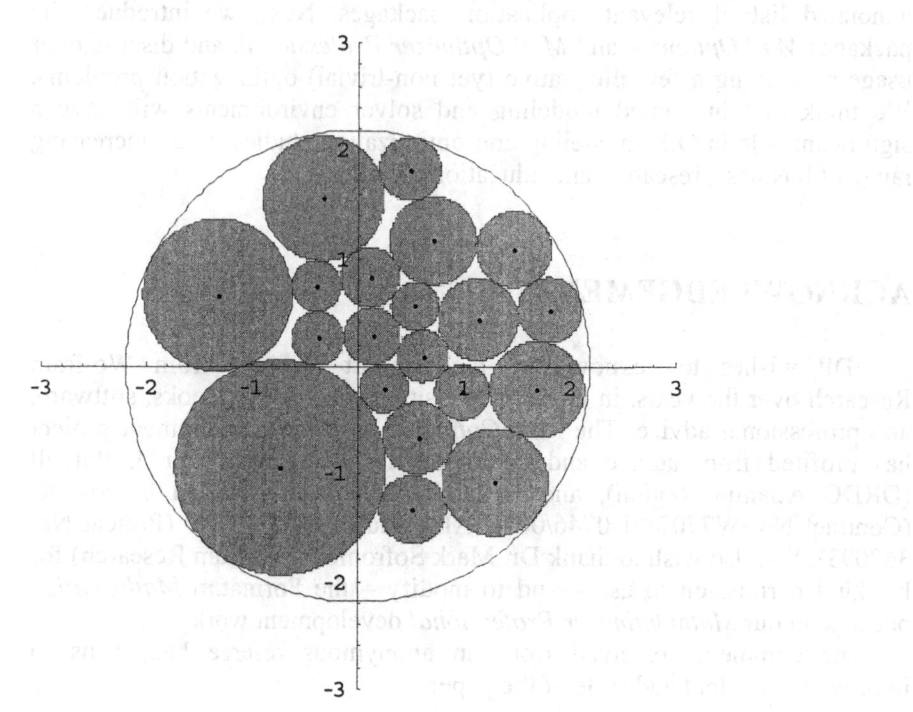

Figure 3. A non-uniform circle packing result for N=20 circles.

Let us remark in this context that in Kampas and Pintér (2002) we have attempted to solve instances of the circle packing problem applying the built-in *Mathematica* function NMinimize, but (in default mode) it could not find a solution of acceptable quality even for the case *N*=5. *MathOptimizer* worked better and found good quality solutions for small configurations (up to *N*=10), but – due to its native *Mathematica* solver functions – solution times are increasing far more rapidly than for *MathOptimizer Professional*. Again, this is just a numerical observation, as opposed to a conclusion: we plan to make a more systematic comparison of global solvers (available for use with *Mathematica*) in the near future, based on detailed numerical tests.

Let us also mention finally that both *MathOptimizer* and *MathOptimizer Professional* are included in a recent peer review of optimization capabilities using *Mathematica* (Cogan, 2003).

6. CONCLUDING REMARKS

This article discusses the potentials of *Mathematica* in Operations Research related modeling and optimization studies. Within this context, we review built-in *Mathematica* optimization functionality and provide an annotated list of relevant application packages. Next, we introduce the packages *MathOptimizer* and *MathOptimizer Professional*, and discuss their usage by solving a few illustrative (yet non-trivial) optimization problems. We think that integrated modeling and solver environments will have a significant role in O.R. modeling and optimization studies, in an increasing range of business, research, and educational contexts.

ACKNOWLEDGEMENTS

JDP wishes to acknowledge the support received from Wolfram Research over the years, in forms of a visiting scholarship, books, software, and professional advice. The *MathOptimizer* software development project has profited from advice and comments by Dr. Christopher J. Purcell (DRDC Atlantic Region), and it has been partially funded by DRDC (Contract No. W7707-01-0746/001/HAL), and by NRC IRAP (Project No. 362093). We also wish to thank Dr. Mark Sofroniou (Wolfram Research) for his kind permission to use – and to modify – the Format.m *Mathematica* package in our *MathOptimizer Professional* development work.

The comments received from an anonymous referee helped us to improve the content and style of the paper.

REFERENCES

Bahder, T.B. (1995) *Mathematica for Scientists and Engineers.* Addison-Wesley, Reading, MA.

Benson, H.P. and Sun, E. (2000) LGO – Versatile Tool for Global Optimization. *ORMS Today* 27 (5) 52-55. See also http://www.lionhrtpub.com/orms/orms-10-00/swr.html.

Bhatti, M.A. (2000) *Practical Optimization Methods With Mathematica Applications.* Springer, New York.

Frontline Systems and Pintér Consulting Services (2001) *Premium Solver Platform – LGO Global Solver Engine for Excel.* Published by Frontline Systems, Inc., Incline Village, NV. See also http://www.solver.com/xlslgoeng.htm.

Cogan, B. (2003) How to get the best out of optimization software. *Scientific Computing World* 71 (2003) 67-68. See http://www.scientific-computing.com/scwjulaug03review_optimisation.html.

Gass, R. (1998) *Mathematica for Scientists and Engineers: Using Mathematica to do Science.* Prentice Hall, Englewood Cliffs, NJ.

Hollis, S. (2003) *A Mathematica Companion for Differential Equations.* Prentice Hall, NJ.

Horst, R. and Pardalos, P.M., eds. (1995) *Handbook of Global Optimization, Vol. 1.* Kluwer Academic Publishers, Dordrecht.

Hu, Y. (2003) Solving large linear optimization problems. Lecture presented at the *2003 Mathematica Developer Conference*, Champaign, IL.

Isenor, G., Pintér, J.D., and Cada, M. (2003) A global optimization approach to laser design. *Optimization and Engineering* 4, 177-196.

Jacob, C. (2001) *Illustrating Evolutionary Computation with Mathematica.* Morgan Kaufmann Publishers, San Francisco, CA.

Kampas, F.J. and Pintér, J.D. (2002) Configuration analysis and design by using optimization tools in *Mathematica. The Mathematica Journal* (to appear).

Kampas, F.J. and Pintér, J.D. (2004a) Generalized circle packings: model formulations and numerical results. *Proceedings of the 2004 International Mathematica Symposium*, Banff, AB.

Kampas, F.J. and Pintér, J.D. (2004b) *Advanced Optimization: Scientific, Engineering, and Economic Applications with Mathematica Examples.* Elsevier, Amsterdam (to appear).

Leyk, Z. (2003) Fast linear algebra in *Mathematica.* Lecture presented at the *2003 Mathematica Developer Conference*, Champaign, IL.

Maeder, R.E. (2000) *Computer Science with Mathematica.* Cambridge University Press, Cambridge, UK.

Mathematica Information Center (2004) http://library.wolfram.com/infocenter/.

Neumaier, A. (2004) Global Optimization. http://www.mat.univie.ac.at/~neum/ glopt.html.

Pardalos, P.M. and Romeijn, H.E., eds. (2002) *Handbook of Global Optimization, Vol. 2.* Kluwer Academic Publishers, Dordrecht.

Pemmaraju, S. and Skiena, S. (2003) *Computational Discrete Mathematics: Combinatorics and Graph Theory with Mathematica.* Cambridge University Press, Cambridge, UK.

Pintér, J.D. (1996) *Global Optimization in Action.* Kluwer Academic Publishers, Dordrecht.

Pintér, J.D. (2001a) *Computational Global Optimization in Nonlinear Systems: An Interactive Tutorial.* Lionheart Publishing, Atlanta, GA.

Pintér, J.D. (2001b) Globally optimized spherical point arrangements: Model variants and illustrative results. *Annals of Operations Research* 104, 213-230.

Pintér, J.D. (2002a) Global optimization: software, test problems, and applications; Chapter 15 (pp. 515-569) in: Pardalos and Romeijn, eds. *Handbook of Global Optimization, Vol. 2.*

Pintér, J.D. (2002b) *MathOptimizer – An Advanced Modeling and Optimization System for Mathematica Users. User Guide.* Published and distributed by Pintér Consulting Services, Inc., Halifax, NS, Canada.

Pintér, J.D. (2003a) *GAMS/LGO User Guide.* Published and distributed by the GAMS Development Corporation, Washington, DC. See http://www.gams.com/solvers/lgo.pdf.

Pintér, J.D. (2003b) GAMS/LGO nonlinear solver suite: key features, usage, and numerical performance. (Submitted for publication.) Available for download at http://www.gams.com/solvers/solvers.htm#LGO.

Pintér, J.D. (2003c) Globally optimized calibration of nonlinear models: techniques, software, and applications. *Optimization Methods and Software* 18, 335-355.

Pintér, J.D. (2004a) *LGO – An Integrated Model Development and Solver Environment for Continuous Global Optimization. User Guide.* (Current edition.) Published and distributed by Pintér Consulting Services, Inc., Halifax, NS, Canada.

Pintér (2004b) *The Maple Global Optimization Toolbox.* Published and distributed by Maplesoft, Inc., Waterloo, ON. See http://www.maplesoft.com/products/toolboxes/globaloptimization/index.shtml.

Pintér, J.D. and Kampas, F.J. (2003) *MathOptimizer Professional – An Advanced Modeling and Optimization System for Mathematica Users with an External Solver Link. User Guide.* Published and distributed by Pintér Consulting Services, Inc., Halifax, NS, Canada.

Pintér, J.D. and Kampas, F.J. (2004) Global optimization in *Mathematica* with MathOptimizer Professional. (Submitted for publication.)

Pintér, J.D. and Purcell, C.J. (2003) Optimization of finite element models with *MathOptimizer* and *ModelMaker*. Lecture presented at the *2003 Mathematica Developer Conference*, Champaign, IL.

Pintér, J.D., Holmström, K., Göran, A.O. and Edvall, M.M. (2004) *TOMLAB /LGO User Guide.* Published and distributed by TOMLAB Optimization AB, Västerås, Sweden and Arcata, CA. See http://tomlab.biz/docs/TOMLAB_LGO.pdf.

Purcell, C.J., Dai, N.M. and Xue, L. (2001) Modelling, analysis & prototyping for rapid manufacturing. Lecture presented at the *2001 Mathematica Developer Conference*, Champaign, IL.

Schwalbe, D. and Wagon, S. (1996) *VisualDSolve: Visualizing Differential Equations with Mathematica.* Springer, New York.

Sodhi, M.S. (2003) *Mathematica 5. ORMS Today* 30 (6), 44-47.

Tervo, J., Kolmonen, P., Lyyra-Laitinen, T., Pintér, J.D., and Lahtinen, T. (2003) An optimization-based approach to the multiple static delivery technique in radiation therapy. *Annals of Operations Research* 119, 205-227.

Trefethen, N.L. (2002) A Hundred-dollar, Hundred-digit Challenge. SIAM News 35 (1), p. 3. See also http://www.siam.org/siamnews/01-02/challenge.pdf.

Wolfram, S. (2003) *The Mathematica Book.* (5th Edition.) Wolfram Media, Inc., Champaign, IL.

Wolfram Research (2004) http://www.wolfram.com/.

VERIFICATION OF BUSINESS PROCESS DESIGNS USING MAPS

Eswar Sivaraman
Department of Systems Engineering & Operations Research
George Mason University
Fairfax, VA 22030
esivaram@gmu.edu

Manjunath Kamath
School of Industrial Engineering & Management
Oklahoma State University
Stillwater, OK 74078
mkamath@okstate.edu

Abstract Business processes form the foundation of an enterprise's operations and determine what the business does, and more importantly, how well the business does what it does. A systematic approach to the design of business processes, supported by a formal foundation for the specification and modeling of business processes, is necessary to (i) capture domain knowledge in a format that is transferable across enterprises and (ii) provide a basis for re-design based on needs of efficiency, changes in market requirements, and reproducibility of process templates for multiple products/services. The specification of a business process is often characterized by combinations of concurrency, choice, and asynchronous completion, the mix of which could lead to incorrect designs. This chapter highlights the verification issues that arise in the design of business processes, and outlines research questions of immediate relevance to the growing interest in business process modeling and enterprise automation solutions. We also discuss **MAPS** – a tool for the **M**odeling and **A**nalysis of **P**rocess model**S**, that has been used effectively as a classroom aid to highlight the importance of design verification as a necessary first step in the design of business processes.

Keywords: Business Process Modeling, Design Verification, Control Flow, Resource Requirements, Enterprise Modeling

1. Verification of Business Process Designs: Motivation & Relevance

A business process is an ordered sequence of tasks/activities involving people, materials, energy, equipment, and information, designed to achieve some specific business outcome [3]. The standard approach to designing and implementing business processes is to rely on a domain expert to develop a process configuration that is subsequently "tuned-up"; however, there is a subtle question that is missing, namely, "what is the guarantee that the process's configuration is correct?" Problems, if any, in the design of a process, are usually detected by simulating the run-time behavior of a process. Currently, design verification capabilities are merely syntactic checks limited to the modeling formalism and do not provide any intelligent feedback to the process designer about whether the process's design is correct or not, and if not, why? There is much scope for developing qualitative design verification techniques that will precede any quantitative performance analysis, especially, techniques that illustrate the potential for using the static structural definition of a business process to improve its design, without recourse to simulated executions [5, 11]. Additionally, the growing interest in the automatic control and coordination of business processes requires that the process, by design, be correct [4, 13]. This will guarantee that operational errors, if any, can be attributed to data inconsistency, failure of IT infrastructure, etc., and not for anything lacking in the design of the process. This paper highlights, with examples, the verification issues that arise in the design of business processes, and outlines the major solution methodologies for the same. We also discuss a new tool called **MAPS** that allows for the modeling and analysis of business process models, and which adds immensely to harnessing the power of design verification.

The remainder of the paper is organized as follows: Section 2 outlines the basics of business process modeling, and the verification issues therein; Section 3 discusses the modeling and design verification capabilities of **MAPS**. Section 4 concludes this paper with ideas for continuing work to help establish design verification as a necessary first step in the design of business processes.

2. Business Processes: Design Requirements & Correctness Issues

The design of a business process minimally requires the specification of the process's logic, and the specifics of resource requirements for the constituent tasks. The logic of a process captures the flow of control among the tasks, and is (usually) graphically represented as a collection of tasks, and logical interconnections, as in Figure 1, the syntax of which is based on [17, 18].

The control flow model describes the process's logic, and is interpreted as follows – (i) Tasks represent individual activities, either atomic, or compos-

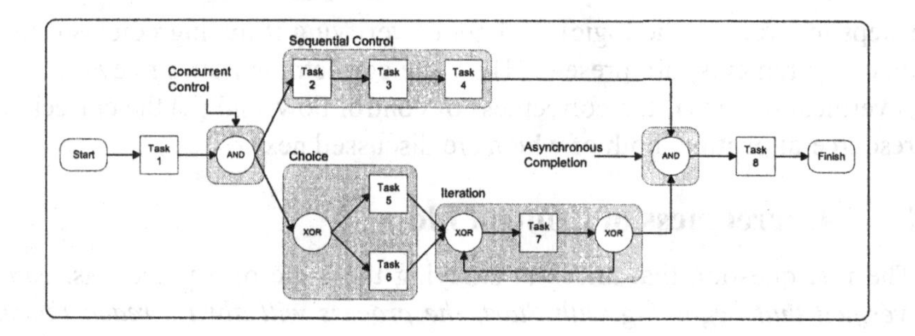

Figure 1. A Sample Control Flow Model

ite (i.e., sub-processes), (ii) AND-Splits (-Joins) model the concurrent creation (asynchronous merging) of several distinct flows of control, and (iii) Exclusive-OR, i.e., XOR-Splits (-Joins) model the creation (or merging) of exactly one of several possible flows of control, and (iv) directed arrows capture the partial and total ordering among the activities. The logic underlying any business process can be captured using just these elements; there is some debate about whether modeling conveniences like inclusive-ORs, n-out-of-m forks and joins, etc., improve modeling capabilities. Such extensions do not really increase the modeling power beyond the four basic elements presented above, and the syntax above is adequate for control flow representation [6].

The resource requirements for the various tasks in a business process is specified as [11]:

- $\mathbf{R} = \{R_1, R_2, \ldots, R_m\}$ is the set of all resources. $\forall R_i \in \mathbf{R}$, $R_i^{\#}$ = number of units of resource R_i.
- $\forall R_i \in \mathbf{R}$, $R_i^{Cap} : \mathbf{T} \rightarrow \mathbb{N}$ is a functional that specifies the number of units of resource R_i *captured* by each task, where $\mathbf{T}$ is the set of all tasks, and $\mathbb{N}$ is the set of non-negative integers.
- $\forall R_i \in \mathbf{R}$, $R_i^{Rel} : \mathbf{T} \rightarrow \mathbb{N}$ is a functional that specifies the number of units of resource R_i *released* by each task.

It is assumed that a resource is a "re-usable element" that is captured by tasks that require it, and released by tasks upon completion, although not necessarily by the same tasks that capture it. Taken together, the specification of control flow and resource requirements is adequate to study the static structural definition of a process, and to verify if the design is correct. Additional details like input-output requirements, infrastructure support, etc., while necessary for completing the process description, are not immediately required for the designer to commence design verification – stated simply, the question of verifying the correctness of a process's design is akin to that of verifying the steps in a recipe (the logic), and not of verifying if the ingredients (input-output requirements) are present. Thus, the correctness issues relevant to design verification are (i) the correctness of control flow, and (ii) the correctness of resource-allocation, both of which are discussed next.

2.1 Correctness of Control Flow

The first question that arises in studying the logic of a process is, *can it be verified that beginning with Start, the process will always reach Finish?* In order that the definition of correctness may be made precise, it remains to understand what counts as an incorrect control flow model – Table 1 illustrates a few examples.

Table 1. Incorrect Control Flow Models – Some Illustrations

Incorrect Control Flow Model	**Discussion**
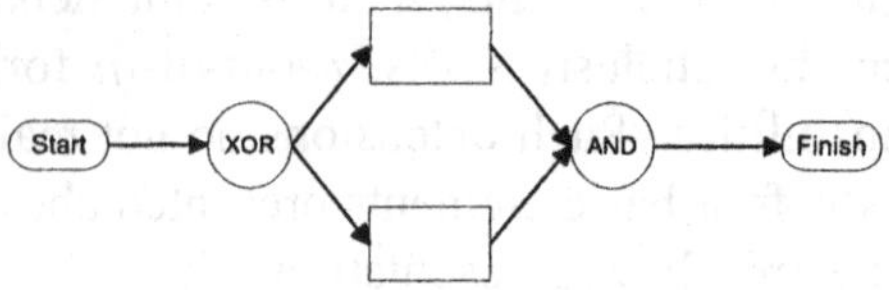 Start XOR AND Finish	An example of deadlock. The process will never terminate, since the AND operand will wait indefinitely for two incoming flows of control, while the XOR creates only one.
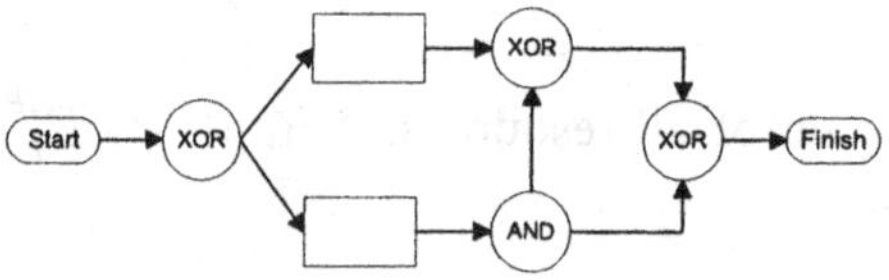 Start XOR XOR AND XOR Finish	A slightly more involved example. If the XOR next to "Start" chooses the top branch, the process will terminate properly; if it chooses the lower branch, the process will terminate twice.
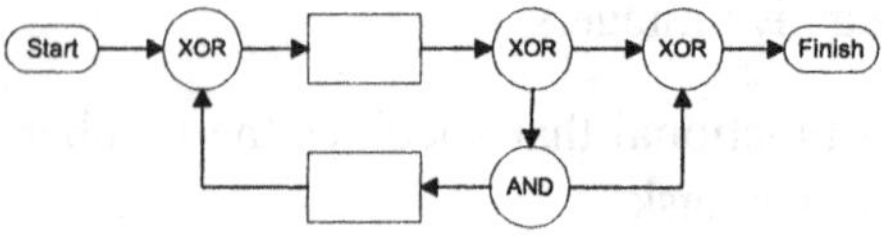 Start XOR XOR XOR Finish AND	A process model with the possibility for infinite repetitions. If the second XOR chooses the branch to its right, the process will terminate properly; if it chooses the lower branch, the process will terminate more than once.

There are three points that clarify themselves in the examples of Table 1, namely:

1 The process must terminate exactly one – this is referred to as *unique termination.*

2 The process must terminate completely, without any residual control flows hanging in the balance – this is referred to as *proper termination.*

3 The bulk of control flow errors arise from interspersing XOR and AND logical operands.

The correctness requirements for control flow can be stated thus: [7]

The Initiation Problem is to determine if there is a sequence of task executions that will lead to the execution of a particular task.

The Termination Problem is to determine if the flow of control in the process will lead to a terminal state.

The question of verifying the correctness of control flow has been shown to be NP-complete [7, 11]. Several restricted versions of the control flow problem have been addressed to date, namely, by requiring that the control flow model be acyclic [14, 8, 9], and that Petri net constructions of the control flow model remain free-choice [1], in order that analysis may be rendered tractable in polynomial time. A more thorough review of the current approaches to control flow verification is presented in [13].

We [11, 12] have developed a new backtracking algorithm (the KORRECTNESS algorithm) for identifying the various process execution scenarios in a control flow model, without imposing any restrictions on the structure of the control flow model. Stated simply, the idea is to identify all paths from *Start* to *Finish*, and to selectively combine them to identify execution scenarios (*meta-paths*) for the process, and to establish that each path does indeed occur in at least one meta-path to establish control flow correctness. Figure 2 illustrates the two valid meta-paths, i.e., valid process execution scenarios for the control flow model of Figure 1.

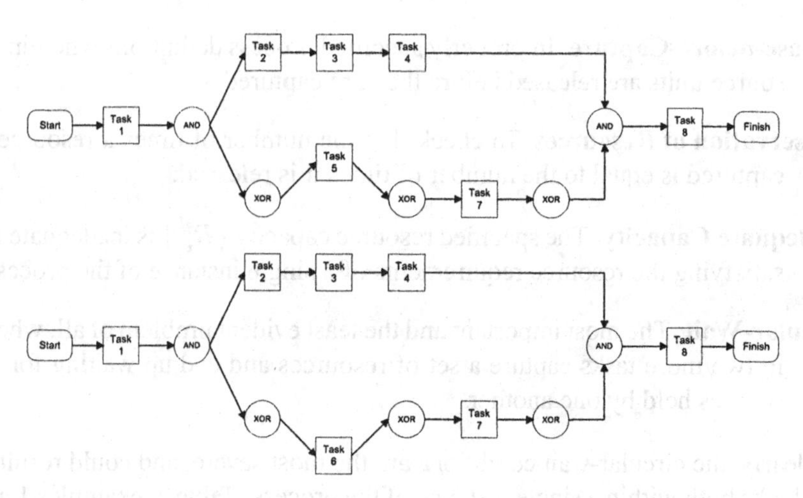

Figure 2. Process Execution Scenarios – An Illustration

The KORRECTNESS algorithm [11, 12] essentially keeps track of counters that are created by each AND-Split and which must subsequently be merged and accounted for, before completion of any instance of the process. While a correct control flow model will present one/more valid meta-paths, where all counters are accounted for, the value of the algorithm lies in its support for diagnostic feedback about sources of control flow errors in control flow models – by identifying the counters that remain to be accounted for in any invalid meta-path, the source of the error can be precisely identified. These ideas for the basis of **MAPS**, a tool for the **M**odeling and **A**nalysis of **P**rocess model**S**, and which is presented in Section 3. A more thorough description of the KORRECTNESS algorithm is contained in [12].

2.2 Correctness of Resource-Allocation

Deadlock arises when tasks that have captured some resources are blocked indefinitely from access to resources held by other tasks. The problems that arise in resource-sharing are best motivated with some examples, as presented in Table 2. The convention followed for these examples is – (i) the capture and release of resources by the tasks is specified with directed arrows entering and leaving the task symbols, and (ii) it is assumed that there is exactly one unit available for each resource, i.e., $R_i^{\#} = 1$.

None of the deadlock situations illustrated in Table 2 are obvious – to this end, it would be of tremendous value to be able to alert the process designer about potential deadlock possibilities that may not be immediately evident from the process's design.

The design problems that may arise from resource-allocation can be categorized into one of:

Release-before-Capture Improperly specified process definitions wherein resource units are released before they are captured.

Conservation of Resources To check that the number of times a resource is captured is equal to the number of times it is released.

Inadequate Capacity The specified resource capacity $\{R_i^{\#}\}$ is inadequate for satisfying the resource requirements of a single instance of the process.

Circular-Wait The most important and the least evident problem of all, wherein two/more tasks capture a set of resources and end up waiting for resources held by one another.

Clearly, the circular-wait conditions are the most severe, and could result in deadlocks both within a single instance of the process (Table 2, examples 1 and 2), and across multiple instances of the same process (Table 2, example 3).

Table 2. Incorrect Resource-Allocation Models – Some Illustrations

Incorrect Resource Model	Discussion
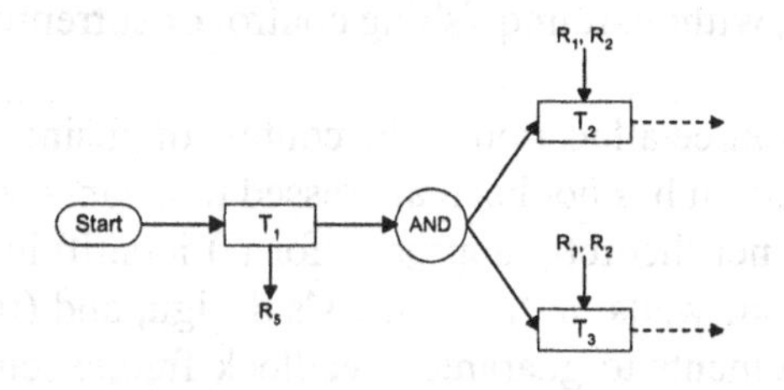	There are two problems, namely, (i) resource R_5 is released before it is captured, and (ii) a potential circular wait could arise if T_2 captures R_1 and T_3 captures R_2 and end up waiting indefinitely for each other to release their resources.
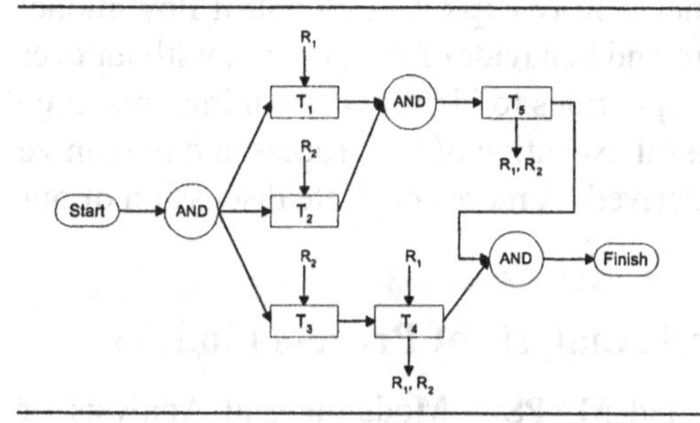	The process will get deadlocked if T_1 captures R_1 and T_3 captures R_2, whereupon neither T_2 nor T_4 can proceed any further, and the resources will never be released – an example of circular-wait.
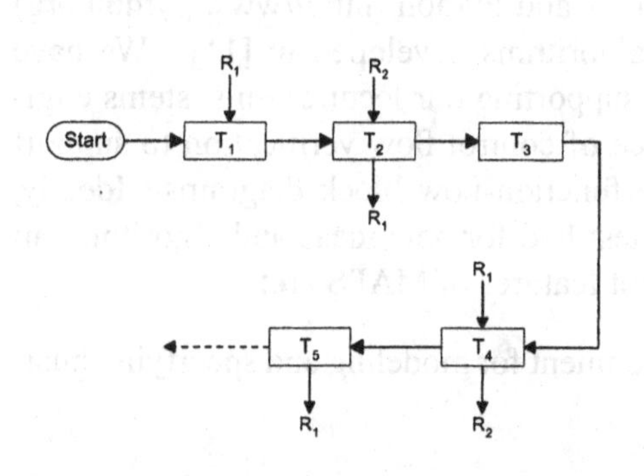	The control flow is very straightforward. However, consider the following incident: T_1 captures R_1 followed by T_2 which captures R_2 and releases R_1, and proceeds to execute T_3; meanwhile, T_1, being enabled, captures R_1 again and starts another *instance* of the process, thereby resulting in deadlock since the previous instance will not release R_2 without R_1 (task T_4), while the second instance will not release R_1 without R_2 (task T_2)– another circular-wait.

The study of deadlock has been motivated primarily by problems in operating systems, and deadlock avoidance in run-time control of flexible manufacturing systems (FMS). In the context of business process modeling, the techniques developed for the former are not applicable, since the process, by its own logic, imposes an ordering of the tasks which is not the case for computing systems where the tasks can be executed in any order. As regards the deadlock avoidance techniques developed for FMSs, they require simulated executions and/or details of the system state advancing through time, making them impractical for basic design verification [16, 19]. The resource-sharing problem in the design of business processes relates to establishing that the sharing of common resources among different tasks, either within a single- or across multiple-instances of the process does not lead to situations wherein two or more tasks compete for resources, without relinquishing control of currently held resources.

The problem of correctness of resource allocation in the context of business process modeling and design verification has not been addressed thus far – we [11] have developed a simple Petri-net theoretic approach for (i) identifying potential deadlocks, especially circular waits, in the process's design, and (ii) computing minimal resource requirements to guarantee deadlock-free execution. Our approach is unique in that it fully exploits the control flow model to gain intuitions about the structure and behavior of the process, without ever requiring any simulations. Also, simple rules to identify the minimal resource requirements that guarantee successful execution of the process and maximize in-process concurrency have been derived. A more complete discussion of our approach is presented in [11].

3. MAPS – Modeling and Analysis of Process modelS

A computerized environment titled **MAPS** – **M**odeling and **A**nalysis of **P**rocess model**S**, implemented in Tcl/Tk and Python (http://www.python.org) has been developed to support the algorithms developed in [11]. We have found **MAPS** to be very effective in supporting our lectures on systems engineering, and to motivate the relevance of control flow verification to support more traditional modeling tools like function-flow block diagrams. Ideally, **MAPS** should grow into a research test-bed for new ideas and algorithms in business process modeling. The salient features of MAPS are:

- A user-friendly graphical environment for modeling and specifying business processes.
- Algorithms for verifying the correctness of control flow and diagnostic feedback about sources of control flow error(s), if any.

3.1 Modeling Interface

The development of a process's design in MAPS begins with the specification of its control flow model, followed by the separate specification of its resource requirements (this is currently under development). Figure 3 presents a screen-shot of MAPS – it illustrates the control flow model of the counter-example presented by Lin *et al.* [8] to show that the algorithms of Sadiq [10] are incomplete; the labels $As\cdot$, $Aj\cdot$ indicate AND-Splits and AND-Joins, re-

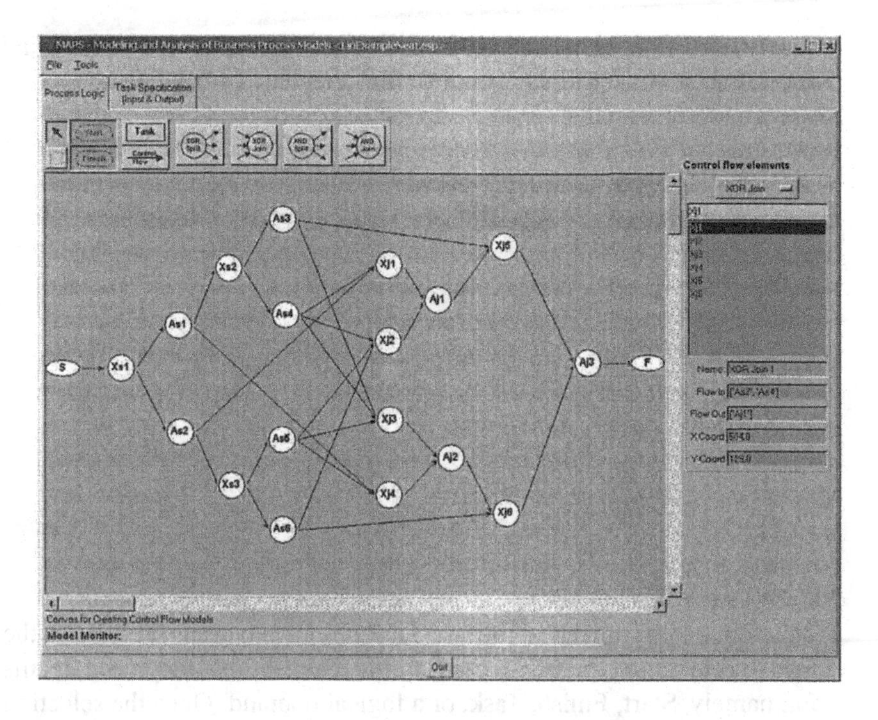

Figure 3. A Sample Screen-shot of MAPS

spectively, and the labels $Xs\cdot$, $Xj\cdot$, indicate, respectively, XOR-Splits, and XOR-Joins.

The major components of the graphical editor are:

1 The stencil that the user uses to select the control flow element being drawn.

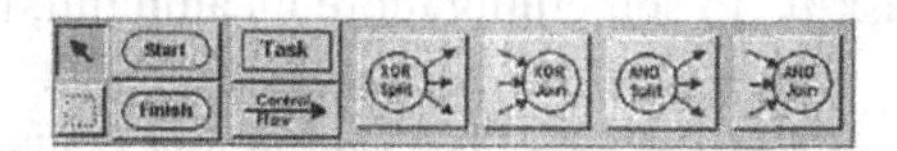

The stencil offers simple click-to-select functionality – the user selects (left-click) the control flow element that needs to be drawn, and then selects a position in the canvas to place them, or, if a control flow arrow is being drawn, the user selects the "source" (*from*) and "destination" (*to*) of the arrow with consecutive clicks inside the face of two elements on the canvas. The model can contain only one "Start" and one "Finish" – the corresponding stencil elements get disabled thereafter, unless those elements are subsequently deleted from the model.

2 The canvas on which the model is drawn – the canvas supports intuitive operations for both movement and deletion of elements in it.

3 The control flow elements' tablet (or frame) that provides an easy and accessible summary of each control flow element's attributes.

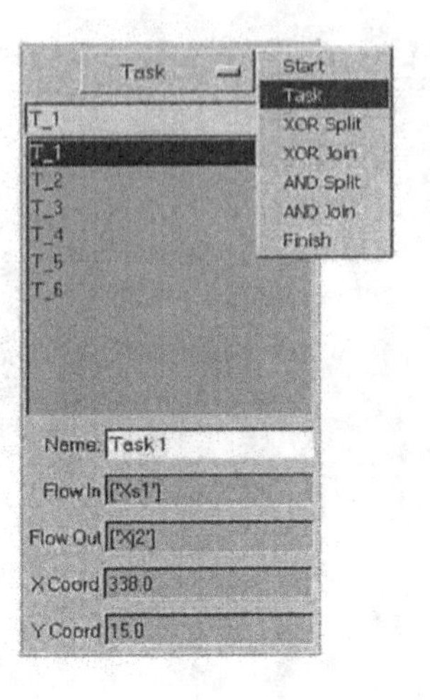

This frame is organized as follows – a drop-down options box allows the user to select the type of the control flow element that they need details for, namely, Start, Finish, Task, or a logical operand. Once the selection is made, say, "Task," a list of all tasks is generated, ordered by their screen IDs, individual selection of which will populate the basic fields beneath to reveal their names, input and output elements, and on-screen graphical coordinates. The screen IDs of the elements are prefixed with a ['S', 'F', 'T_', 'Xs', 'Xj', 'As', 'Aj'] to indicate that they are either a "Start," "Finish," "Task," "XOR-Split," "XOR-Join," "AND-Split," or an "AND-Join." All of the basic fields are fixed and cannot be edited, except for the names of tasks, which can be changed from the program-generated "Task n" to something more meaningful, if needed.

4 The model monitor that provides continuous feedback to the modeler about their actions through short messages, examples of whose outputs are shown in Figure 4.

Model Status: You are currently drawing an arrow starting at Task 1 and ending at XOR Split 1

Delete Alert: You have deleted Xs1

Error: You cannot draw more than one edge leading out of an Task. Please restart all over again.

Error: You cannot draw more than one edge coming into an XOR Split. Please restart all over again.

Delete Alert: Control flow arrow from T_2 to T_3 has been deleted.

Figure 4. Feedback from the Model Monitor in MAPS – Some Examples

> The model monitor will be indispensable when dealing with complex models – it includes several useful features that will guide the user in the construction of the model, and more so, when the user is contemplating the deletion of some elements. Unlike commercial grade software with the luxury of *undos* and such, the user will have to rely on the model monitor to inject the requisite caution in dealing with mouse-clicks, right or left.

In addition to the features listed above, *help balloons* have been programmed to appear liberally across all aspects of the application to clarify the purpose of all four components above. The models thus created can be saved and retrieved with standard $File \rightarrow Save$ operations.

3.2 Analysis and Verification Capabilities of MAPS

The current version of MAPS includes support for verifying the model syntax, and the KORRECTNESS algorithm [12] for control flow verification, both of which are described below.

Syntax Verification. The syntax checks enforced in MAPS ensure that the model does not have any abandoned elements, that it has a "Start" and a "Finish", and that all other elements have properly defined "from" and "to" elements, as illustrated below.

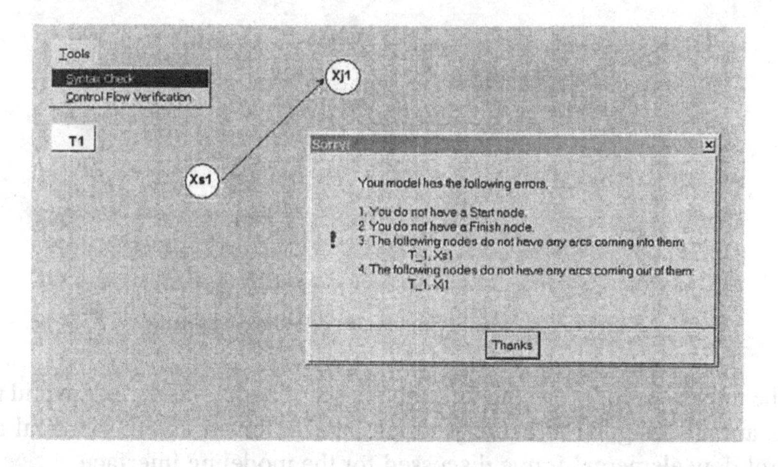

Additionally, should the user attempt to draw, say, two arcs leading into a task (or an AND/XOR-split), or other such basic modeling errors that violate the definition of the elements, the model monitor will alert the modeler to the same, thereby allowing for immediate model checks as well – the syntax verification capabilities in MAPS are dynamic and work constantly during the development of the model.

Control Flow Verification. The verification of control flow correctness follows the KORRECTNESS algorithm, in that it proceeds by generating the set of $S - F$ paths, the set of valid meta-paths, and the set of invalid meta-paths, if any. Figure 5 shows a snap-shot of the application interface after the KORRECTNESS algorithm has been applied to the control flow model shown in Figure 3 – note that the results of the control flow verification procedure will open up on a new page titled "Korrectness Results".

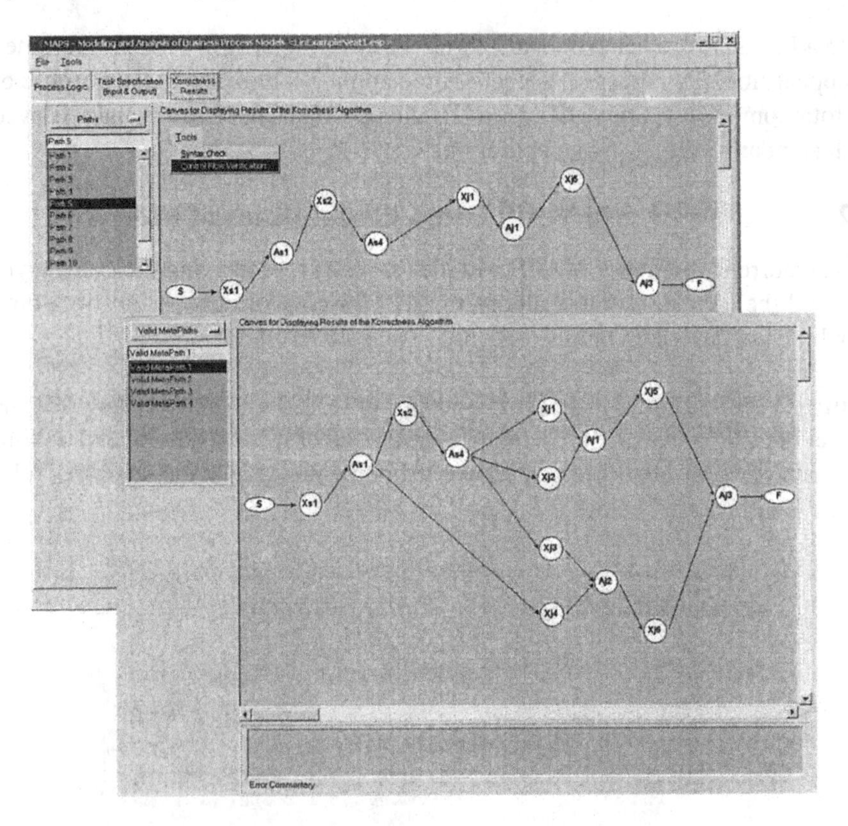

Figure 5. Identifying the Set of Valid Meta-paths

The interfaces for browsing through the set of paths, valid, and invalid meta-paths are all designed to be very simple, and follow a layout identical to the control flow elements' frame discussed for the modeling interface.

Figure 6 illustrates a snap-shot of the application interface after the KORRECTNESS algorithm has been applied to an incorrect control flow model – it illustrates one of several invalid meta-paths in the process.

Much like the "model monitor" in the modeling interface, the "Error Commentary" provides feedback about the source of the control flow error – the meta-path in Figure 6 is invalid because one of the three control flows required

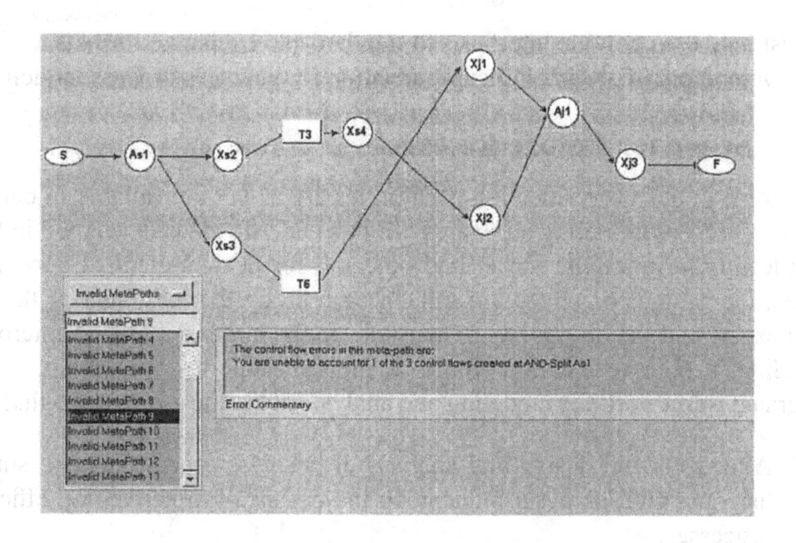

Figure 6. Identifying the Set of Valid Meta-paths

at AND-Split As_1 is missing. This "error commentary" feature of MAPS in unique in providing precise reasons as to the failure of a control flow design – what is missing however, is a way to automate the correction of these control flow errors, or at the very least, to give suggestions to fix the same.

Summary. This section outlined the use of **MAPS**, a computerized environment that was developed to support the algorithms developed in [11]. MAPS has been designed to be easily extendible with new functionalities, either in graphical modeling, or algorithmic support. We have found **MAPS** to be an excellent tool for classroom instruction – there are as yet no comparable tools for modeling and analysis of business process models, that work with just the control flow model's syntax. Two other verification tools currently in use are Woflan [15] and FlowMake [10] – the former relies on Petri net formalizations of the control flow model, and the latter works with graph-theoretic reductions. Woflan relies on free-choice Petri net constructions to make analysis tractable, but is not restricted to only free-choice constructions; it offers an impressive array of diagnostic checks that support design correction efforts. While Woflan reports outputs using Petri-net theoretic constructs, **MAPS** works with the syntax of the control flow model allowing for easy communication between the domain expert and the business process modeler/analyst.

MAPS is an ongoing work-in-progress, and currently includes an editor for creating, storing, and retrieving control flow models, and analysis support to establish control flow correctness. We are presently working to incorporate algorithms for verifying the correctness of resource requirements for a business process, and also provide interfaces to standardized exchange formats to aid in the development of an ontological foundation for business process modeling.

4. Scope for Future Research

Automatic design verification capabilities are as yet unavailable in current business process modeling softwares. To this end, the ideas developed in [11] significantly advance the power and potential for development of good processes, the designs of which are influenced both by the judgment of the domain expert and the objectivity of analysis. In the larger context of enterprise modeling, design verification would be indispensable toward developing an integrated framework for modeling and analysis of business processes that [5]:

1 Allows for the design and analysis of business processes to be simultaneous, with analysis influencing the design of effective and efficient processes.

2 Clarifies ambiguities in the domain expert's interpretation, experience, and expectations of the business process through immediate qualitative analysis.

3 Provides a seamless, almost invisible translation between the description of the business process, and the formalism that feeds the underlying analysis.

4 Provides linkages to other analysis techniques that can be used to derive summary metrics about the run-time performance of the process.

The following ideas merit further inquiry:

1. **Automation of Business Process Redesign** Suppose the process's design, i.e., control flow and/or resource allocation, is incorrect; can the redesign of business processes to eliminate design errors be automated? More specifically, can human intuition be replaced with algorithmic deduction? Currently, diagnostic checking is limited to identification; can it be extended to include elimination? While domain knowledge cannot be replaced, can control-flow design decisions be abstracted away from the context of the process?

2. **Automatic Reconfiguration of Business Processes** Suppose the process's design, i.e., control flow and resource-allocation, is correct; is it possible to suggest approaches to reconfigure or "optimize" the process's design, based on the nature of the resource-allocation requirements and the precedence-order relationships that are imposed by the logic of the

process? This would require a more precise understanding of the expectations of "optimality" in business process designs – stated simply, the domain expert has identified a particular configuration for the process; can it be improved?

3. **Standards for Business Process Specification** To develop a formalism for modeling and specification of business processes that blends the ease of modeling intuition with the rigor required for design verification. Ideally, a formalism that capitalizes on the transparency of XML, the intuitions of graphical models, and the support of underlying design verification techniques would greatly enhance the value-addition of enterprise automation.

The question raised in (3) has already been initiated and is actively being pursued by the Business Process Management Initiative (`www.bpmi.org`). The questions raised in (1) and (2) are more fundamental and as yet unexplored; to allow for a computer to suggest a better process design is very intriguing, and such an ability would lend new meaning to "automation" in business process automation. The extent of the questions' appeal is surpassed only by the vagueness of their answers.

Acknowledgments

This research was supported, in part, by the National Science Foundation (grant #DMI-0075588) under the Scalable Enterprise Systems initiative. The authors would like to thank two anonymous referees for their review and comments.

References

[1] Aalst, W.M.P. (2000). "Workflow Verification: Finding Control-Flow Errors Using Petri-Net Based Techniques". In Aalst, W.M.P., Desel, J., and Oberweis, A., editors, *Business Process Management – Models, Techniques, and Empirical Studies*, volume 1806 of *Lecture Notes in Computer Science*, pages 161–183. Springer-Verlag.

[2] Arkin, A. (2003). "Business Process Modeling Language". Technical Report, Business Process Management Initiative, `http://www.bpmi.org/library.esp`.

[3] CSC Corporation (2002). "The Emergence of Business Process Management". Computer Sciences Corporation White Paper, `http://www.bpmi.org/library.esp`.

[4] Cichocki, A., Helal, A., Rusinkiewicz, M., and Woelk, D. (1998). *Workflow and Process Automation: Concepts and Technology*. Kluwer Academic Publishers.

[5] Dalal, N.P., Kamath, M., Kolarik, W.J., and Sivaraman, E. (2004). "Toward an Integrated Framework for Modeling Enterprise Processes". *Comm. of the ACM*, 47(3):83–87.

[6] Kiepuszewski, B., Hofstede, A.H.M., and Aalst, W.M.P. (2004). "Fundamentals of Control Flow in Workflows". *Acta Informatica*, 39(3):143–209.

[7] Hofstede, A.H.M., Orlowska, M.E., and Rajapakse, J. (1998). "Verification Problems in Conceptual Workflow Specifications". *Data & Knowledge Engineering*, 24(3):239–256.

[8] Lin, H., Zhao, H., Li, H., and Chen, Z. (2002). "A Novel Graph Reduction Algorithm to Identify Structural Conflicts". In *Proceedings of the 35th Annual Hawaii International Conference on System Science (HICSS-35'02)*. IEEE Computer Society Press.

[9] Sadiq, W. and Orlowska, M.E. (2000). "Analyzing Process Models using Graph Reduction Techniques". *Information Systems*, 25(2):117–134.

[10] Sadiq, W. (2001). *On Verification Issues in Conceptual Modeling of Workflow Processes*. PhD Dissertation, Department of Computer Science and Electrical Engineering, The University of Queensland, Australia.

[11] Sivaraman, E. (2003). *Formal Techniques for Analyzing Business Process Models*. PhD Dissertation, School of Industrial Engineering & Management, Oklahoma State University, Stillwater, OK, USA.

[12] Sivaraman, E. and Kamath, M. (2004). "An Algorithm for Verifying the Correctness of Control Flow in Business Process Models". Under review for publication in the *INFORMS J. of Computing*.

[13] Sivaraman, E. and Kamath, M. (2004). "Verification Issues in the Design of Business Processes: A Review". Under review for publication in *IIE Transactions*.

[14] Straub, P. and Hurtado, C.L. (1995). "The Simple Control Property of Business Process Models". In *Proceedings of the XV Conference of the Chilean Computer Society*, Arica, Chile, Oct. 30-Nov. 3.

[15] Verbeek, H.M.W., Baasten, T. and Aalst, W.M.P. (2001). "Diagnosing Workflow Processes using Woflan". *The Computer Journal*, 44(4):246–279.

[16] Viswanadham, N., Narahari, Y. and Johnson, T.J. (1990). "Deadlock Prevention and Deadlock Avoidance in Flexible Manufacturing Systems". *IEEE Trans. on Robotics and Automation*, 6:713–723.

[17] WfMC (1995). *The Workflow Reference Model*. published by the Workflow Management Coalition, Document WFMC TC00-1003, `http://www.wfmc.org/standards/docs.htm`.

[18] WfMC (1999). *Terminology & Glossary*. published by the Workflow Management Coalition, Document WFMC TC-1011, `http://www.wfmc.org/standards/docs.htm`.

[19] Zhou, M., Jeng. M. and Fanti, M.P. (editors) (2004). Special Issues on Deadlock Resolution in Computer-Integrated Systems. *IEEE Trans. on SMC – Part A*, 34(1).

ALPS: A FRAMEWORK FOR IMPLEMENTING PARALLEL TREE SEARCH ALGORITHMS

Yan Xu
Operations R & D, SAS Institute Inc., Cary NC 27513
Yan.Xu@sas.com

Ted K. Ralphs
Department of Industrial and Systems Engineering, Lehigh University, Bethlehem PA 18015
tkralphs@lehigh.edu

Laszlo Ladányi
Department of Mathematical Sciences, IBM T. J. Watson Research Center, Yorktown Heights NY 10598
ladanyi@us.ibm.com

Matthew J. Saltzman
Department of Mathematical Sciences, Clemson University, Clemson SC 29634
mjs@clemson.edu

Abstract ALPS is a framework for implementing and parallelizing tree search algorithms. It employs a number of features to improve scalability and is designed specifically to support the implementation of *data intensive* algorithms, in which large amounts of *knowledge* are generated and must be maintained and shared during the search. Implementing such algorithms in a scalable manner is challenging both because of storage requirements and because of communications overhead incurred in the sharing of data. In this abstract, we describe the design of ALPS and how the design addresses these challenges. We present two sample applications built with ALPS and preliminary computational results.

Keywords: Parallel Algorithm, Integer Programming, Software, Branch and Bound

1. Introduction

Tree search algorithms are a general class in which the nodes of a directed, acyclic graph are systematically searched in order to locate one or more *goal nodes.* In most cases, the graph to be searched is not known a priori, but is constructed dynamically based on information discovered during the search process. We assume the graph has a unique *root node* with no incoming arcs, which is the first node to be examined. In this case, the search order uniquely determines a rooted tree called the *search tree.* Although tree search algorithms are easy to parallelize in principle, the absence of a priori knowledge of the shape of the tree and the need to effectively share information generated during the search makes such parallelization challenging and scalability difficult to achieve. In [Ralphs et al., 2003] and [Ralphs et al., 2004], we examined the issues surrounding parallelization of tree search algorithms and presented a high-level description of a class hierarchy for implementing such algorithms. In this abstract, we follow up on those works by presenting further details of the search handling layer of the proposed hierarchy, called the Abstract Library for Parallel Search (ALPS), which will soon have its first public release.

A variety of existing software frameworks are based on tree search. For mixed-integer programming—the application area we are most interested in—most packages employ a sophisticated variant of branch and bound. Among the offerings for solving generic mixed-integer programs are bc-opt [Cordier et al., 1999], FATCOP [Chen and Ferris, 2001], MIPO [Balas et al., 1996], PARINO [Linderoth, 1998], SIP [Martin, 1998], SBB [Forrest, 2004], GLPK [Makhorin, 2004], and bonsaiG [Hafer, 1999]. Of this list, FATCOP and PARINO are parallel codes. Commercial offerings include ILOG's CPLEX, IBM's OSL (soon to be discontinued), and Dash's XPRESS. Generic frameworks that facilitate extensive user customization of the underlying algorithm include SYMPHONY [Ralphs, 2004], ABACUS [Jünger and Thienel, 2001], BCP [Ladányi and Ralphs, 2001], and MINTO [Nemhauser et al., 1994], of which SYMPHONY and BCP are parallel codes. Other frameworks for parallel branch and bound include BoB [Benchouche et al., 1996], PICO [Eckstein et al., 2000], PPBB-Lib [Tschoke and Polzer, 1998], and PUBB [Shinano et al., 1995]. Good overviews and taxonomies of parallel branch and bound are provided in both [Gendron and Crainic, 1994] and [Trienekens and Bruin, 1992]. Eckstein et al. [Eckstein et al., 2000] also provides a good overview of the implementation of parallel branch and bound. A substantial number of papers have been written specifically about the application of parallel branch and bound to discrete optimization problems, including [Bixby et al., 1995; Correa and Ferreira, 1995; Grama and Kumar, 1995; Mitra et al., 1997].

The goal of the ALPS project is to build on the best existing methodologies while addressing their shortcomings to produce a framework that is more general and extensible than any of the current options. As such, we provide support for the implementation of a range of algorithms that existing frameworks are not general enough to handle. Our design is centered around the abstract notion of *knowledge generation and sharing*, which is very general and central to implementing scalable versions of today's most sophisticated tree search algorithms. Such algorithms are inherently *data-intensive*, i.e., they generate large amounts of knowledge as a by-product of the search. This knowledge must be organized, stored, and shared efficiently. ALPS provides explicit support for these procedures and allows for user-defined knowledge types, making it easy to create derivative frameworks for a wide range of specific classes of algorithms. While our own experience is in developing algorithms for solving mixed-integer linear programs, we have in mind to develop a number of additional layers providing support for tree search algorithms in other areas, such as global optimization. Although we present limited computational results, we want to emphasize that this research is ongoing and that the results are intended merely to illustrate the challenges we still face. The main goal of the paper is to describe the framework itself. ALPS is being developed in association with the Computational Infrastructure for Operations Research (COIN-OR) Foundation [Lougee-Heimer, 2003], which will host the code.

1.1 Tree Search Algorithms

In a tree search algorithm, each node in the search graph has associated data, called its *description*, that can be used to determine if it is a goal node, and if it has any successors. To specify such an algorithm, four main elements are required. The *fathoming rule* determines whether a node has successors that need to be explored. The *branching method* specifies how to generate the descriptions of a node's successors. The *processing method* determines whether a node is a goal node and whether it has any successors. The *search strategy* specifies the processing order of the candidate nodes.

Each node has an associated *status*, which is one of: `candidate` (available for processing), `active` (currently being processed), `fathomed` (processed and has no successors), or `processed` (not fathomed, hence has successors). The search consists of repeatedly selecting a candidate node (initially, the root node), processing it, and then either fathoming or branching. The nodes are chosen according to *priorities* assigned during processing.

Variants of tree search algorithms are widely applied in areas such as discrete optimization, global optimization, stochastic programming, artificial intelligence, game playing, theorem proving, and constraint programming. One of the most common variants in discrete optimization is *branch and bound*, originally suggested by Land and Doig [Land and Doig, 1960]. In branch and bound, branching consists of partitioning the feasible set into subsets. Processing consists of computing a bound on the objective function value, usually by solving a relaxation. A node can be fathomed if (1) the solution to the relaxation is in the original feasible set (in which case, the best such solution seen so far is recorded as the *incumbent*), (2) the objective value of the solution to the relaxation exceeds the value of the incumbent, or (3) the subset is proved to be empty.

1.2 Parallelizing Tree Search

In principle, tree search algorithms are easy to parallelize. Sophisticated variants, however, involve the generation and sharing of large amounts of *knowledge*, i.e., information helpful in guiding the search and improving the effectiveness of node processing. Inefficiencies in the mechanisms by which knowledge is maintained and shared result in *parallel overhead*, which is additional work performed in the parallel algorithm that would not have been performed in the sequential one. The goal of any parallel implementation is to limit this overhead as much as possible.

We assume a simple model of parallel computation in which there are N processors with access to their own local memory and complete connectivity with other processors. We further assume that there is exactly one process per processor at all times, though this process might be multi-threaded. The main sources of parallel overhead for tree search algorithms are:

- *Communication Overhead*: time spent actively sending or receiving knowledge.
- *Idle Time*: time spent waiting for knowledge to be transferred from another processor (including *task starvation*, when the processor is waiting for more work to do).
- *Redundant Work*: time spent performing unnecessary work, usually due to a lack of appropriate global knowledge.
- *Ramp-Up/Ramp-Down*: idle time at the beginning/end of the algorithm during which there is not enough work for all processors.

The effectiveness of the knowledge-sharing mechanism is the main factor affecting this overhead. The sources of overhead listed above highlight the tradeoff between centralized storage and decision making, which incurs increased communication and idle time, and decentralized storage and decision making, which increases performance of redundant work. Achieving the proper balance is the challenge we face. *Scalability* is a measure of how well this balance is achieved, i.e., how well an algorithm takes advantage of increased computing resources, primarily additional processors. Our measure of scalability is the rate of increase in overhead as additional processors are made available. A parallel algorithm is considered scalable if this rate is near linear. An excellent general introduction to the analysis of parallel scalability is provided in [Kumar and Gupta, 1994].

2. Implementation

2.1 Knowledge Sharing

In [Ralphs et al., 2004], building on ideas in [Trienekens and Bruin, 1992], we proposed a tree search methodology driven by the concept of knowledge discovery and sharing. We briefly review the concepts from the earlier work here. The design of ALPS is predicated on the idea that all information required to carry out a tree search can be represented as knowledge that is generated dynamically and stored in various local *knowledge pools* (KPs), which share that knowledge when needed. A single processor can host multiple KPs that store different types of knowledge and are managed by a *knowledge broker* (KB). Examples of knowledge generated while solving mixed-integer programs include feasible solutions, search-tree nodes, and valid inequalities.

The KB associated with a KP may field two types of requests on its behalf: (1) new knowledge to be inserted into the KP or (2) a request for relevant knowledge to be extracted from the KP, where "relevant" is defined for each category of knowledge with respect to data provided by the requesting process. A KP may also choose to "push" certain knowledge to another KP, even though no specific request has been made.

The most fundamental knowledge generated during the search is the descriptions of the search-tree nodes themselves. The node descriptions are stored in KPs called *node pools.* The node pools collectively contain the list of candidate nodes. The tradeoff between centralization and decentralization of knowledge is most evident in the mechanism for sharing node descriptions among the processors, known as *load balancing.* Effective load balancing reduces both idle time associated with task starvation and performance of redundant work. Load balancing methods have been studied extensively [Fonlupt et al., 1998; Henrich, 1993; Kumar et al., 1994; Laursen, 1994; Sanders, 1998; Sinha and Kalé, 1993], but many of the suggested schemes are not suited for our framework. The simplest approach is a *master-worker* design that stores all node descriptions in a single, central node pool. This makes work distribution easy, but incurs high communication costs. This is the approach we have taken in our previous frameworks, SYMPHONY and BCP. It works well for small numbers of processors, but does not scale well, as the central node pool inevitably becomes a computational and communications bottleneck.

2.2 The Master-Hub-Worker Paradigm

To overcome the drawbacks of the master-worker approach, ALPS employs a *master-hub-worker* paradigm, in which a layer of "middle management" is inserted between the master process and the worker processes. In this scheme, a *cluster* consists of a hub, which is responsible for managing a fixed number of workers. As the number of processes increases, we simply add more hubs and more clusters of workers. This scheme is similar to one implemented by Eckstein et al. in the PICO framework [Eckstein et al., 2000], except that PICO does not have the concept of a master. This decentralized approach maintains many of the advantages of global decision making while reducing overhead and moving some computational burden from the master process to the hubs. This burden is then further shifted from the hubs to the workers by increasing the task granularity, as described below. Cluster size is computed based on the number of hubs and the number of processors, which are set by the user at run time.

The basic unit of work in our design is a *subtree.* Each worker is capable of processing an entire subtree autonomously and has access to all of the methods needed to manage a tree search. Designating a subtree as the fundamental unit of work helps to minimize memory requirements by enabling the use of efficient data structures for storing subtrees using a differencing scheme similar to that used in both SYMPHONY and BCP. In this scheme, node descriptions are not stored explicitly, but rather as differences from their predecessors' descriptions. This increased granularity also reduces idle time due to task starvation, but, without proper load balancing, may increase the performance of redundant work.

2.3 Load Balancing

Recall that each node has an associated priority that can be thought of as indicating the node's "quality," i.e., the probability that the node or one of its successors is a goal node. In assessing the distribution of work to the processors, we need to consider not only *quantity*, but also *quality*. ALPS employs a three-tiered load balancing scheme, consisting of *static*, *intra-cluster dynamic*, and *inter-cluster dynamic* load balancing. Static load balancing, or *mapping*, takes place during the initial phase of the algorithm. The first task is to generate a group of successors of the root node and distribute them to the workers to initialize their local node pools. ALPS uses a *two-level root initialization* scheme, a generalization of the *root initialization* scheme of [Henrich, 1993]. During static load balancing, the master creates and distributes a user-specified number of nodes for hubs. The hubs in turn create a user-specified number of successors for their workers, then the workers initialize their subtree pools and begin.

Time spent performing static load balancing is the main source of ramp-up, which can be significant when node processing times are large. The problem of reducing ramp-up has long been recognized as a challenging one [Gendron and Crainic, 1994; Borbeau et al., 2000; Eckstein et al., 2000]. Two-level root initialization reduces ramp-up by parallelizing the root initialization process itself. Implementation of two-level root initialization is straightforward, but our experience has shown that it can work quite well if the number of nodes distributed to each worker is large enough and node processing times are short.

Inside a cluster, the hub manages dynamic load balancing. Intra-cluster load balancing is initiated when an individual worker reports to the hub that its workload is below a given threshold. Upon receiving the request, the hub asks its most loaded worker to donate a subtree to the requesting worker. In addition, the hub periodically checks the qualities of the workloads of its workers. If it finds that the qualities are unbalanced, the hub asks the workers with the most high priority nodes to share their workload with the workers that have fewer such nodes.

The master is responsible for balancing the workload among hubs, which periodically report their workload information to the master. The master has a roughly accurate global view of the system load and the load of each cluster at all times. If either the quantity or quality of work is unbalanced among the clusters, the master identifies pairs of *donors* and *receivers*. Donors are clusters whose workloads are greater than the average workload of all clusters by a given factor. Receivers are the clusters whose workloads are smaller than the average workload by a given factor. Donors and receivers are paired and each donor sends a subtree to its paired receiver.

A unique aspect of our load balancing scheme is that it takes account of the differencing scheme for storing subtrees. In order to prevent subtrees from becoming too fractured for efficient storage using differencing, we try at all times to ensure that the search-tree nodes are distributed in a way such that the nodes stored together locally constitute connected subtrees of the search tree. This means the tree structure must be taken into account when sharing nodes during the load balancing. Candidate nodes that constitute the leaves of a subtree are grouped, and the entire subtree is shared, rather than just the nodes themselves. To achieve this, each subtree is assigned a priority level, defined as the average priorities of a given number of its best nodes. During load balancing, the donor chooses the best subtree in its subtree pool and sends it to the receiver. If a donor does not have any subtrees in its subtree pool, it splits the subtree that it is currently exploring into two parts and sends one of them to the receiver. In this way, differencing can still be used effectively.

2.4 Task Management

Because each process hosts a KB and several KPs, it is necessary to have a scheme for enabling multi-tasking. In order to maintain maximum portability and to assert control over task scheduling, we have implemented our own simple version of threading. ALPS processes are message driven—each process devotes one thread to listening for and responding to messages at all times. Other threads are devoted to performing computation as scheduled. Because each processor's KB controls the communication to and from the process, it also controls task scheduling. The KB receives external messages, forwards them to the appropriate local KP if needed, and forwards all locally generated messages to the appropriate remote KB. When not listening for messages, the KB schedules the execution of computational tasks by the local KPs. The KB decides when and for how long to process each task.

3. Class Structure

ALPS consists of a library of C++ classes from which can be derived specialized classes that define various tree search algorithms. Figure 1 shows the ALPS class hierarchy. Each block represents a C++ class, whose name is listed in the block. The lines ending with triangles represent inheritance relationships. For example, the `AlpsSolutionPool`, `AlpsSubtreePool` and `AlpsNodePool` classes are derived from the class `AlpsKnowledgePool`. The lines ending with diamonds represent associative relationships. For instance, `AlpsKnowledge` contains as a data member a pointer to an instance of `AlpsEncoded`. ALPS is comprised of just three main base classes and a number of derived and auxiliary classes. These classes support the core concept of knowledge sharing and

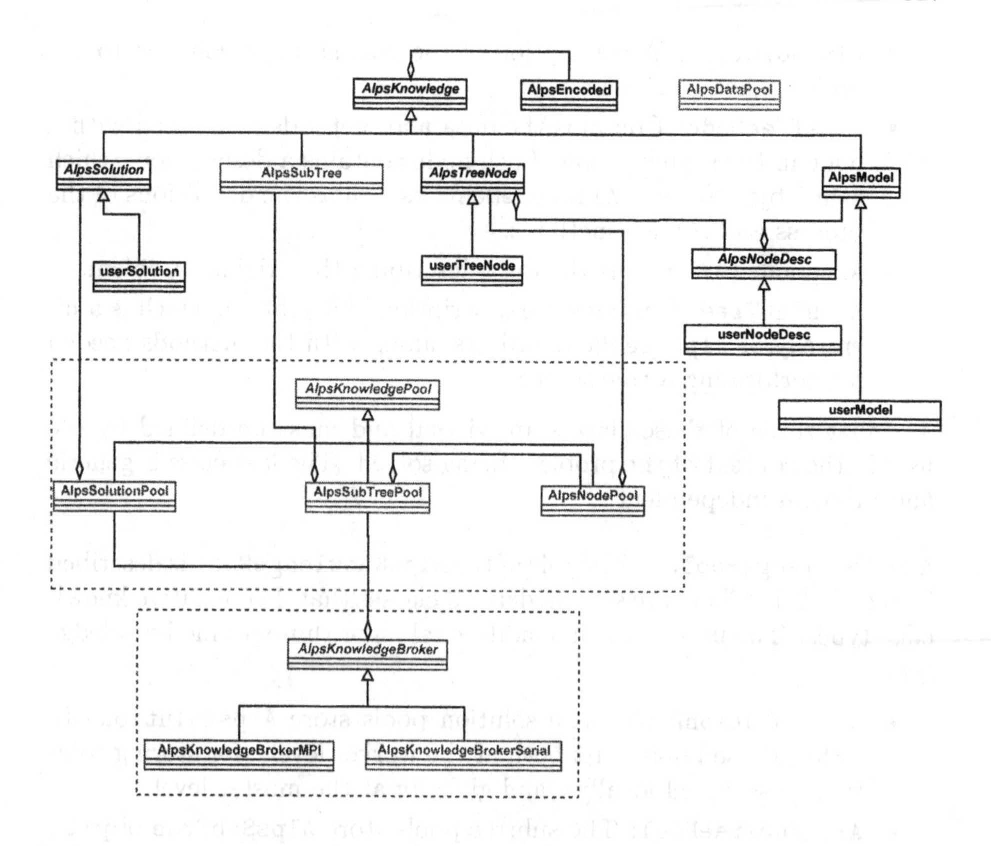

Figure 1. The ALPS class hierarchy.

are described in the paragraphs below. The classes named `UserXXX` in the figure are those that must be defined by the user to develop a new application. Two examples are described in Section 4.

`AlpsKnowledge`. This is the virtual base class for any type of information that must be shared or moved from one process to another. `AlpsEncoded` is an associated class that contains the encoded or packed form of an `AlpsKnowledge` object. The packed form contains the data needed to describe an object of a particular type in the form of a character string. This representation typically takes much less memory than the object itself; hence, it is appropriate both for storage of knowledge and for communication of knowledge between processors. The packed form is also independent of type, which allows ALPS to deal with user-defined knowledge types. Finally, duplicate objects can be quickly identified by hashing their packed forms. ALPS has the following four native knowledge types:

- `AlpsSolution`: A description of the goal state or solution to the problem being solved.
- `AlpsTreeNode`: Contains the data and methods associated with a node in the search graph. Each node contains a description, which is an object of type `AlpsNodeDesc`, as well as the definitions of the process and branch methods.
- `AlpsModel`: Contains the data describing the original problem.
- `AlpsSubTree`: Contains the description of a subtree, which is a hierarchy of `AlpsTreeNode` objects, along with the methods needed for performing a tree search.

The first three of these classes are virtual and must be defined by the user in the context of the problem being solved. The last class is generic and problem-independent.

`AlpsKnowledgePool`. The role of the `AlpsKnowledgePool` is described in Section 2.1. There are several derived classes that define native knowledge types. The user can define additional algorithm-specific knowledge types.

- `AlpsSolutionPool`: The solution pools store `AlpsSolution` objects. These pools exist both at the worker level—for storing solutions discovered locally—and globally at the master level.
- `AlpsSubTreePool`: The subtree pools store `AlpsSubTree` objects. These pools exist at the hub level for storing subtrees that still contain unprocessed nodes.
- `AlpsNodePool`: The node pools store `AlpsTreeNode` objects. These pools contain the queues of candidate nodes associated with the subtrees as they are being searched.

`AlpsKnowledgeBroker`. This class encapsulates the communication protocol. The KB is the driver for each processor and is responsible for sending, receiving, and routing all data that resides on that processor. Each KP must be registered so that the KB knows how to route each specific type of knowledge when it arrives and where to route requests for specific types of knowledge from other KBs. This is the only class whose implementation depends on the communication protocol. Currently, the protocols supported are a serial layer and an MPI [Gropp et al., 1999] layer.

- `AlpsKnowledgeBrokerMPI`: A KB for multiprocessor execution via the MPI message-passing interface.
- `AlpsKnowledgeBrokerSerial`: A KB for uniprocessor execution.

```
#include Alps.h
#include AlpsUser.h     // User-derived classes

int main(int argc, char* argv[])
{
    UserModel model;
    UserParams userPar;

#if defined(SERIAL)
    AlpsKnowledgeBrokerSerial broker(argc, argv, model, userPar);
#elif defined(PARALLEL_MPI)
    AlpsKnowledgeBrokerMPI broker(argc, argv, model, userPar);
#endif
    broker.registerClass("MODEL", new UserModel);
    broker.registerClass("SOLUTION", new UserSolution);
    broker.registerClass("NODE", new UserTreeNode);
    broker.search();
    broker.printResult();
    return 0;
}
```

Figure 2. Sample main function.

4. Applications and Preliminary Results

Developing an application with ALPS consists mainly of implementing derived classes, and writing the `main()` function. As described in Section 3, the user must derive algorithm-specific classes from the base classes `AlpsModel`, `AlpsTreeNode`, `AlpsNodeDesc`, and `AlpsSolution`. The user may also want to define algorithm-specific parameters by deriving a class from `AlpsParameterSet`, or he may even want to define new types of knowledge. A sample code for `main()` is shown in Figure 2.

4.1 Knapsack Solver

The binary knapsack problem is to select from a set of items a subset with the maximum total profit and not exceeding a given total weight. The profit is additive. By deriving classes `KnapModel`, `KnapTreeNode`, `KnapNodeDesc`, `KnapSolution` and `KnapParameterSet`, we have developed a solver for the binary knapsack problem employing a very simple branch and bound algorithm. The nodes of the search tree are described by subproblems obtained by fixing a subset of the items in the global set to be either in or out of the selected subset. The branching procedure consists of selecting an item and requiring it to be in the selected subset in one successor node and not in the other. Processing consists

N	Wall-clock	Ramp-up	Idle	Speedup	Efficiency	Nodes
1	1335	–	–	–	–	254 k
4	296	0%	2.9%	4.5	1.13	85 m
8	160	0%	2.6%	8.3	1.04	85 m
16	94	0%	7.8%	14.2	0.89	85 m
32	53	0%	7.9%	26.3	0.83	85 m

Table 1. Overall results on four knapsack instances.

of solving the knapsack problem without binary constraints (subject to the items that are fixed) to obtain a lower bound, which is then used to determine the node's priority (lower is better). Fathoming occurs when the solution to the relaxation is feasible to the binary problem or the lower bound exceeds the value of the incumbent. The search strategy is to choose the candidate node with the lowest lower bound (best first).

To illustrate the performance of the solver, we randomly generated four difficult knapsack instances using the method described in [Martello and Toth, 1990]. These results are not meant to be comprehensive. Clearly, further testing on a much larger scale is needed and complete performance results will be reported in a full paper to follow. Testing was conducted on a Beowulf cluster with 48 dual processor nodes. Each node has two 1.0-GHz Pentium III processors and 512 megabytes of RAM. The operating system was Red Hat Linux 7.2. The message-passing library used was LAM/MPI. Five trials were run for each instance, with two hubs employed when the number of processors was eight or more. Table 1 shows the number of processors used (N), the wall-clock running time (in seconds), the percentage idle time, the speedup (ratio of the sequential and parallel running times), the parallel efficiency (ratio of the speedup to the number of processors), and the number of nodes enumerated. The efficiency approximates the percentage of running time devoted to useful work and should ideally be near one. Efficiencies significantly below one indicate the presence of overhead. We used SBB [Forrest, 2004] to produce the sequential running times for comparison. Because our solver does not employ advanced techniques such as dynamic cut generation or primal heuristics, we disabled these capabilities with SBB as well. SBB still generated many fewer search-tree nodes due to its use of strong branching. Nonetheless, the comparison provides a useful baseline. From Table 1, we see that the speedup is near linear. Ramp-up time is negligible, but idle time still leaves room for improvement. The number of nodes enumerated is not increasing, which indicates that the performance of redundant work is not a problem.

Problem	N	Wall-clock	Ramp-up	Idle	Speedup	Eff	Nodes
gesa3	1	1626	–	–	–	–	403
gesa3	4	614	9.8%	0	2.6	0.66	445
gesa3	8	269	35.1%	0.2%	6.0	0.76	337
gesa3	16	161	49.1%	0.1%	10.1	0.63	247
blend2	1	1565	–	–	–	–	2339
blend2	4	258	12.8%	0	6.1	1.53	1019
blend2	8	213	14.0%	0.2%	7.3	0.92	717
blend2	16	129	34.1%	0	12.1	0.76	980
fixnet6	1	2716	–	–	–	–	2729
fixnet6	4	703	1.0%	0	3.9	0.98	3598
fixnet6	8	626	3.0%	0.2%	4.3	0.54	4703
fixnet6	16	376	4.6%	0	7.2	0.45	6570
cap6000	1	4287	–	–	–	–	6129
cap6000	4	1344	0.2%	0	3.2	0.80	9551
cap6000	8	1012	0.3%	0	4.2	0.53	12363
cap6000	16	640	1.2%	0.2%	6.7	0.42	14121

Table 2. Computational results of sample MILP problems.

4.2 Mixed-integer Linear Program Solver

For the knapsack solver, node processing times were negligible and good feasible solutions were discovered early in the solution process, which makes scalability relatively easy to achieve. As a more stringent test, we have developed a generic solver for mixed-integer linear programs (MILPs) called ALPS Branch and Cut (ABC), employing a straightforward branch and cut algorithm with cuts generated using the COIN-OR Cut Generation Library [Lougee-Heimer, 2003]. ABC consists of the classes `AbcModel`, `AbcTreeNode`, `AbcNodeDesc`, `AbcSolution`, and `AbcParameterSet`. The search strategy is best first. Strong branching is used to choose the variables to be branched on. ABC also uses the SBB rounding heuristic as a primal heuristic.

We tested ABC using four problems: *gesa3*, *blend2*, *fixnet6*, and *cap6000* from MIPLIB3 [Bixby et al., 1998]. As above, these results are meant to be illustrative, not comprehensive. As with the knapsack example, two hubs were used when the number of processes was eight or more. The results are summarized in Table 2.

From Table 2, we see that for generic MILPs, parallel efficiency is not as easy to achieve. However, the source of overhead is quite problem dependent. For *gesa3* and *blend2*, ramp-up is a major problem, due to large node processing time near the top of the tree. Neither *gesa3* nor *blend2* exhibits signs of the performance of redundant work. Also, as the number of processors increases, the number of search nodes decreases. This is primarily due to the fact that good feasible solutions are found early in the search process. For *fixnet6* and *cap6000*, ramp-up is not a problem, but the number of nodes processed increases when the number of processes increases, indicating the presence of redundant work. For these problems, good feasible solutions are not found until much later in the search process. These results illustrate the challenges that we still face in improving scalability. We discuss prospects for the future in the final section.

5. Summary and Future Work

In this paper, we have described the main features of the ALPS framework. Two applications were developed to test ALPS. The limited computational results highlight the challenges we still face in achieving scalability. The preliminary results obtained for ABC highlight the two most difficult scalability issues to address for MILP—reduction of ramp-up time and elimination of redundant work. Controlling ramp-up time is the most difficult of these. Attempts to branch early in order to produce successors more quickly have thus far been unsuccessful. A number of other ideas have been suggested in the literature. Two that we are currently exploring are (1) using a branching procedure that creates a large number of successors instead of just the current two, and (2) utilizing the processors idle during ramp-up in order to find a good initial feasible solution, thereby helping to eliminate redundant work. The first approach seems unlikely to be successful, but the second one may hold the key. This approach is also being explored by Eckstein et al. in the context of PICO. As for eliminating redundant work, this can be done by fine-tuning our load balancing strategies, which are currently relatively unsophisticated, to ensure a better distribution of high-priority work.

In future work, we will continue to improve the performance of ALPS by refining our methods of reducing parallel overhead as discussed above. Also, we will continue development of the Branch, Constrain, and Price Software (BiCePS) library, the data handling layer for solving mathematical programs that we are building on top of ALPS. BiCePS will introduce dynamically generated cuts and variables as new types of knowledge and support the implementation of parallel branch and bound algorithms in which the bounds are obtained by Lagrangian relaxation. Finally, we will build the BiCePS Linear Integer Solver (BLIS) on top of BiCePS. BLIS will be a LP-based branch, cut, and price solver for MILPS, like ABC, but with user customization features akin to SYMPHONY and BCP.

Acknowledgments. This research was partially supported through NSF grant ACI-0102687 and the IBM Faculty Partnership Program.

References

Balas, E., Ceria, S., and Cornuéjols, G. (1996). Mixed 0-1 programming by lift-and-project in a branch-and-cut framework. *Management Science*, 42:1229–1246.

Benchouche, M., Cung, V.-D., Dowaji, S., Cun, B. L., Mautor, T., and Roucairol, C. (1996). Building a parallel branch and bound library. In *Solving Combinatorial Optimization Problems in Parallel*. Springer, Berlin.

Bixby, R., Ceria, S., McZeal, C., and Savelsbergh, M. (1998). An updated mixed integer programming library: MIPLIB 3. Technical Report TR98-03, Department of Computational and Applied Mathematics, Rice University.

Bixby, R., Cook, W., Cox, A., , and Lee, E. (1995). Parallel mixed integer programming. Research Monograph CRPC-TR95554, Rice University Center for Research on Parallel Computation.

Borbeau, B., Crainic, T., and Gendron, B. (2000). Branch-and-bound parallelization strategies applied to a depot location and container fleet management problem. *Parallel Computing*, 26:27–46.

Chen, Q. and Ferris, M. C. (2001). FATCOP: A fault tolerant Condor-PVM mixed integer program solver. *SIAM Journal on Optimization*, 11:1019–1036.

Cordier, C., Marchand, H., Laundy, R., and Wolsey, L. A. (1999). bc-opt: A branch-and-cut code for mixed integer programs. *Mathematical Programming*, 86:335–353.

Correa, R. and Ferreira, A. (1995). Parallel best-first branch and bound in discrete optimization: A framework. Technical Report 95-03, Center for Discrete Mathematics and Theoretical Computer Science.

Eckstein, J., Phillips, C. A., and Hart, W. E. (2000). Pico: An object-oriented framework for parallel branch and bound. Technical Report RRR 40-2000, Rutgers University.

Fonlupt, C., Marquet, P., and Dekeyser, J. (1998). Data-parallel load balancing strategies. *Parallel Computing*, 24:1665–1684.

Forrest, J. (2004). Simple branch and bound. Available from http://www.coin-or.org.

Gendron, B. and Crainic, T. (1994). Parallel branch and bound algorithms: Survey and synthesis. *Operations Research*, 42:1042–1066.

Grama, A. and Kumar, V. (1995). Parallel search algorithms for discrete optimization problems. *ORSA Journal on Computing*, 7:365–385.

Gropp, W., Lusk, E., and Skjellum, A. (1999). *Using MPI*. MIT Press, Cambridge, MA, USA, 2nd edition.

Hafer, L. (1999). bonsaiG: Algorithms and design. Technical Report SFU-CMPTTR 1999-06, Simon Frazer University Computer Science.

Henrich, D. (1993). Initialization of parallel branch-and-bound algorithms. In *Second International Workshop on Parallel Processing for Artificial Intelligence(PPAI-93)*.

Jünger, M. and Thienel, S. (2001). The abacus system for branch and cut and price algorithms in integer programming and combinatorial optimization. *Software Practice and Experience*, 30:1325–1352.

Kumar, V., Grama, A. Y., and Vempaty, N. R. (1994). Scalable load balancing techniques for parallel computers. *Journal of Parallel and Distributed Computing*, 22:60–79.

Kumar, V. and Gupta, A. (1994). Analyzing scalability of parallel algorithms and architectures. *Journal of Parallel and Distributed Computing*, 22:379–391.

Ladányi, L. and Ralphs, T. (2001). *COIN/BCP User's Manual*. Available from http://www.coin-or.org.

Land, A. H. and Doig, A. G. (1960). An automatic method for solving discrete programming problems. *Econometrica*, 28:497–520.

Laursen, P. S. (May, 1994). Can parallel branch and bound without communication be effective? *SIAM Journal on Optimization*, 4:33–33.

Linderoth, J. (1998). *Topics in Parallel Integer Optimization*. PhD thesis, School of Industrial and Systems Engineering, Georgia Institute of Technology, Atlanta, GA.

Lougee-Heimer, R. (2003). The Common Optimization INterface for Operations Research. *IBM Journal of Research and Development*, 47:57–66.

Makhorin, A. (2004). Introduction to GLPK. Available from http://www.gnu.org/software/glpk/glpk.html.

Martello, S. and Toth, P. (1990). *Knapsack Problems: algorithms and computer implementation*. John Wiley & Sons, Inc., USA, 1st edition.

Martin, A. (1998). Integer programs with block structure. Habilitation Thesis, Technical University of Berlin, Berlin, Germany.

Mitra, G., Hai, I., and Hajian, M. (1997). A distributed processing algorithm for solving integer programs using a cluster of workstations. *Parallel Computing*, 23:733–753.

Nemhauser, G. L., Savelsbergh, M. W. P., and Sigismondi, G. S. (1994). Minto, a mixed integer optimizer. *Operations Research Letters*, 15:47–58.

Ralphs, T. (2004). *SYMPHONY Version 4.0 User's Manual*. Available from `http://www.branchandcut.org/SYMPHONY`.

Ralphs, T., Ladányi, L., and Saltzman, M. J. (2003). Parallel branch, cut, and price for large-scale discrete optimization. *Mathematical Programming*, 98:253–280.

Ralphs, T., Ladányi, L., and Saltzman, M. J. (2004). A library hierarchy for implementing scalable parallel search algorithms. *The Journal of Supercomputing*, 28:215–234.

Sanders, P. (1998). Tree shaped computations as a model for parallel applications. In *ALV'98 Workshop on application based load balancing*, pages 123–132.

Shinano, Y., Harada, K., and Hirabayashi, R. (1995). A generalized utility for parallel branch and bound algorithms. In *Proceedings of the 1995 Seventh Symposium on Parallel and Distributed Processing*, pages 392–401, Los Alamitos, CA. IEEE Computer Society Press.

Sinha, A. and Kalé, L. V. (1993). A load balancing strategy for prioritized execution of tasks. In *Seventh International Parallel Processing Symposium*, pages 230–237, Newport Beach, CA.

Trienekens, H. W. J. M. and Bruin, A. d. (1992). Towards a taxonomy of parallel branch and bound algorithms. Report EUR-CS-92-01, Erasmus University, Rotterdam.

Tschoke, S. and Polzer, T. (1998). *Portable Parallel Branch and Bound Library User Manual: Library Version 2.0*. Department of Computer Science, University of Paderborn.

VI

CLASSIFICATION, CLUSTERING, AND RANKING

TABU SEARCH ENHANCED MARKOV BLANKET CLASSIFIER FOR HIGH DIMENSIONAL DATA SETS

Xue Bai[1,2] and Rema Padman[1]

[1] *The H. John Heinz III School of Public Policy and Management*
Carnegie Mellon University, Pittsburgh PA 15213
USA

[2] *Center for Automated Learning and Discovery*
School of Computer Science
Carnegie Mellon University, Pittsburgh PA 15213
USA

{xbai,rpadman}@andrew.cmu.edu

Abstract Data sets with many discrete variables and relatively few cases arise in health care, ecommerce, information security, and many other domains. Learning effective and efficient prediction models from such data sets is a challenging task. In this paper, we propose a Tabu Search enhanced Markov Blanket (TS/MB) procedure to learn a graphical Markov Blanket classifier from data. The TS/MB procedure is based on the use of restricted neighborhoods in a general Bayesian Network constrained by the Markov condition, called Markov Equivalent Neighborhoods. Computational results from a real world data set drawn from the health care domain indicate that the TS/MB procedure converges fast, is able to find a parsimonious model with substantially fewer predictor variables than in the full data set, has comparable or better prediction performance when compared against several machine learning methods, and provides insight into possible causal relations among the variables.

Keywords: Tabu Search, Markov Blanket, Bayesian Networks

Introduction

The deployment of comprehensive information systems and online databases has made extremely large collections of real-time data readily available. In many domains such as genetics, clinical diagnoses, direct marketing, finance, and on-line business, data sets arise with thousands of variables and a small ratio of cases to variables. Such data present dimensional difficulties for classification of a target variable (Berry and Linoff, 1997), and identification of critical predictor variables. Furthermore, they pose even greater challenges in the determination of actual influence, i.e., causal relationships between the target variable and predictor variables. The problem of identifying essential variables is critical to the success of decision support systems and knowledge discovery tools due to the impact of the number of variables on the speed of computation, the quality of decisions, operational costs, and understandability and user acceptance of the decision model. For example, in medical diagnosis (Cooper et al., 1992), the elimination of redundant tests may reduce the risks to patients and lower healthcare costs. In this study, we address this problem of efficiently identifying a small subset of predictor variables from among a large number, and estimating the causal relationship between the selected variables and the target variable, using *Markov Blanket* (MB) and *Tabu Search* (TS) approaches.

The Markov Blanket of a variable Y, (MB(Y)), by definition, is any set of variables such that Y is conditionally independent of all the other variables given $MB(Y)$. A *Markov Blanket Directed Acyclic Graph* (MB DAG) is the Directed Acyclic Graph of that set. When the parameters of the MB DAG are estimated, the result is a *Bayesian Network,* defined in the next section. Recent research by the machine learning community has sought to identify the Markov Blanket of a target variable by filtering variables using statistical decisions for conditional independence and using the MB predictors as the input features of a classifier. However, learning MB DAG classifiers from data is still an open problem (Chickering, 2002). There are several challenges: the problem of learning the graphical structure is NP hard (Chickering et al., 2003); selecting associations in the presence of limited data is quite unreliable; and the presence of multiple local optima in the space of possible structures makes the learning process difficult. Given these challenges, learning the Markov Blanket instead of the complete Bayesian Network allows us to limit the number of associations that we want to consider.

In this paper, we propose a Tabu Search enhanced Markov Blanket procedure that finds a parsimonious MB DAG. This two-stage algorithm generates an MB DAG in the first stage as a starting solution; in the second stage, the Tabu Search metaheuristic strategy is applied to improve the effectiveness of the MB DAG as a classifier, with conventional Bayesian updating. Classification using the Markov Blanket of a target variable in a Bayesian Network has important properties: it specifies a statistically efficient prediction of the probability distribution of a variable from the smallest subset of variables; it provides accuracy while avoiding overfitting due to redundant variables; and it provides both a classifier and some insight into causal relations between a reduced set of predictors and the target variable.The TS/MB procedure proposed in this paper allows us to move through the search space of Markov Blanket structures quickly and escape from local optima, thus learning a more robust structure.

Metaheuristic search methods such as genetic algorithms (Holland, 1975), Artificial Neural Networks (Freeman and Skapura, 1991), simulated annealing (Metropolis et al., 1953, Johnson et al., 1989), Tabu search (Glover, 1997), and others have been applied to machine learning and data mining methods such as decision trees and Bayesian Networks (Sreerama et al., 1994, Harwood and Scheines, 2002) with significant success in finding good solutions and accelerating convergence in the learning process. Recent applications include a Genetic Algorithm based approach to building accurate decision trees in the marketing domain (Fu et al., 2004), Neural Networks applied to Hybrid Intelligent Systems for Stock Market Analysis (Abraham et al., 2001) and hill-climbing heuristics to model a reinforcement learning algorithm for learning to control partially-observable Markov decision processes (Moll et al., 2000).

This paper is organized as follows: Section 2 provides some background and literature review. Section 3 presents our proposed method and examines several relevant issues such as move selection, neighborhood structure, and evaluation metric for our specific problem. Section 4 details an outline of the algorithm and experimental design. Section 5 reports computational results. Section 6 presents our conclusions and directions for future research.

1. Representation and Background Knowledge

Bayesian Networks and Markov Blankets

DEFINITION 1 *A Bayesian Network for a set of variables* $X = \{X_1, ..., X_n\}$ *consists of: (i) a network structure* S *that encodes a set of conditional independence assertions among variables in* X*; (ii) a set* $P = \{p_1, ..., p_n\}$ *of local conditional probability distributions associated with each node and its parents.*

In Definition 1, S is a directed acyclic graph (DAG) which, along with P, entails a joint probability distribution p over the nodes.

A Bayesian Network is a graphical representation of the joint probability distribution of a set of random variables. It also has a causal interpretation: a directed edge from one variable to another, $X \rightarrow Y$, represents the claim that X is a direct cause of Y with respect to other variables in the DAG (Spirtes et al., 2000, Pearl, 2000).

DEFINITION 2 P *satisfies the Markov condition for* S *if every node* X_i *in* S *is independent of its non-descendants, conditional on its parents.*

The Markov Condition implies that the joint distribution p can be factorized as a product of conditional probabilities, by specifying the distribution of each node conditional on its parents (Pearl, 2000). In particular, for a given structure

S, the joint probability distribution for X can be written as

$$p(X) = \prod_{i=1}^{n} p_i(X_i|pa_i) \ , \tag{1}$$

where pa_i denotes the set of parents of X_i.

DEFINITION 3 *Given the set of variables X and target variable Y, a Markov Blanket (MB) for Y is a subset Q of variables in X such that Y is independent of $X \backslash Q$, conditional on the variables in Q.*

Given a Bayesian Network (S, P), the Markov Blanket for Y consists of pa_Y, the set of parents of Y; ch_Y, the set of children of Y; and $pa\, ch_Y$, the set of parents of children of Y.

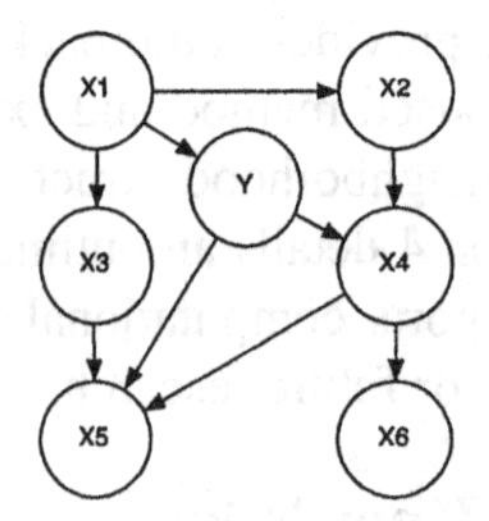

Figure 1. Bayesian Network (S, P).

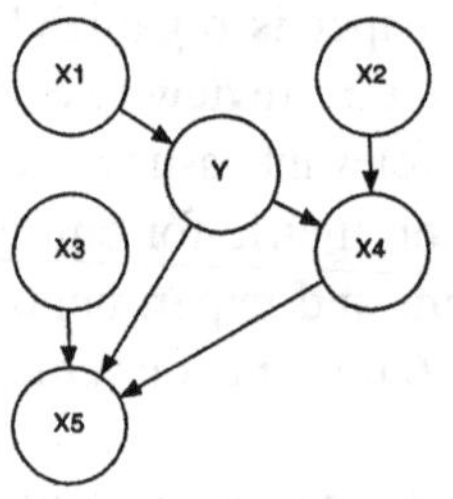

Figure 2. Markov Blanket for Y.

For example, consider the two DAGs in Figure 1 and 2, above. The factorization of p entailed by the Bayesian Network (S, P) is

$$\begin{aligned} p(Y, X_1, ..., X_6) &= C \cdot p(Y|X_1) \cdot p(X_4|X_2, Y) \cdot p(X_5|X_3, X_4, Y) \cdot \\ &\quad \cdot p(X_2|X_1) \cdot p(X_3|X_1) \cdot p(X_6|X_4) \ , \end{aligned} \tag{2}$$

where C is a normalizing constant. The factorization of the conditional probability $p(Y|X_1, ..., X_6)$ entailed by the Markov Blanket for Y corresponds to the product of those (local) factors in equation (2) that contain the term Y.

$$p(Y|X_1, ..., X_6) = C' \cdot p(Y|X_1) \cdot p(X_4|X_2, Y) \cdot p(X_5|X_3, X_4, Y) \ , \tag{3}$$

where C' is a different normalizing constant.

There are a few recent studies that have applied Markov Blanket classifiers and compared their accuracy against alternative machine learning methods using data sets with few variables or based on limiting assumptions that affect the scalability of the methods (Koller and Sahami, 1996, Margaritis and Thrun, 1999, Tsamardinos and Aliferis, 2002). Theoretically correct Bayesian algorithms (Chickering, 2002) for finding DAGs are now known, but have not been applied to the problem of finding MBs for data sets with large numbers of variables.

DEFINITION 4 *MB DAGs that have the same Markov factorization for the target node are Markov equivalent, and the set of all MB DAGs that are Markov equivalent belong to the same Markov equivalence class.*

For example, the MB DAGs shown in Figure 3 are in the same *Markov equivalence class*.

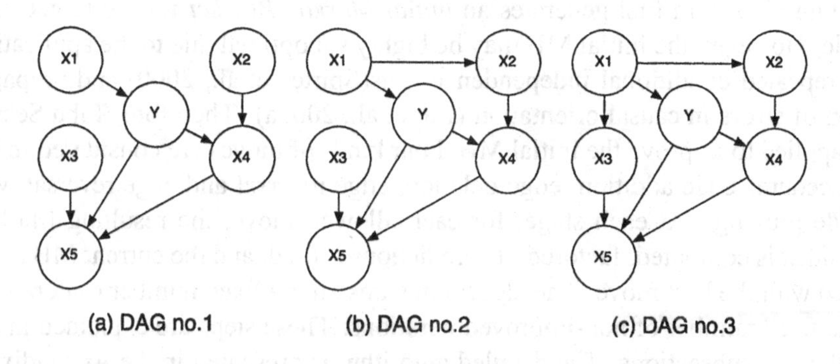

Figure 3. Markov Equivalent MB DAGs for Y.

The conditional probability distribution for MB(Y) is the same for (a), (b), and (c) in Figure 3, which is:

$$p(Y|X_1, ..., X_6) = C' \cdot p(Y|X_1) \cdot p(X_4|X_2, Y) \cdot p(X_5|X_3, X_4, Y) \quad , \quad (4)$$

where C' is a constant.

Basic Tabu Search Heuristic

Tabu Search is a powerful meta-heuristic strategy that helps local search heuristics to explore the solution space by guiding them out of local optima (Glover, 1997). Its strategic use of memory and responsive exploration is based on selected concepts that cut across the fields of artificial intelligence and operations research. It has been applied successfully to a wide variety of continuous and combinatorial optimization problems (Johnson and McGeoch, 1997, Toth and Vigo, 2003), capable of reducing the complexity of the search process and accelerating the rate of convergence. In its simplest form, Tabu Search starts with a feasible solution and chooses the *best move* according to an evaluation function while taking steps to ensure that the method does not re-visit a solution previously generated. This is accomplished by introducing *tabu restrictions* on possible moves to discourage the reversal and in some cases repetition of selected moves. The tabu list that contains these forbidden move attributes is known as the short term memory function. It operates by modifying the search trajectory to exclude moves leading to new solutions that contain attributes (or attribute mixes) belonging to solutions previously visited within a time horizon governed by the short term memory. Intermediate and long-term memory functions may also be incorporated to intensify and diversify the search.

2. Tabu Search Enhanced Markov Blanket Algorithm

Our algorithm first generates an *initial Markov Blanket* for the target variable. However, the initial MB may be highly suboptimal due to the application of repeated conditional independence tests(Spirtes et al., 2000) and propagation of errors in causal orientation (Bai et al., 2004a). Therefore, Tabu Search is applied to improve the initial MB. Four kinds of moves are considered in the procedure: edge addition, edge deletion, edge reversal and edge reversal with node pruning. At each stage, for each allowed move, the resulting Markov Blanket is computed, factored, its predictions scored, and the current MB modified with the best move. The algorithm stops after a fixed number of iterations or a fixed number of non-improved iterations. These steps are explained in the following subsections. The detailed algorithm is presented in the Appendix.

Creation of the Initial Solution

The *InitialSolution* procedure starts from an empty graph *G*, identifies the associated nodes for the target node through independence and conditional independence (CI) tests. This is then repeated for the associated nodes of each node adjacent to the target, removing edges based on CI tests among those associates. We use χ^2 statistic to test for statistical independence. After this step, variables adjacent to the target are identified, which are the parents or children of the target. We call this set *PC*. The next step is to generate *PCPC*, the parents and children of the variables in *PC*, again removing edges based on CI tests among the associates. The procedure then prunes away all the nodes that are not in *PC* or *PCPC*, and starts orienting edges. The edge orientation rules used are: *Collider* Orientation Rule (Spirtes et al., 2000), *Meek's Rules* (Spirtes and Meek, 1995), and two other rules, all based on causal reasoning theory (Pearl, 2000, Spirtes et al., 2000)[1]. Finally, redundant, undirected, and bidirected edges are pruned to avoid cycles. The output of *InitialSolution* is an initial Markov Blanket, used as the input to the *TabuSearch* procedure.

Markov Equivalent Neighborhoods and Choice of Moves

Our algorithm uses the set of logical Markov equivalence classes as the set of possible states in the search space instead of searching over the space of DAGs. The representation scheme is MB DAGs. The set of operators are the feasible moves, which transforms the current MB DAG from the Markov equivalence class it belongs to, to an MB DAG which belongs to *another* Markov equivalence class. Thus the neighborhood for any state is the set of *new* Markov equivalence classes that can be constructed via one feasible move. We call this a Markov Equivalent Neighborhood.

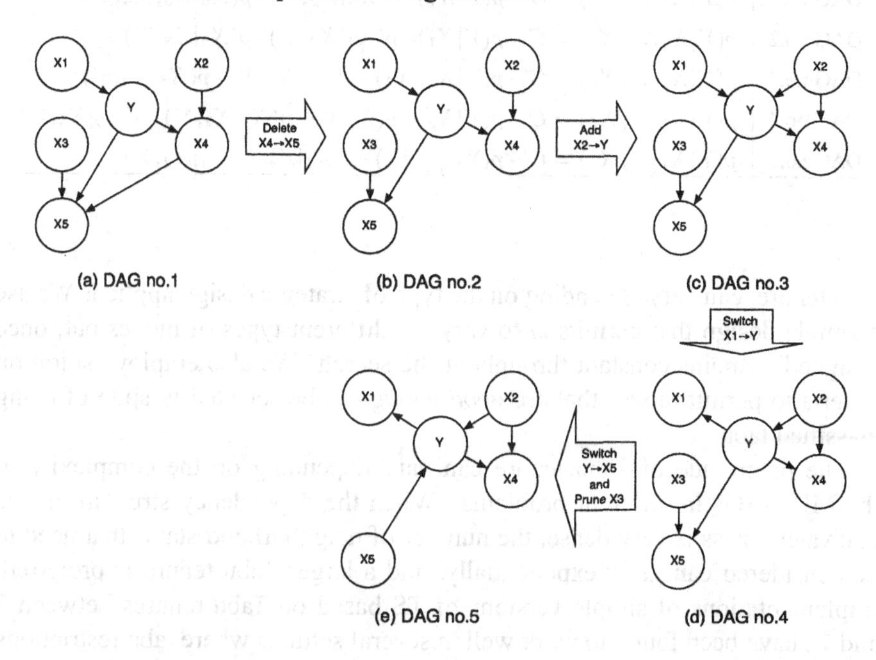

Figure 4. An Example of Moves in Tabu Search Enhanced Markov Blanket Procedure

We allow the following kinds of moves, as illustrated in Figure 4: edge addition ((b) to (c)), edge deletion ((a) to (b)), edge reversal ((c) to (d)), and edge reversal with node pruning ((d) to (e)). Moves that yield Markov Blankets *within* the same Markov equivalent class or moves that result in cyclic graphs are not valid moves. At each stage, for each allowed move, the resulting Markov Blanket is constructed, the conditional probability for the target node is factored and computed, its prediction is scored, and the current MB is modified with the best move.

Table 1 lists the corresponding scores of each move in Figure 4.

Tabu List and Tabu Tenure

In our implementation, the tabu list keeps a record of m previous moves, or more precisely, of the attributes of the current solution that are changed by these moves. By reference to this record, new solutions are classified tabu if they contain attributes of previous solutions encountered within the m move horizon. A move is tabu if it leads to a tabu solution. The value of m, called the

Table 1. Scores of Moves in Tabu Search Enhanced Markov Blanket Procedure in Figure 4

DAGs	Score Computation
DAG no.1	$p(Y\|X_1, ..., X_5) = C' \cdot p(Y\|X_1) \cdot p(X_4\|X_2, Y) \cdot p(X_5\|X_3, X_4, Y)$
DAG no.2	$p(Y\|X_1, ..., X_5) = C' \cdot p(Y\|X_1) \cdot p(X_4\|X_2, Y) \cdot p(X_5\|X_3, Y)$
DAG no.3	$p(Y\|X_1, ..., X_5) = C' \cdot p(Y\|X_1, X_2) \cdot p(X_4\|X_2, Y) \cdot p(X_5\|X_3, Y)$
DAG no.4	$p(Y\|X_1, ..., X_5) = C' \cdot p(Y\|X_2) \cdot p(X_1\|Y) \cdot p(X_4\|X_2, Y) \cdot p(X_5\|X_3, Y)$
DAG no.5	$p(Y\|X_1, ..., X_5) = C' \cdot p(Y\|X_2, X_5) \cdot p(X_1\|Y) \cdot p(X_4\|X_2, Y)$

tabu tenure, can vary depending on the type of strategic design applied. We use a simple design that permits *m* to vary for different types of moves but, once assigned, remains constant throughout the search. We also employ aspiration criteria to permit moves that are *good enough* to be selected in spite of being classified tabu.

The magnitude of Tabu tenure can vary depending on the complexity of the MB DAGs in different problems. When the dependency structure of the equivalent class is very dense, the number of neighborhood states that need to be considered can grow exponentially, and a larger Tabu tenure is preferred. Implementations of simple versions of TS based on Tabu tenures between 7 and 12 have been found to work well in several settings where tabu restrictions rule out a non-trivial portion of the otherwise available moves (Glover, 1997). Our setting appears to be one of this character, and in our experiments, we use a static Tabu tenure of 7 because the structure of the MB DAGs is not complex. Considering the computational cost of each type of move, it is reasonable to assign larger value of Tabu tenure to moves that are computationally more expensive. Such moves include edge reversals that involve pruning nodes that are no longer present in the resulting MB DAG, or edge reversals that result in significant changes in the parent-child relations in the resulting MD DAG. It is possible to optimize these parameters in future research and to replace the use of a static tenure by a dynamic tenure.

Evaluation Criteria

Prediction accuracy is widely used in the Machine Learning community for comparison of classification effectiveness. However, accuracy estimation is not the most appropriate metric when cost of misclassification or class distributions are not specified precisely (Provost et al., 1998). For example, wrong prediction in the diagnosis and treatment of a seriously ill patient has different consequences than incorrect prediction of the patronage of an online consumer.

The quality metric AUC, the area under the Receiver Operator Characteristic (ROC) curve, takes into account the costs of the different kinds of misclassification (Hanley and McNeil, 1983). An ROC curve is a plot of true-positive rate and false positive rate in binary classification problems as a function of the variation in the classification threshold. This metric has gained popularity among statisticians for evaluating diagnosis tests, and has also been adopted by the machine learning community for general binary classification evaluation. ROC curves are similar to the precision/recall curves used in information retrieval as well as lift curves used in marketing communities. AUC ranges from 0 to 100 percent. The higher the AUC is, the better the quality of the classifier.

We evaluate our algorithm using both AUC and prediction accuracy to score each move. In experiments where AUC is used as the scoring criterion, the procedure calculates AUC for every neighborhood move and identifies the move with the highest AUC as the best move, and similarly with prediction accuracy.

Intensification and Diversification

We alternate between intensification and diversification in the TS/MB procedure. The intensification process focuses on convergence to the best Markov Blanket DAG in a local Markov equivalent neighborhood. The diversification process attempts to explore MB structures that are far from the current neighborhood. The distance of two Markov equivalence classes can be roughly understood as the difference in the MB structures and the resulting Markov factorizations. In our experiments, we seed different starting solutions by altering the alpha level of the independence tests and by altering the edge orientation rules in the InitialSolution procedure. By doing this, we generate starting solutions with a variety of independence structures and complexity.

3. Experimental Design

Data

We tested our algorithm on Prostate cancer (PCA) data set (Adam et al., 2002). PCA data set has been widely used in medical informatics research (Tsamardinos and Aliferis, 2002). The task is to diagnose prostate cancer from analysis of mass spectrometry signal peaks obtained from human sera. 326 serum samples from 167 PCA patients, 77 patients with benign prostate hyperplasia (BPH), and 82 age-matched unaffected healthy men are used for learning and prediction. Peak detection was performed using Ciphergen SELDI software versions 3.0 and 3.0.5. Powerful peaks in discriminating normal versus PCA, normal versus BPH, and BPH versus PCA were selected as features for classification. After the clustering and peak alignment process, 779 peaks were identified. Table 2 summarizes the characteristics of PCA data set.

Table 2. Characteristics of PCA Data Set

Problem Type	Variables	#Samples	Variable Types	Target Variable Type
Mass-Spec Diagnosis	779	326	discretized	binary

Design of Experimental Parameters

The parameters in our experiments are: data-splits, scoring criteria, starting solution structure, the depth of conditional independence search (*d*), and significance level (α). We split the data in two ways: 90 percent for training/10 percent for testing and 80 percent for training/20 percent for testing. The applied scoring criteria are AUC and prediction accuracy. By orienting the edges differently, we create two different types of Markov Blanket structures for the starting solution, as shown in Figure 5 and 6.

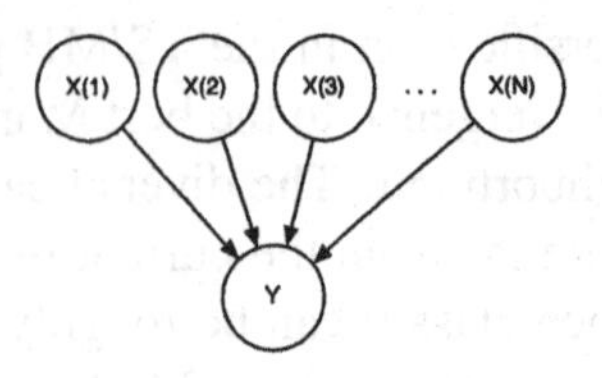

Figure 5. Starting solution I

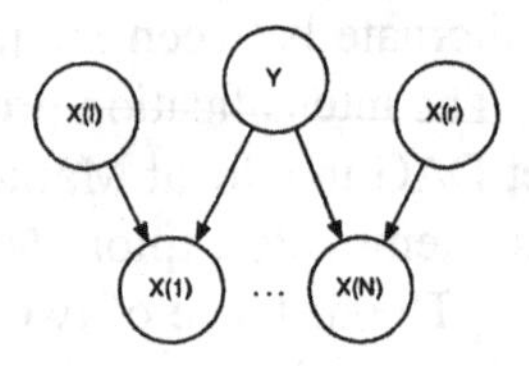

Figure 6. Starting solution II

At the end of the *InitialSolution* stage, the algorithm may terminate with some undirected or bidirected edges incident on the target variable *Y*. Before Tabu Search is applied, these edges need to be oriented. Variations in the orientation of the undirected edges result in different starting solutions. In Structure I, all the undirected edges associated with the target variable Y are oriented as parents of *Y*, i.e., *Y* does not have any children. In Structure II, all the undirected edges associated with the target variable *Y* are oriented as children of *Y*, with possible parents of their own.

The depth of conditional independence search (*d*) and the significance level (α) are usually exogenous variables that the user has to provide. The depth of search specifies the density or connectivity, i.e., the presumptive maximum number of parents each node can have, of the primary graphical structure, before the edge orientation step. The alpha level is the threshold for the statistical independence tests. The smaller alpha is, the stricter the tests are and the stronger is the dependence between two nodes with an existing edge. In our experiments, we fixed the depth of conditional independence search (*d*) at 3, which means the maximum number of parents assumed for each node in the primary graph is 3.

There are thus a total of 32 configurations of parameter combinations (Table 3). Each configuration is cross validated. For example, in a 5-fold cross validation scheme, the data set is divided into 5 subsets. In each run, one of the 5 subsets is used as the test set and the other 4 subsets are assembled to form a training set. Then the average error across all 5 trials is computed. The advantage of this method is that it matters less how the data gets divided. Every data point gets to be in a test set exactly once, and gets to be in a training set 4 times. The mean and the variance of the estimated error are reduced as the number of folds increases. In that sense, 5-fold cross validation yields more conservative estimates than 10-fold cross validation.

Table 3. Experimental Parameter Configurations

Parameters	Data-splits (train/test)	Scoring Criteria	Starting Solution	Alpha
Configurations	90%/10%	AUC, Accuracy	Type I, Type II	0.001,0.005
	80%/20%			0.01,0.05

We use a nested, stratified cross-validation scheme (Weiss and Kulikowski, 1991). In the inner layer, the procedure trains and optimizes the Markov Blanket on training data for each parameter configuration. The configuration that yields the best MB according to the scoring criterion is chosen as the best configuration. The outer layer of cross-validation estimates the performance of the optimized Markov Blanket classifier on the testing data. We report both the AUC and prediction accuracy on the testing set to evaluate the classification performance of the generated models.

4. Computational Results

Table 4 and Table 5 present the classification accuracy and AUC for different combinations of the experimental parameters. The results are averaged over cross-validation runs. The number in parentheses is the standard error.

As shown in Table 4 and Table 5, the optimal parameter configurations for best prediction accuracy (259/67, AUC or Accuracy, Structure I, 0.05) is different from the optimal parameter configurations for best AUC (293/33, AUC or Accuracy, Structure I, 0.05). In real world applications, users can choose the evaluation criterion they deem appropriate for the application. As shown in Figures 7 and 8, the resulting best fitting MB DAGs are also different.

In our experiments, the scoring criterion used in Tabu Search does not impact the classification performance, both in terms of prediction accuracy and AUC. Setting the alpha value at 0.05 generates the best result for both accu-

Table 4. Classification Accuracy on the Testing Set (%)

Data-splits (train/test)	Scoring Criteria	Starting Solution	Alpha(α) (α=0.001)	(α=0.005)	(α= 0.01)	(α=0.05)
293/33	AUC	I	83.8 (1.3)	87.1 (1.2)	83.8 (1.5)	90.3 (1.3)
		II	83.8 (1.4)	87.1 (1.0)	74.1 (3.5)	90.3 (1.0)
	Accuracy	I	83.8 (2.2)	87.1 (1.0)	83.8 (2.0)	90.3 (1.7)
		II	83.8 (2.3)	87.1 (1.1)	74.1 (4.9)	83.8 (3.5)
259/67	AUC	I	90.3 (1.2)	88.7 (1.2)	83.8 (2.2)	93.5 (1.0)
		II	61.6 (28.3)	88.7 (1.2)	83.8 (2.2)	91.9 (1.2)
	Accuracy	I	90.3 (1.2)	88.7 (1.3)	90.3 (1.0)	93.5 (1.0)
		II	90.3 (1.2)	88.7 (1.3)	90.3 (1.0)	91.9 (1.9)

Table 5. Classification AUC on the Testing Set (%)

Data-splits (train/test)	Scoring Criteria	Starting Solution	Alpha(α) (α=0.001)	(α=0.005)	(α= 0.01)	(α=0.05)
293/33	AUC	I	97.1 (1.0)	95.0 (2.1)	94.5 (1.6)	98.3 (0.8)
		II	97.1 (1.0)	94.1 (1.1)	94. (1.0)	96.2 (1.2)
	Accuracy	I	97.1 (1.1)	95.0 (1.9)	94.5 (1.3)	98.3 (0.8)
		II	97.1 (0.9)	95.0 (2.1)	94.1 (1.9)	83.7 (1.8)
259/67	AUC	I	96.9 (1.2)	96.0 (0.9)	94.0 (0.8)	96.3 (1.1)
		II	69.3 (24.0)	96.0 (1.0)	94.0 (1.1)	96.5 (1.9)
	Accuracy	I	96.9 (1.2)	96.0 (0.9)	95.8 (0.9)	95.0 (0.8)
		II	96.9 (1.2)	96.0 (1.3)	97.6 (1.2)	96.5 (1.2)

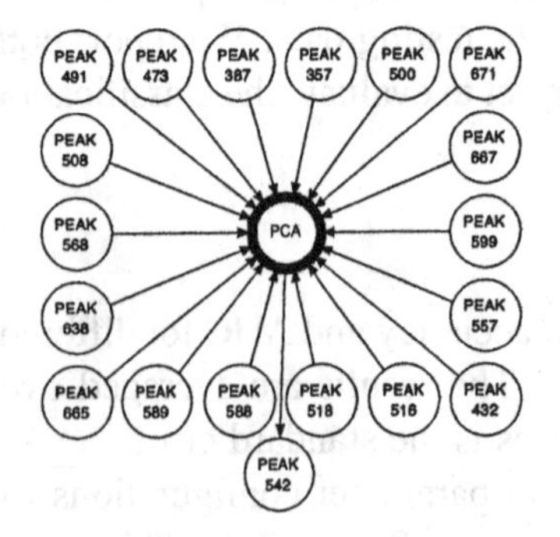

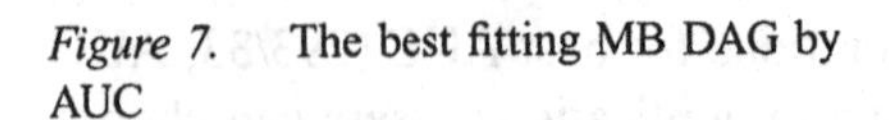

Figure 7. The best fitting MB DAG by AUC

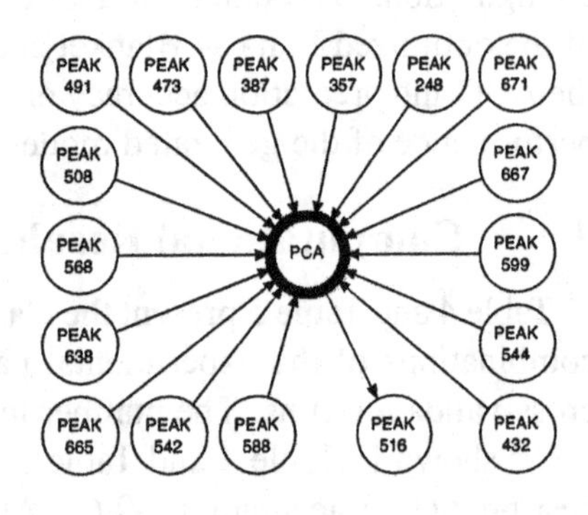

Figure 8. The best fitting MB DAG by accuracy.

racy and AUC. The reason may be that a larger alpha value imposes fewer constraints on accepting dependence in the independence tests. This allows the algorithm to generate more complex MB structures and test more edges. Furthermore, the Structure I initial solution seems to be a better configuration under both evaluation criteria. All the directed edges are robust over almost all cross validation runs, with very small variation. However, more extensive and systematic experiments on larger data sets are necessary to explore these relationships further.

In Tables 6, 7 and 8 we present the average best-fitting classification results when compared against several state-of-the-art classifiers in three different ways. Comparison I presents the results when using the full set of variables as input. Comparison II uses the same number of variables as identified by TS/MB as input for all the other classifiers, selected by information gain criterion. Comparison III uses the exact same variables as identified by TS/MB as input variables.

We report both the AUC and prediction accuracy on the testing set as well as the size of reduction in the set of variables. The size reduction was evaluated based on the fraction of variables in the resulting models. All metrics (variable size reduction, AUC, and accuracy) were averaged over cross-validation splits.

Table 6. Average Performance - Comparison I

Method	AUC (%)	Accuracy (%)	#Original Variables	#Predictor Variables	Size Reduction	C. V. Folds
TS/ MB	98.3	90.3	779	19	39.5	5
naive Bayes	97.5	89.3	779	779	1	5
SVM	97.1	98.5	779	779	1	5
Voted Perceptron	73.9	58.0	779	779	1	5
Maximum Entropy	97.4	98.8	779	779	1	5
Poisson Learner	97.4	88.3	779	779	1	5
K-NN	96.3	88.6	779	779	1	5
Logistic Regression	94.7	90.2	779	779	1	5

For Comparison I, we use the full data set as the input for each method. As shown in Table 6, the TS/MB classifier produces a substantially smaller variable set. In terms of AUC, the TS/MB classifier consistently yields the best results; on accuracy, even with the smaller number of variables employed, it produces results comparable to the average performance of the state-of-the-art methods. Moreover, our algorithm identifies the 19 most discriminatory peaks out of 779 peaks that were identified by SELDI software program and the follow up clustering and peak alignment processes.

One might be interested to see how well these methods will do using the same number of features. In order to do this comparison, we selected the same number of features as the TS/MB classifier yields with the highest information gain score, and used these as the input for the six alternative classifiers. Note that information gain is just a scoring criterion that ranks the features, but does not tell us how many of them we should use [2].

Table 7 presents the results for Comparison II. The TS/MB classifier dominates other methods both in terms of AUC and accuracy. This is possible because the TS/MB classifier encodes and takes advantage of dependencies among predictors which all other methods fail to incorporate. For problems where the independence assumption is not adequate, our algorithm is highly effective, and gives more robust predictions. In Table 8, we compare the results by using exact the same variables as were selected by the TS/MB classifier as the input variables for the other six methods. Similar to Comparison II, the TS/MB substantially outperforms all the competitors.

Table 7. Average Performance - Comparison II

Method	AUC (%)	Accuracy (%)	#Original Variables	#Predictor Variables	Size Reduction	C. V. Folds
TS/MB	98.3	90.3	779	19	39.5	5
naive Bayes	67.5	63.2	779	19	39.5	5
SVM	63.3	62.0	779	19	39.5	5
Voted Perceptron	65.2	59.2	779	19	39.5	5
Maximum Entropy	64.7	64.1	779	19	39.5	5
Poisson Learner	60.5	55.2	779	19	39.5	5
K-NN	65.6	58.6	779	19	39.5	5
Logistic Regression	73.6	56.4	779	19	39.5	5

Table 8. Average Performance - Comparison III

Method	AUC (%)	Accuracy (%)	#Original Variables	#Predictor Variables	Size Reduction	C. V. Folds
TS/MB	98.3	90.2	779	19	39.5	5
naive Bayes	77.4	69.4	779	19	39.5	5
SVM	72.6	69.9	779	19	39.5	5
Voted Perceptron	75.4	67.8	779	19	39.5	5
Maximum Entropy	74.9	70.9	779	19	39.5	5
Poisson Learner	78.5	69.4	779	19	39.5	5
K-NN	72.1	65.3	779	19	39.5	5
Logistic Regression	98.1	90.0	779	19	39.5	5

5. Conclusions

On average, the TS/MB classifier reduces the set of predictor variables by at least an order of magnitude from the full set of variables, in some cases to a sufficiently small set for entry into hand calculators or paper and pencil decision procedures in clinical and marketing decision settings. At the same time, when compared to the state-of-the-art classification methods, the TS/MB classifier procedures excellent classification results, especially in real world applications where the cost of misclassification has significant implications. Moreover, the algorithm generates a graphical structure that represents the relationships between the variables and provides additional insight into causal discovery. These experiments, as well as more results we have obtained on data sets from other domains, such as health care, Internet marketing (Bai et al., 2004a) and sentiment extraction (Bai et al., 2004b), suggest that for problems where the ratio of samples to the number of the variables is small or the independence assumption is not appropriate, the two-stage MB classifier is superior in terms of the prediction performance, effectiveness in identifying critical predictors, and robustness.

It is possible that different Markov Blanket graphical structures consistent with the TS/MB classifier output would give slightly different classification results. Because any undirected and bi-directed edges are deleted after the edge orientation step, these deletions might be suboptimal decisions. Tabu Search iteratively investigates alternative orientations and further edge additions to minimize the extent of sub-optimality. On the other hand, theoretically, two DAGs that have the same Markov factorization are Markov equivalent. In our case, two Markov Blankets can be Markov equivalent even if some edges are oriented in a different, but statistically non-differentiable way. This research can be extended to address the interesting problem of simultaneously building classifiers for all variables in a large variable data set, or the problem of discovering a causal model for all variables in such data. Future research could be the exploration of heuristic approaches to global causal discovery problems.

Acknowledgments

The authors thank Professor Fred Glover of the University of Colorado at Boulder and Professors Clark Glymour, Peter Spirtes, and Joseph Ramsey of Carnegie Mellon University for valuable suggestions and comments.

Appendix: The Algorithm

InitialMBsearch (Data D, Target Y, Depth d, Significance α)

1 L_Y = **findAdjacencies** $(Y, \{X_1, ..., X_N\}, d, \alpha)$

2 **for** $X_i \in L_Y$

2.1. L_{X_i} = **findAdjacencies** $(X_i, \{X_1, ..., X_N\} \backslash X_i, d, \alpha)$

3 G = **prune** $(Y \cup L_Y \cup_i L_{X_i})$
4 **orient** (G)
5 *MB* = **prune** *(G)*
6 *MB* = **fineTune** *(MB)*
7 *MBDag* = **prune** *(MB)*
8 **return** *MBDag*

TabuSearch (Data D, Target Y)
1 **init** *(bestSolution = currentSolution = MBDag, bestScore = 0, ...)*
2 **repeat until** *(bestScore does not improve for k consecutive iterations)*
2.1. *form candidateMoves for currentSolution*
2.2. **find** *bestMove among candidateMoves according to function* **score**
2.3. **update** *currentSolution by applying bestMove*
2.4. **add** *bestMove to tabuList // not re-considered in the next t iterations*
2.5. **if** *(bestScore <* **score** *(bestMove))*
2.5.1. **update** *bestSolution and bestScore by applying bestMove*
3 **return** *bestSolution // MBDag*

findAdjacencies (Node Y, List of Nodes L, Depth d, Significance α)
1 $A_Y := \{X_i \in L$: X_i is dependent of Y at level $\alpha\}$
2 **for** $X_i \in A_Y$ and **for** all distinct subsets $S \subset \{A_Y \backslash X_i\}^d$
2.1. **if** X_i is independent of Y given S at level α, **then** remove X_i from A_Y
3 **for** $X_i \in A_Y$
3.1. $A_{X_i} := \{X_j \in L$: X_j is dependent of X_i at level $\alpha, j \neq i\}$
3.2. **for** all distinct subsets $S \subset \{A_{X_i}\}^d$
3.2.1. **if** X_i is independent of Y given S at level α, **then** remove X_i from A_Y
4 **return** A_Y

orient (Graph G)
1 **for** each triple of vertices (X, Y, Z) in G
1.1. **if** pair (X, Y) and (Y, Z) are adjacent, pair (X, Z) are not adjacent (i.e. pattern $X - Y - Z$), and **if** $\neg(X \perp Z \mid Y)$, **then** orient $X - Y - Z$ as $X \rightarrow Y \leftarrow Z$ **(Collider Orientation Rule)**
1.2. **if** $X \rightarrow Y$ and Y, Z are adjacent, X, Z are not adjacent (i.e.pattern $X \rightarrow Y - Z$), **and** there is no arrow into Y, **then** orient $Y - Z$ as $Y \rightarrow Z$ (**Meek's Rule 1**).
2 **if** $\exists$(directed path from X to Y), **and** $\exists$ (undirected edge between X and Y), **then** orient $X - Y$ as$X \rightarrow Y$ (**Meek's Rule 2)**
3 **for** any undirected edge connected to X (i.e.$X - Y$)
3.1. **if** $\exists$ (Z, W) s.t. Z adjacent to X, W is adjacent to X, and$W \rightarrow Y \leftarrow Z$, **then** orient $X - Y$ as $X \leftarrow Y$ (**Meek's Rule 3)**
4 **return** G

fineTune (Graph G, Target Y)

1 **for** any X in G s. t. $X \leftrightarrow Y$ or $X - Y$, reorient this edge as $X \rightarrow Y$

2 **for** any X, Z in G s. t. $Z - X \rightarrow Y$ ($Z \notin \{Y\} \cup PC$), remove the edge $Z - X$

3 **return** G

prune (Graph G,Target Y)

1 **for** any X , $X \notin \{Y\} \cup ParentChild(Y) \cup ParentChild(ParentChild(Y))$

1.1. remove X and all the associated edges

2 **return** G

Notes

1. Details are presented in the algorithm in the Appendix.
2. The software was kindly provided by the Text Learning Group at Carnegie Mellon University.

References

Abraham, A., Nath, B., and Mahanti, P.K. (2001). Hybrid intelligent systems for stock market analysis. *Computational Science*, pages 337–345.

Adam, B.L., Qu, Y., Davis, J.W., Ward, M.D., Clements, M.A., Cazares, L.H., Semmes, O.J., Schellhammer, P.F., Yasui, Y., Feng, Z., and Wright, G.L.Jr. (2002). Serum protein fingerprinting coupled with a pattern-matching algorithm distinguishes prostate cancer from benign prostate hyperplasia and healthy men. *Cancer Research*, 62:3609–3614.

Bai, X., Glymour, C., Padman, R., Spirtis, P., and Ramsey, J. (2004a). Mb fan search classifier for large data sets with few cases. Technical Report CMU-CALD-04-102, School of Computer Science, Carnegie Mellon University.

Bai, X., Padman, R., and Airoldi, E. (2004b). Sentiment extraction from unstructured text using tabu search-enhanced markov blanket. In *Proceedings of KDD Workshop on Mining for and from the Semantic Web (MSWKDD)*. Springer-Verlag Germany.

Berry, M.J.A. and Linoff, G.S. (1997). *Data Mining Techniques: For Marketing, Sales, and Customer Support*. John Wiley and Sons.

Chickering, D.M. (2002). Learning equivalence classes of bayesian-network structures. *Journal of Machine Learning Research*, 3:507–554.

Chickering, D.M., Meek, C., and Heckerman, D. (2003). Large-sample learning of bayesian networks is np-hard. In *Proceedings of Nineteenth Conference on Uncertainty in Artificial Intelligence*, pages 124–133. Morgan Kaufmann.

Cooper, G.F., Aliferis, C.F., Aronis, J., Buchanan, B.G., Caruana, R., Fine, M.J., Glymour, C., Gordon, G., Hanusa, B.H., Janosky, J.E., Meek, C., Mitchell, T., Richardson, T., and Spirtes, P. (1992). An evaluation of machine-learning methods for predicting pneumonia mortality. *Artificial Intelligence in Medicine*, 9:107–139.

Freeman, J. and Skapura, D. (1991). *Neural Networks*. Addison-Wesley.

Fu, Z., Golden, B.L., Lele, S., Raghavan, S., and Wasil, E.A. (2004). A genetic algorithm-based approach for building accurate decision trees. *INFORMS Journal on Computing*, 15:3–22.

Glover, F. (1997). *Tabu Search*. Kluwer Academic Publishers.

Hanley, J.A. and McNeil, B.J. (1983). A method of comparing the areas under receiver operating characteristic curves derived from the same cases. *Radiology*, 148:839–843.

Harwood, S. and Scheines, R. (2002). Genetic algorithm search over causal models. Technical Report CMU-PHIL-131, Department of Philosophy, Carnegie Mellon University.

Holland, J. H. (1975). *Adaptation in Natural and Artificial Systems*. University of Michigan Press.

Johnson, D.S., Aragon, C.R., McGeoch, L.A., and Schevon, C. (1989). Optimization by simulated annealing: an experimental evaluation. *Part I, graph partitioning, Operations Research*, 37:6:865–892.

Johnson, D.S. and McGeoch, L.A. (1997). The traveling salesman problem: A case study in local optimization. In E.H.L., Aarts and J.K., Lenstra, editors, *Local Search in Combinatorial Optimization*, pages 215–310. John Wiley and Sons.

Koller, D. and Sahami, M. (1996). Towards optimal feature selection. In *Proceedings of the Thirteenth International Conference on Machine Learning*, pages 284–292. Morgan Kaufmann.

Margaritis, D. and Thrun, S. (1999). Bayesian network induction via local neighborhoods. In *Advances in Neural Information Processing System*.

Metropolis, N., Rosenbluth, A., Rosenbluth, M., Teller, A., and Teller, E. (1953). Equation of state calculations by fast computing machines. *Journal. Chemical Physics*, 21-6:1087–1092.

Moll, R., Perkins, T.J., and Barto, A.G. (2000). Machine learning for subproblem selection. In *Proceedings 17th International Conf. on Machine Learning*, pages 615–622. Morgan Kaufmann, San Francisco, CA.

Pearl, J. (2000). *Causality: Models, Reasoning, and Inference*. Cambridge University Press.

Provost, F., Fawcett, T., and Kohavi, R. (1998). The case against accuracy estimation for comparing induction algorithms. In *Proceedings of the Fifteenth International Conference on Machine Learning*, pages 445–453.

Spirtes, P., Glymour, C., and Scheines, R. (2000). *Causation, Prediction, and Search*. MIT Press.

Spirtes, P. and Meek, C. (1995). Learning bayesian networks with discrete variables from data. In *Proceedings of the First International Conference on Knowledge Discovery and Data Mining*, pages 294–299. AAAI Press.

Sreerama, K. M., Kasif, S., and Salzberg, S. (1994). A system for induction of oblique decision trees. *Journal of Artificial Intelligence Research*, 2:1–32.

Toth, P. and Vigo, D. (2003). The granular tabu search and its application to the vehicle routing problem. *INFORMS Journal on Computing*, 15:4:334–346.

Tsamardinos, I. and Aliferis, C.F. (2002). Algorithms for large-scale local causal discovery in the presence of small sample or large causal neighborhoods. Technical Report DSL-02-08, Vanderbilt University.

Weiss, S.M. and Kulikowski, C.A. (1991). *Computer Systems That Learn*. Morgan Kaufmann.

DANCE MUSIC CLASSIFICATION USING INNER METRIC ANALYSIS

A Computational Approach and Case Study Using 101 Latin American Dances and National Anthems

Elaine Chew, Anja Volk (Fleischer) and Chia-Ying Lee*
University of Southern California Viterbi School of Engineering
Integrated Media Systems Center
Epstein Department of Industrial and Systems Engineering
3715 McClintock Avenue GER240 MC:0193, Los Angeles CA90089-0193, USA.
{echew, avolk, leechiay}@usc.edu

Abstract This paper introduces a method for music genre classification using a computational model for Inner Metric Analysis. Prior classification methods focussing on temporal features utilize tempo (speed) and meter (periodicity) patterns and are unable to distinguish between pieces in the same tempo and meter. Inner Metric Analysis reveals not only the periodicity patterns in the music, but also the accent patterns peculiar to each musical genre. These accent patterns tend to correspond to perceptual groupings of the notes. We propose an algorithm that uses Inner Metric Analysis to map note onset information to an accent profile that can then be compared to template profiles generated from rhythm patterns typical of each genre. The music is classified as being from the genre whose accent profile is most highly correlated with the sample profile. The method has a computational complexity of $O(n^2)$, where n is the length of the query excerpt. We report and analyze the results of the algorithm when applied to Latin American dance music and national anthems that are in the same meter (4/4) and have similar tempo ranges. We evaluate the efficacy of the algorithm when using two variants on the model for Inner Metric Analysis: the metric weight model and the spectral weight model. We find that the correct genre is either the top rank choice or a close second rank choice in almost 80% of the test pieces.

Keywords: Music information processing, genre classification, rhythmic similarity.

*The research has been funded in part by the Integrated Media Systems Center, an NSF ERC, Cooperative Agreement No.EEC-9529152 and USC's WISE program. Any opinions, findings and conclusions or recommendations expressed in this material are those of the authors and do not necessarily reflect those of NSF or WISE.

Introduction

This paper proposes a method for genre classification by automatic extraction of rhythmic features from digital music. Accurate genre classification is critical to the efficient retrieval of music information. Previous research in genre classification using temporal information utilized tempo (speed) and meter (periodicity) as distinguishing features. A problem results when such classification methods encounter music of different genres that have the same tempo and meter, not an infrequent occurrence. We propose an algorithm for automatic classification that is capable of making subtle distinctions between different musical genres that can be in the same meter and tempo. The algorithm can be used either alone or as a add-on to existing classification methods. We test the algorithm using selected Latin American dances – the tango, the rumba, the bossa nova and the merengue – and national anthems. All test pieces have the same 4/4 meter, that is to say, they exhibit a periodicity pattern that cycles every four beats, and are typically played in a moderate tempo. Classification techniques that use tempo and periodicity information would not be able to distinguish between these pieces. The difference between these dances lie in the placement of strong and weak pulses within the four-beat framework. The entire test set consists of 101 pieces. We tested two versions of the algorithm that compute the accent profiles using two variations on the Inner Metric Analysis model [2, 5, 7], namely, the metric weight model and the spectral weight model. The correct genre was ranked either first or second in close to 80% of the test pieces. The metric weight model ranked the correct genre first in 57 of the test pieces, and within the top two rankings in 78 of the test pieces. The corresponding results for the spectral weight model were 71 and 79 respectively.

The perception of musical rhythm is one of the most basic human perceptual facilities. Music consists of a succession of sounds that can be represented as sequences of events that vary over time. The human ear possesses the innate ability to group these events into beats, units of time marking the rhythmic pulse of the music; the beats, in turn, coalesce into larger groups that determine the higher level metrical structures that are marked by patterns of strong (accented) and weak beats. Several other researchers have used temporal information to perform genre classification. This body of work use predominantly the beat and tempo (speed) information to differentiate between pieces from distinct genres. For example, Tzanetakis and Cook [12] use a combination of tempi, pitch histograms and sound texture to perform genre classification in audio files. They use beat histograms to show the predominant tempi in a musical audio excerpt. Tzanetakis and Cook reported a 61% accuracy rate in distinguishing between ten classes of music ranging from classical to metal, with classical and jazz having four and six subclasses respectively. Dixon,

Pampalk and Widmer [1] used a combination of tempo ranges and periodicity patterns (meter) to perform genre classification. They reported an 80% accuracy rate in distinguishing between fourteen styles of ballroom dance music. Gouyon and Herreraa [4] report a high degree of success (between 81 and 95%) in distinguishing between duple and triple meters. Other work on the use of periodicity patterns in similarity assessment includes Foote, Cooper and Nam's results on using self-similarity to extract periodicity patterns [3]. The use of periodicity patterns enabled finer grain distinction between dance music in different meters and improved upon tempo-only approaches to classification in the temporal domain. However, these methods that use only periodicity and tempo information encounter problems classifying music that are from different genres but exhibit the same periodicity and tempo characteristics. Much information is lost by not considering the structures encoded in the temporal patterns exhibited in the pieces themselves.

Our method focuses on extracting and comparing the grouping patterns induced by the note onsets present in the music. We employ the computational method of the Inner Metric Analysis described in [2], [5] and [7] to extract these temporal groupings from note onset time information and map them to time series of impulses that chart the relative strength of the pulse at each point in time. The method uses persistent and regular pulses at all grid levels and phase shifts to induce the metrical patterns present in the piece, and can be applied to both melodic and polyphonic music. We use tempo-invariant numeric input to bypass the problems introduced by performance tempo fluctuations and focus on the problem of perceptual grouping of pulses as exhibited in the piece. We compare the resulting accent profile for each piece of music to that of template rhythms typical of each musical genre by calculating the correlation coefficient between the two time series. Note that periodicity information will emerge in the resulting profiles. The music is classified as being from the genre whose template produces the highest correlation value. Distinct from some previous approaches, our classification method does not require pitch or tempo information.

The paper is organized as follows: Section 1 introduces the computational model for Inner Metric Analysis. Section 2 describes the algorithm for quantifying rhythmic similarity and presents the prototypical rhythm templates and details of the test set. Section 3 reports our computational results. Sections 3.2 and 4 presents a discussion of the results and conclusions respectively. A quick overview of the rhythm notation used in this paper can be found at *www-rcf.usc.edu/~echew/papers/ICS2005*.

1. Inner Metric Analysis

We use the models for Inner Metric Analysis described by Volk in [2] and [13] to map note onsets to numeric time series that reflect their accent strength. In traditional Western music notation, the grouping of beats is prescribed by a time signature in the musical score that indicates the period of the cycle and the unit of the beat. We call the time structure imposed by the time signature the *outer meter* . Each time signature imposes a particular accent pattern on the notes in each cycle that forms the rhythmic feature of the piece. Another type of time structure can be induced from the notes of the piece, by the groupings that arise out of the note content itself – we call this the *inner meter* . The two types of metrical structures may not always coincide. While the outer meter usually remains constant in a piece of music, the inner meter can shift dynamically according to the local grouping patterns. The correspondence, or lack thereof, between the *outer* and *inner meter* led to Volk's definition of metric coherence [2, 13], which is said to exist if the the patterns in the inner meter exhibit the same period and phase as that of the outer meter. The investigation of a wide range of compositions of different styles and epochs concerning the occurrence of metric coherence has proven that it serves as an adequate description of the relation of the metric structure expressed by the notes of a piece and the outer meter. In this section, we give an overview of two methods for computing the inner meter from note onset information only. In Sections 1.1 and 1.2, we show how the inner metric structure can be generated by means of the *metric* and *spectral* weights respectively. More detailed descriptions of the basic concepts of Inner Metric Analysis and a software implementation can be found in [5], [2], [7] and [11].

1.1 Metric Weights

The objective of the model for metrical coherence is the mapping of a numeric weight to each note. The main concept behind the metric weight is the idea of a *local meter* , a maximal and successive chain of equally spaced events. If X denotes the set of all note onsets in a piece of music, then a local meter is defined as a maximal subset:

$$m(s, d, k) = \{s + id, i = 0, \ldots, k\} \subset X,$$

where s denotes the starting point or first onset, d the fixed distance between the onsets of m or period, k the length of m. The local meter, $m(s, d, k)$, is maximal in that there does not exist another local meter $m' \subset X$ s.t. $m \subset m'$.

Figure 1 shows all local meters for a short excerpt from "The Girl From Ipanema" by Antonio Carlos Jobim, a well known example of the bossa nova genre. The discs on the first line below the notes (labeled X) show all onsets in the excerpt. There are three local meters present in this excerpt, namely,

$m(6,2,2)$, $m(6,4,2)$ and $m(6,8,2)$. These three local meters are indicated as successions of dark discs (•) in the second (A), third (B) and fourth (C) lines respectively. The triangles (△) show the extensions of the local meters, to be discussed in the next section.

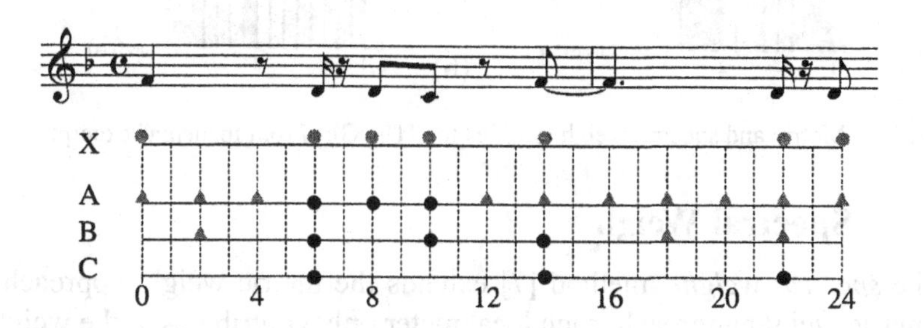

Figure 1. Local Meters in excerpt from "The Girl From Ipanema."

The intuition behind the computation of the metric weight is that longer successions of events should contribute more weight to an onset in its chain than shorter ones. The more stably established in a chain of successive and regular events, the more significant the onset. Hence, the *metric weight* of an onset, $o \in X$, is the sum of a power function of the lengths of all local meters of size at least ℓ, of which o is an element. Let $M(\ell)$ be the set of all local meters of size at least ℓ, that is to say, $M(\ell) = \{m(s,d,k) : k \geq \ell\}$. The general metric weight of an onset, $o \in X$, is as follows:

$$W_{\ell,p}(o) = \sum_{\{m \in M(\ell) : o \in m\}} k^p. \tag{1}$$

For the purposes of this paper, we shall use the metric weight function when $\ell = 2$ and $p = 2$. For example, the second note in the excerpt in Figure 1 is a member of three local meters – $m(6,2,2)$, $m(6,4,2)$ and $m(6,8,2)$ – each of which has a length of two segments. Hence, the metric weight of this second note is $2^2 + 2^2 + 2^2 = 12$. Whereas, the third note is only a member of one local meter, $m(6,2,2)$, and its metric weight is $2^2 = 4$. Figure 2(a) shows the metric weight profile for all notes in the excerpt of "The Girl From Ipanema" shown in Figure 1. The higher the line in the graph, the greater the corresponding weight.

Note that if n is the length of the excerpt to be analyzed, all local meters at a given grid level d, $m(*,d,*)$, can be located in $O(n)$ time. There are n/ℓ such grid levels that are of interest. Hence, the computational complexity of the method for generating the metric weights is $O(n^2)$. The same is true for the spectral weight method described in the section to follow.

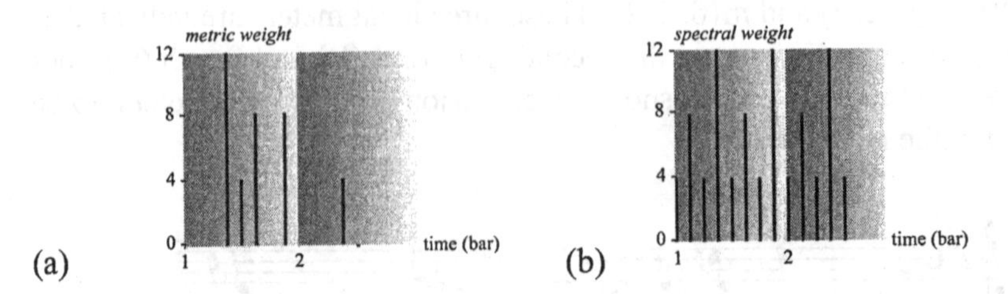

Figure 2. Metric and spectral weight profiles for "The Girl From Ipanema" excerpt.

1.2 Spectral Weights

The *spectral weight* method [7] extends the metric weight approach. In the metric weight approach, each local meter only contributes to the weight of onsets that belong to that local meter. In the spectral weight approach, each local meter contributes to the weight of onsets in its extension throughout the piece. The spectral weight exists not only for note events, but also for silence events (rests), on the other hand, it is not as sensitive to local changes in the inner meter as the metric weight.

Let the *extension* of a local meter be $ext(m(s,d,k)) = \{s + id, \forall i\}$. In Figure 1, the union of all symbols (circles, •, and triangles, △) on each line (A, B and C) shows the elements in the extensions of the corresponding local meters – $ext(m(6,2,2))$, $ext(m(6,4,2))$ and $ext(m(6,8,2))$. Each local meter, $m(s,d,k)$, contributes to the spectral weight of the events in its extension, $t \in ext(m(s,d,k))$. The spectral weight is defined as:

$$SW_{\ell,p}(t) = \sum_{\{m \in M(\ell): t \in ext(m)\}} k^p. \tag{2}$$

The metric and spectral weights for the "Girl From Ipanema" excerpt are given in Figures 2(a) and (b) respectively. While the metric weight profile does not reveal a regular grouping for this short example, the spectral weights can be separated into three hierarchical pulse levels, corresponding to the strongest pulses at the fourth and eighth elements in the bar, a moderate pulse on the second and sixth elements in the bar, and the underlying pulse on the beat.

2. Measuring Rhythmic Similarity

The comparison algorithm is based on the assumption that rhythms that are similar will generate similar distributions of strong and weak accent patterns on the time line as represented by their metric and spectral weight profiles. Section 2.1 outlines our method for genre classification. The test corpus is discussed in Section 2.2 and the prototypical rhythms for each class of music are presented in Section 2.3. An example of the classification process in action is given in Section 2.4. The actual classification results for the test corpus will be presented in Section 3.

2.1 Quantifying Similarity

The core classification process involves two stages: the pre-processing of template rhythms representative of each class of dance music, and the classification of musical samples. In the first stage, we use template rhythms that are known to be typical of each genre to generate metric or spectral weight profiles that are representative of the class of music. The template rhythms are lifted from "The Complete Book of the World's Dance Rhythms" by Kleon Raptakis [9] and augmented with examples gleaned from the CD "The Fabulous Ballroom Collection" [6].

In the calculation of the metric and spectral weights, local meters must be at least of length ℓ to contribute to the weight of an event. As a result, the first (and last) few events in a sample are assigned low weights that may not correspond to their true accent strengths. To minimize this edge effect and to get a stable accent profile, each rhythm pattern is repeated four times and a sample profile from the middle of this test set is singled out as the prototype. In general, the profiles for the second and third repetitions are highly similar. In this paper, we use the profile of the second repetition. The following steps are applied to each template rhythm:

Stage 1: Pre-processing of template rhythms
Step 1: Repeat the template rhythm four times.
Step 2: Calculate its metric/spectral weight.
Step 3: Extract metric/spectral weight profile of 2nd repetition.
Step 4: Normalize all weights to fall between 0 and MAX (=10).

The template rhythms and their accent profiles will be described in Section 2.3.

The second stage involves the comparison of each sample's metric or spectral weight profile to the corresponding profiles of the prototypical rhythms. A short melodic fragment is excerpted manually near the beginning of the test sample and treated in the same fashion as the template rhythms (repeated four times, and middle of accent profile excerpted for comparison). We found that using short melodic fragments from a single instrument works well enough for classification purposes. The constant and regular succession of onsets typical of many Latin American pieces results in a regular onsets that do not reveal much accent differentiation when one considers the union of onsets from all instruments. The classification stage consists of the following steps:

Stage 2: Classification of music sample
Step 1: Obtain melodic fragment near beginning of sample.
Step 2: Repeat the rhythm pattern four times.
Step 3: Calculate its metric/spectral weight.
Step 4: Extract metric/spectral weight profile of 2nd repetition.
Step 5: Normalize all weights to fall between 0 and MAX (=10).
Step 6: Extend (by repetition) both the sample profile and the template profile to the lowest common multiple of sample and template lengths.
Step 7: Compute the correlation coefficient between the (extended) sample and template profiles (details to follow).
Step 8: Assign piece to genre with highest correlation.

Suppose that we wish to compare two accent profiles of the same length, $\{x_i : i = 1, \ldots, N\}$ and $\{y_i : i = 1, \ldots, N\}$, where N is large enough to represent all events in the accent profile. $x_i = 0$ if there is no accent weight at that point in time; the same is true for y_i. We use the correlation coefficient formula: $r = \sum x_i y_i / \sqrt{\sum x_i^2} \sqrt{\sum y_i^2}$.

2.2 Test Corpus

Our test corpus consists of twenty-two tangos, seventeen rumbas, twenty-five bossa novas, seventeen merengues and twenty national anthems from Latin America. All rhythms in the test corpus are of the same periodicity, 4/4 time, cycling after four beats; and, all except the merengue are typically played in a moderate tempo. Hence, the distinguishing features lie in the rhythm patterns themselves and the instrumentation, and not the periodicity or tempi.

The sources for our tango, rumba, bossa nova and merengue test data are listed online at *www-rcf.usc.edu/~echew/papers/ICS2005/appendix.html*. The source for the anthems was Reed and Bristow's "National Anthems of the World" [10]. We use as input to the classification process note material from a melodic fragment near the beginning of each piece. These melodic fragments were extracted manually and the duration sequences encoded in a text format. In the case of tangos, merengues, bossa novas and anthems, fragments of the melodies were excerpted as input. In the case of the rumba, the excerpt generally came from the bass instrument.

2.3 Prototypical Rhythms

We present here the prototypical rhythms for the tango, rumba, bossa nova and merengue. These rhythms were collected from the "Complete Book of the World's Dance Rhythms" [9] and augmented by templates manually transcribed from the CD "The Fabulous Ballroom Collection" [6].

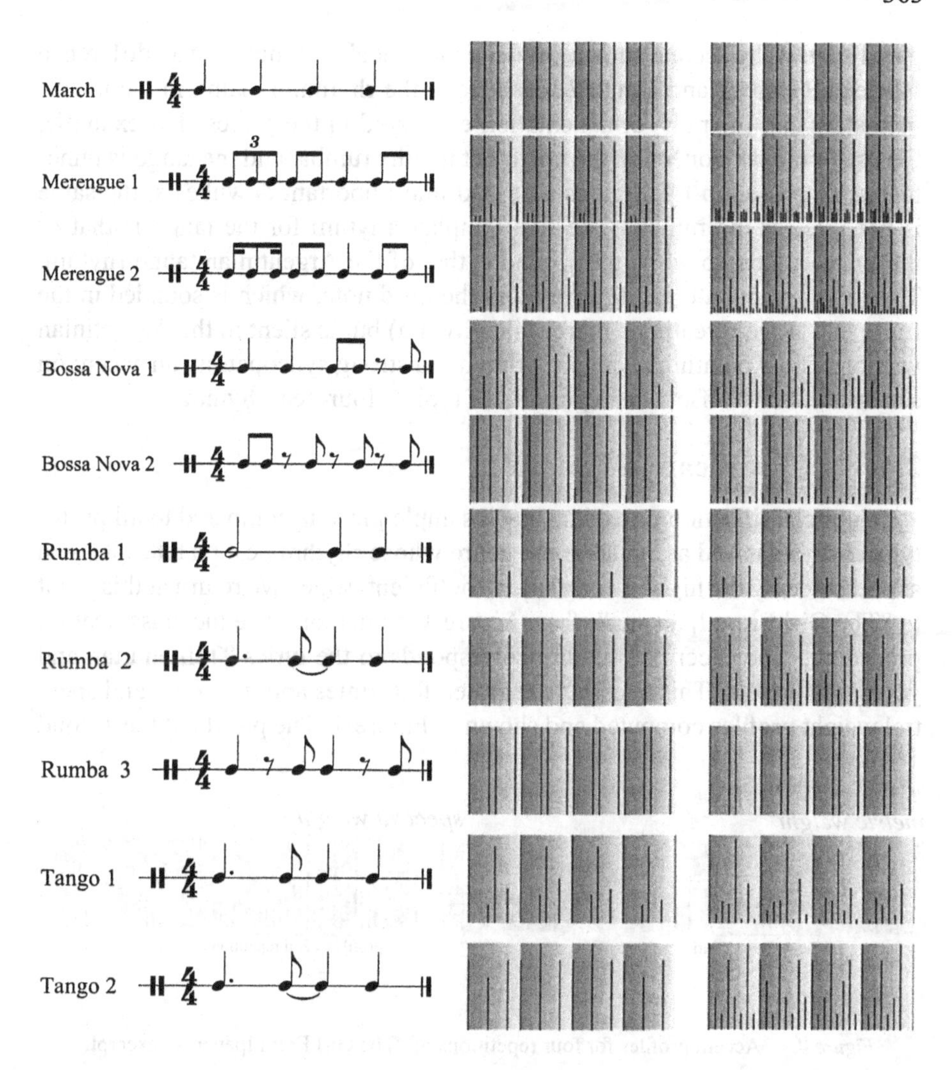

Figure 3. Prototypical rhythms with their respective metric (left chart) and spectral (chart on right) weight profiles.

Figure 3 documents the list of template rhythms for the musical genres under consideration. Each profile is derived from four repetitions of the rhythm shown. The first template is for the four-beat March rhythm typical of many national anthems. The typical merengue rhythms show a busy flow of notes. The bossa nova rhythms are characterized by eighth notes on the off-beats. The phase shift of the accent pattern to the off beats is particularly apparent in the profiles for the second bossa nova rhythm. Rumba 1 and Rumba 3 have strong accents on the first and third beats, as reflected in their metric and spectral weight profiles. Rumba 3 has an equally strong accent on the second (syncopated) note. The second rumba rhythm is identical to Tango 2. The difference between Rumba 2 and Tango 2 lies beyond the rhythmic realm, in the instrumentation and the pitch ornamentation employed in the pieces. For example, the use of the bandoneon in the tango but not the rumba, and the tango is punctuated by glissando-like figures that lead into important downbeats, the same is not true for the rumba. The first template rhythm for the tango is that of the European tango while the second is that of the Argentinian tango rhythm. The two rhythm patterns differ only in the third note, which is sounded in the European tango (creating a march-like rhythm) but is silent in the Argentinian version. The Argentinian tango rhythm has a strong syncopation, an accent on a weak beat that offsets the regularity of typical four-beat rhythm.

2.4 Classification Example

In our classification procedure, each sample piece is compared to all prototypical rhythms and assigned to the genre whose rhythm best matches its own as measured by the highest correlation coefficient value. We return at this point to "The Girl From Ipanema" (from Figure 1) to demonstrate the classification procedure. The excerpted rhythm corresponds to the lyrics "Tall and tan and young and love-". This excerpt is repeated four times and its metric and spectral weight profiles computed and shown in Figure 4. The profile of the second

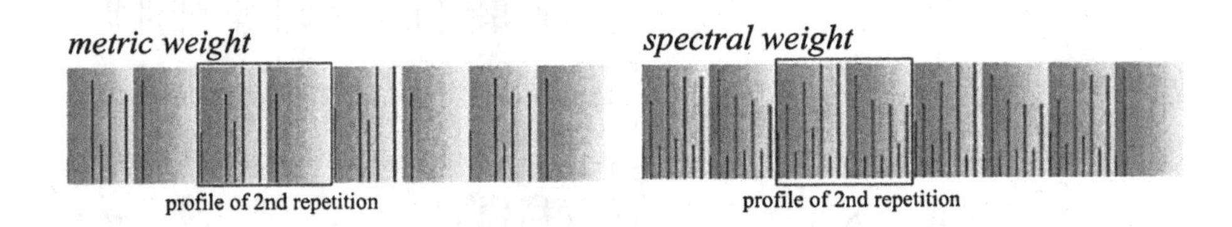

Figure 4. Accent profiles for four repetitions of "The Girl From Ipanema" excerpt.

repetition is excerpted for comparison to all rhythm templates (see Section 2.3). We shall refer to this as the Ipanema profile. Each rhythm template (the profile of the second repetition of each template rhythm in Figure 3) is repeated twice so that it is of the same length as the Ipanema example. We then compute the correlation coefficient between the Ipanema profile and the extended rhythm templates. The coefficient correlation values for all rhythm templates using both metric and spectral weights are presented in Table 1 (where A = anthem, B = bossa nova, M = merengue, R = rumba and T=tango).

Both the metric weight and the spectral weight approaches rank the Bossa nova 2 rhythm highest, followed by the Rumba 3 rhythm. Consider the Bossa nova 2 rhythm in Figure 3. Both the Bossa nova 2 template and the Ipanema

Table 1. Correlation values for comparison of Ipanema profile to rhythm templates.

	B2	R3	B1	T1	R1	M2	M1	T2	R2	A
metric wt	0.62	0.55	0.49	0.41	0.36	0.34	0.34	0.31	0.31	0.26
spectral wt	0.62	0.61	0.44	0.33	0.37	0.45	0.54	0.31	0.31	0.30

profiles (metric and spectral weights) display prominent weights on the off beats, thus identifying the Ipanema rhythm as being from the bossa nova genre.

3. Computational Results

This section reports our computational results using the method outlined in Section 2.1, the test corpus detailed in Section 2.2 and the rhythm templates described in Section 2.3. Section 3.1 describes the experiments and the classification results and Section 3.2 provides a discussion of the results.

3.1 Classification Experiments and Results

We performed separate classification tests using the metric weight model and the spectral weight model. The summary statistics on the correlation coefficient values are reported in Table 2.

Table 2. Summary statistics for correlation coefficient values.

Rhythm profile method	***Highest corr coeff*** *average (std dev)*	***Lowest corr coeff*** *average (std dev)*
metric weight	0.7676 (0.1784)	0.2058 (0.1338)
spectral weight	0.7209 (0.1804)	0.2456 (0.1268)

We first examine the rank one assignment results, that is to say, those with the highest correlation coefficient scores. The left and right charts and tables in Figure 5 document the rank one classification results using the metric and spectral weight models respectively. The bar chart shows the classification results for each test sample class. Each table is a transformed confusion matrix that maps directly to the segments in its corresponding bar chart. The numbers in bold indicate the correct assignments. In infrequent cases of ambiguity, a piece is considered to have been correctly classified if the rhythm template with which its melodic fragment achieved the highest correlation coefficient value is one of the prototypical rhythms of its genre. Consider the identical rhythm templates: Rumba 2 and Tango 2. If a rumba matches this rhythm template, then it is considered to have been classified correctly.

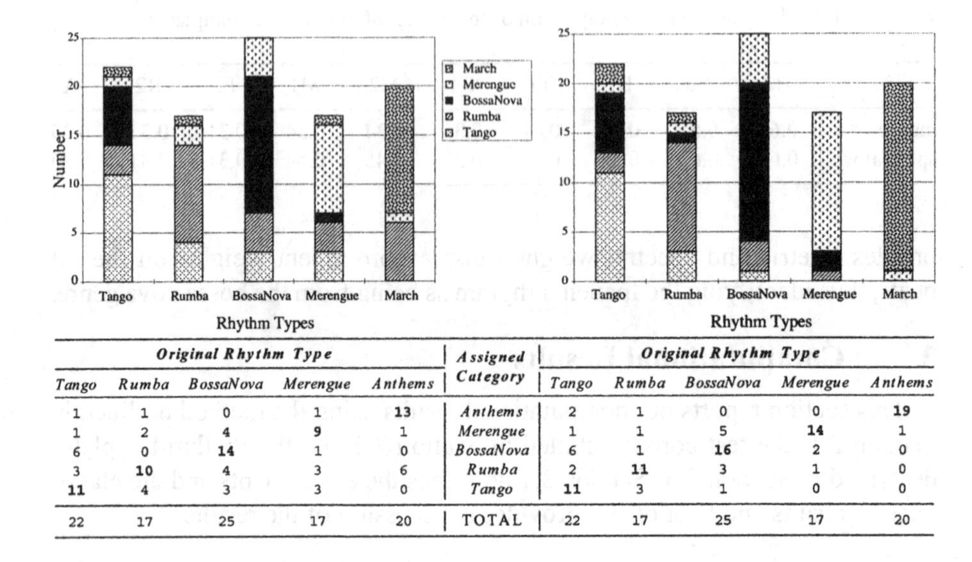

Original Rhythm Type					Assigned	Original Rhythm Type				
Tango	*Rumba*	*BossaNova*	*Merengue*	*Anthems*	*Category*	*Tango*	*Rumba*	*BossaNova*	*Merengue*	*Anthems*
1	1	0	1	**13**	*Anthems*	2	1	0	0	**19**
1	2	4	**9**	1	*Merengue*	1	1	5	**14**	1
6	0	**14**	1	0	*BossaNova*	6	1	**16**	2	0
3	**10**	4	3	6	*Rumba*	2	**11**	3	1	0
11	4	3	3	0	*Tango*	**11**	3	1	0	0
22	17	25	17	20	TOTAL	22	17	25	17	20

Figure 5. Rank one categorization results using metric (left) and spectral (right) weight models.

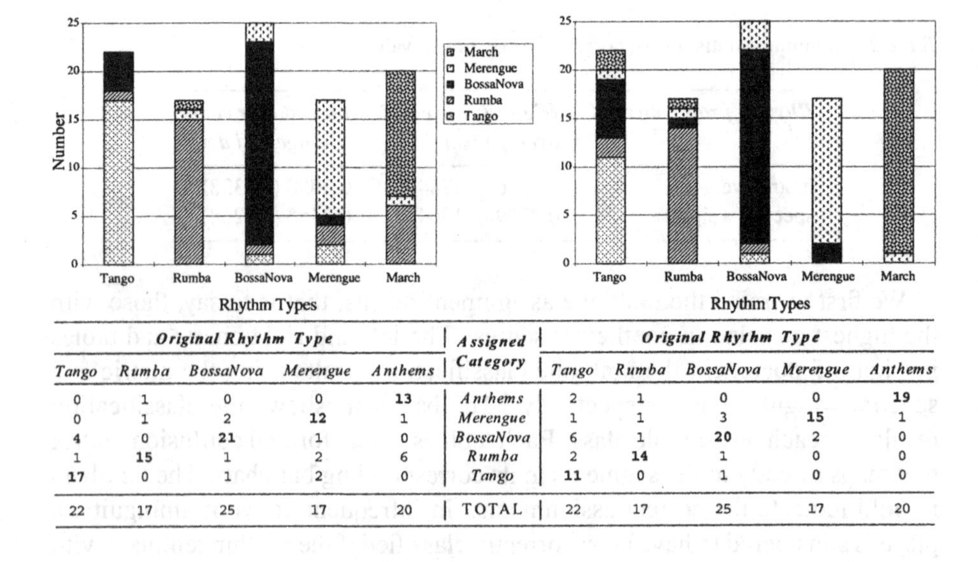

Original Rhythm Type					Assigned	Original Rhythm Type				
Tango	*Rumba*	*BossaNova*	*Merengue*	*Anthems*	*Category*	*Tango*	*Rumba*	*BossaNova*	*Merengue*	*Anthems*
0	1	0	0	**13**	*Anthems*	2	1	0	0	**19**
0	1	2	**12**	1	*Merengue*	1	1	3	**15**	1
4	0	**21**	1	0	*BossaNova*	6	1	**20**	2	0
1	**15**	1	2	6	*Rumba*	2	**14**	1	0	0
17	0	1	2	0	*Tango*	**11**	0	1	0	0
22	17	25	17	20	TOTAL	22	17	25	17	20

Figure 6. Rank one and two (within 0.05) categorization results using metric (left) and spectral (right) weight models.

Sometimes, a piece may be misclassified based on the top ranked assignment, but a close second rank competitor may be the correct genre assignment. Hence, we next take into account second rank results that are correct and close to the top rank classifications. If the correlation coefficient corresponding to the rank two classification of the sample is within, say, 0.05 of that of the rank one classification, then the second classification is considered one of the top choices. For this second scenario, the figures and tables in Figure 6 document the corresponding classification results using the metric and spectral weight models. Finally, Table 3 summarizes the classification results for each method.

Table 3. Summary of classification results.

Experiment	***Proportion correct: only rank one assignments***	***Proportion correct: rank one & two assignments (rank two within 0.05 of one)***
metric weight	0.5644 (57 out of 101)	0.7723 (78 out of 101)
spectral weight	0.7030 (71 out of 101)	0.7822 (79 out of 101)

3.2 Discussion and Analysis

In this section, we evaluate the results of our approach for fine-grain differentiation of the musical genres employing the Inner Metric Analysis model presented in Section 1.

Performance Summary. In general, the method using the spectral weights accurately classified more of the test pieces than the one using metric weights. Consider the top-ranked assignment results presented in Figure 5. The spectral weight approach provided more accurate classifications – 70.3% correct versus the metric weight's 56.4% – showing a marked improvement in the accurate classification of the merengue and march (anthem) samples. For example, the six anthems that were incorrectly classified as rumbas (Rumba 1) using the metric weight approach (see chart on left in Figure 5) were adjusted to the right category in the spectral weight approach (chart on right in same figure). The metric weight approach performed markedly better than the spectral weight method only in assigning the correct classification of tango within the first and second choices (see the two tables at the bottom of Figure 6).

In a significant portion of cases for the metric weight approach where the first rank answer was incorrect, a close second rank answer was the correct one. The same phenomenon was observed to a lesser degree for the spectral weights approach. Compare the metric weight experiment results documented in the left charts in Figures 5 and 6. In 21 out of the 44 incorrect genre assignments for the dance pieces, the correct assignment was the rank two answer whose correlation coefficient was within 0.05 of the rank one answer. For example, all four rumba pieces misclassified as tangos (Tango 1) had as second highest scoring template a rumba rhythm (Rumba 1 and Rumba 2). In fact, when considering both the highest and close second highest correlation coefficient scores, the performance of both the metric weight and spectral weight methods were comparable as shown in Table 3.

Table 4a. Most frequent classification errors using metric weight method.

Original type	*Probable orig.type (#)*	*Misclass. type (#)*
Anthem	Ma(6)	R1(6)
Tango	T1(4)	B1(4)
Rumba	R1(3), R2(1)	T1(4)
BossaNova	B1(1), B2(3)	R3(4)
BossaNova	B1(3)	M2(3)
Tango	T1(1), T2(1)	B2(2)
Merengue	M2(2)	T2(2)
Anthem	Ma(1)	M2(1)
BossaNova	B1(1)	M1(1)
Merengue	M1(1)	T1(1)

Table 4b. Most frequent classification errors using spectral weight method.

Original orig.type	*Probable type (#)*	*Misclass. type (#)*
Tango	T1(2), T2(2)	B1(4)
BossaNova	B1(4)	M1(4)
Rumba	R2(3)	T1(3)
Tango	T1(2)	B2(2)
Merengue	M1(2)	B2(2)
BossaNova	B1(1)	M2(1)
Anthem	Ma(1)	R1(1)

LEGEND: B = bossa nova, Ma = march
M = merengue, R = rumba, T = tango

Most Frequent Errors. We next consider the most frequent classification errors. Referring to Figure 5, the largest segment in each bar in the graph corresponds to the correct assignment, while the second largest segment represents the most frequent classification error. The details of these most frequent classification errors for the metric weight and spectral weight methods are summarized in Tables 4a and 4b respectively. Each table lists the most frequent erroneous assignment for each genre as well as their probable correct type. The probably type is the template in the correct genre that has the highest correlation with the sample's accent profile. For example, of the four rumbas misclassified as Tango1 in the metric weight table, three matched the Rumba1 (R1) rhythm and one the Rumba2 (R2) template. In order to assess the degree of error for misclassified samples, we present the similarity matrix for the template rhythms in Table 5. The tables show the similarity matrix for the template rhythms computed using the metric and spectral weights respectively.

The most frequent error using the metric weight profiles was the misclassification of anthems (marches) as rumbas (Rumba 1, R1, rhythm). This is not a surprising outcome since one might expect the R1 rhythm template to be a plausible march rhythms. In his experiments on pulse salience and metrical ac-

Table 5. Similarity matrices for rhythm templates.

	metric weight method										spectral weight method									
	T1	T2	R1	R2	R3	B1	B2	M1	M2	Ma	T1	T2	R1	R2	R3	B1	B2	M1	M2	Ma
T1	x	.67	.97	.67	.78	.76	.18	.82	.79	.78	x	.99	.62	.99	.51	.19	.42	.70	.63	.58
T2		x	.48	1.0	.60	.32	.37	.61	.61	.53		x	.55	1.0	.49	.17	.45	.69	.61	.51
R1			x	.48	.69	.79	.06	.80	.77	.80			x	.55	.58	.49	.23	.47	.55	.86
R2				x	.60	.32	.37	.61	.61	.53				x	.49	.17	.45	.69	.61	.51
R3					x	.71	.54	.64	.64	.50					x	.62	.66	.88	.74	.46
B1						x	.33	.82	.81	.81						x	.52	.64	.64	.70
B2							x	.18	.20	.04							x	.66	.51	.18
M1								x	.99	.98								x	.84	.55
M2									x	.97									x	.64
Ma										x										x

cent, Parncutt used the R1 rhythm to represent a march [8]. The second most frequent error was the misclassification of tangos as bossa novas (B1), rumbas as tangos (T1), and bossa novas as rumbas (R3). Consider the accent profile of the T1 and the B1 rhythms in Figure 3: they both show strong accents on the first and third beats, a high degree of correspondence with the outer meter, which may explain the algorithm's confusion between the T1 and B1 classification. The correlation between the R1 and T1 rhythms is high (0.97 as shown in the left similarity matrix in Table 5); and, both the B2 and R3 rhythms show strong accents on the weak beats (see Figure 3).

The most frequent error using the spectral weight profiles was the misclassification of tangos as bossa novas (B1) and bossa novas as merengues (M1). As mentioned in the above paragraph, both the tango and B1 rhythms show strong accents on the first and third beats. Both the B1 and M1 rhythms exhibit strong accents on beats 1, 2 and 3.

4. Conclusions

We have presented an $O(n^2)$ method for music genre classification using a computational model for Inner Metric Analysis. We have shown that the inner metrical patterns of the music provide valuable information that can make fine distinctions between pieces in the same tempo and meter. We tested the algorithm on 101 Latin American dances and national anthems and ranked the correct genre within the top two choices in close to 80% of the test pieces.

The current implementation of the algorithm utilized manual preprocessing of the data to extract a melodic fragment for analysis. Future implementations could automate this process, introduce random selection of input data or classification by majority vote from all tracks, and explore the sensitivity of the algorithm to the choice of input fragments. While we have used score-generated MIDI input to encode musical information in order to validate the approach using data that is as precise as possible, further work needs to be done to modify the algorithm to perform classification under uncertainty, for example, irregularity of beats as would be the case in performed music.

In conclusion, the classification method we have presented can discriminate between musical genres based on accent patterns in the same meter (periodicity). Such differentiation by accent patterns provided by Inner Metric Analysis can be used either as a standalone technique or in conjunction with tempo information as a method for genre classification that employs a more complete set of temporal features.

References

[1] Dixon, Simon, Pampalk, Elias and Widmer, Gerhard (2003). "Classification of Dance Music by Periodicity Patterns," in Proceedings of the 4th International Conference on Music Information Retrieval, Baltimore, MD, p. 159-165.

[2] Fleischer, Anja (2002). "A Model of Metric Coherence," in Proceedings of the 2nd International Conference on Understanding and Creating Music, Caserta, Italy.

[3] Foote, Jonathan, Cooper, Matthew and Nam, Unjung (2002). "Audio Retrieval by Rhythmic Similarity," in Proceedings of the 2002 International Conference on Music Information Retrieval at IRCAM, Paris.

[4] Gouyon, Fabien, and Herrera, Perfecto (2003). "Determination of the meter of musical audio signals: Seeking recurrences in beat segment descriptors" in Proceedings of the 114th Convention of the Audio Engineering Society, Amsterdam, The Netherlands, 2003.

[5] Mazzola, Guerino (2002). *The Topos of Music.* Basel: Birkhauser.

[6] Murray, Arthur (1998). *The Fabulous Ballroom Collection.* RCA CD No.63136 (UPC: 90266313624)

[7] Nestke, Andreas and Noll, Thomas (2001). "Inner Metric Analysis',' in Haluska, Jan (ed.) Music and Mathematics. Bratislava: Tatra Mountains Mathematical Publications.

[8] Parncutt, Richard (1994). "A Perceptual Model of Pulse Salience and Metrical Accent in Musical Rhythms," Music Perception, 1994, Vol. 11, No. 4, p. 409-464.

[9] Raptakis, Kleon (1966). *The Complete Book of the World's Dance Rhythms.* Astoria, N.Y.: K. Raptakis Publications.

[10] Reed, William L. and Bristow, Michael J. (1987). *National Anthems of the World.* 7th ed. ISBN: 0713719621. London; New York: Blandford Press.

[11] Volk, Anja and Noll, Thomas (2004). Rubato, Rubettes and related Java-Tools: A User's Guide. Manuscript.

[12] Tzanetakis, George and Cook, Perry (2002). "Musical Genre Classification of Audio Signals," IEEE Transactions on Speech and Audio Processing, 10(5), July 2002.

[13] Volk, Anja (2004). Metric Investigations in Brahms' Symphonies. In Mazzola, Guerino and Thomas Noll (ed): Perspectives of Mathematical and Computer-Aided Music Theory, epOs Music, Osnabruck.

ASSESSING CLUSTER QUALITY USING MULTIPLE MEASURES - A DECISION TREE BASED APPROACH

Kweku-Muata Osei-Bryson
Department of Information Systems & The Informaton Systems Research Institute, Virginia Commonwealth University, Richmond, VA 23284

Abstract: Clustering is a popular data mining technique, with applications in many areas. Although there are many clustering algorithms, none of them is superior on all datasets. Typically these clustering algorithms while providing summary statistics on the generated set of clusters do not provide easily interpretable detailed descriptions of the set of clusters that are generated. Further for a given dataset, different algorithms may give different sets of clusters, and so it is never clear which algorithm and which parameter settings is the most appropriate. In this paper we propose the use of a decision tree (DT) based approach that involves the use of multiple performance measures for indirectly assessing cluster quality in order to determine the most appropriate set of clusters.

Key words: Cluster Quality; Decision Tree; Performance Measures; Multi-Criteria Decision Analysis

1. INTRODUCTION

Clustering (or segmentation) is a popular data mining technique (e.g. Cristofor and Simovici, 2002; Dhillon, 2001; Ben-Dor and Yakhini, 1999; Huang, 1997; Fisher, 1997; Benfield and Raftery, 1992) that attempts to partition a dataset into a meaningful set of mutually exclusive clusters (or segments). Within the context of data mining, clustering is considered to be a form of unsupervised learning since there is no target variable to guide the learning process. There are numerous algorithms available for doing clustering. They may be categorized in various ways such as: hierarchical (e.g. Murtagh, 1983; Ward, 1963) or partitional (e.g. Mc Queen, 1967), deterministic or probabilistic (e.g. Bock, 1996), hard or fuzzy (e.g. Bezdek, 1981; Dave, 1992).

Typically, these clustering algorithms while providing summary statistics on the generated set of clusters (e.g. mean of each variable, distance between clusters), do not provide easily interpretable detailed descriptions of the set of clusters that are generated. Further for a given dataset, different algorithms may give different sets of clusters, and so it is never clear which algorithm and which parameter settings (e.g. number of clusters) is the most appropriate. For as noted by Jain et al. (1999) "There is no clustering technique that is universally applicable in uncovering the variety of structures present in multidimensional data sets". They thus raised the questions: "How is the output of a clustering algorithm evaluated? What characterizes a 'good' clustering result and a 'poor' one?" Ankerst et al. (1999) also commented that "Most of the recent research related to the task of clustering has been directed towards efficiency. The more serious problem, however, is effectivity, i.e. the quality or usefulness of the result". In this paper we focus on the evaluation of the quality of the output of the clustering process.

It might seem that the best approach for identifying the most appropriate partitioning of a given dataset should involve the use of multiple clustering algorithms and parameter settings. One major difficulty is that given the volume of the output data that is to be compared and bounds on human mental processing such a comparison could be a challenging task. This task is made no easier by the fact that it is not even meaningful to compare some output statistics. For example, one algorithm may use distance that is based on the L-2 norm while the other algorithm may use distance that is based on the L-1 norm. Given this challenge the user may be tempted to use a single algorithm with specified ranges of values for some parameters (e.g. number of clusters) and accept the generated set of clusters as the most appropriate one. Difficulties with this approach include how to select the algorithm, and the fact that these algorithms are not optimal algorithms so cannot guarantee that the value that they select from a given range is optimal. This suggests seems that the best approach would involve the application of multiple clustering algorithms with different parameter settings and a non-taxing approach for comparing the various sets of clusters that would not be generated by these algorithms. In this paper we propose the use of a decision tree (DT) based approach that involves the use of multiple performance measures for assessing cluster quality in order to determine the most appropriate set of clusters.

It is fairly well known that while clustering is a form of unsupervised learning, that application of a clustering algorithm results in a data driven labeling (i.e. the cluster identifier) of the dataset, and as such decision tree induction could be used to provide a detailed description of the set of clusters. Thus some data mining software (e.g. SAS Enterprise Miner) automatically generates a profile tree, which is equivalent to a decision tree in which the target variable is the cluster identifier, and the entire dataset is used for training. Thus what we are proposing is not entirely new. What differs here is that we are proposing that the traditional approach to DT induction be applied to generate the corresponding DTs (e.g. after clustering but before DT induction, doing a partitioning of dataset into training, validation and test subsets based on stratified sampling) and to use a multi-criteria approach to select the best DT and thus the best set of clusters. The latter phase is similar to that proposed by Osei-Bryson (2004).

Our work here differs from that of other researchers who have considered the use of DT induction in clustering. For example, Liu et al. (2000) recently proposed using DTs for generating clusters. While their work focused on an approach for generating clusters, our work here focuses on evaluating sets of clusters that could have been generated by any clustering algorithm, including that proposed by Liu et al.

2. OVERVIEW ON RELATED CONCEPTS

Given that our approach for evaluating cluster output involves the use of DT induction and multi-criteria decision making, in this section we present an overview of concepts from each area that is relevant to our approach.

2.1 Decision Tree Induction

A DT is a tree structure representation of the given decision problem such that each non-leaf node is associated with one of the decision variables, each branch from a non-leaf node is associated with a subset of the values of the corresponding decision variable, and each leaf node is associated with an IF-THEN rule that can be used to predict the value of the target variable. A DT is thus an explanatory as well as a predictive model.

The generation of a DT includes partitioning the model dataset into either two parts (i.e. training and validation) or three parts (i.e. training, validation and test). There are two major phases of the DT generation process: the *growth phase* and the *pruning phase* (e.g. Kim and Koehler, 1995). The *growth phase* involves inducting a DT from the *training* dataset. The *pruning phase* aims to generalize the DT that was generated in the *growth phase* in order to avoid over fitting. Therefore in the *pruning phase*, the DT is evaluated against the *validation* dataset in order to generate a subtree of the DT generated in the *growth phase* that has the lowest error rate against the validation dataset. It follows that this DT is not independent of the *training* dataset or the *validation* dataset. Since the *test* dataset was not used for generation or post-pruning of the DT, the accuracy rate of the DT on this third partition of the original dataset is used to estimate the generalization accuracy of the DT.

Many DM software packages (e.g. C5.0, SAS Enterprise Miner, IBM Intelligent Miner) provide facilities that make the generation of DTs a relatively easy task. However, it is known that for a given dataset, the use of different splitting methods and other parameter settings might result in different DTs, and so it may be necessary to generate multiple DTs in order to get the most appropriate one. Most previous approaches to comparing DTs have focused on a single performance measure, typically some measure of accuracy, although it is usually acknowledged that multiple factors are important for evaluating DTs (e.g. Bohanec and Bratko, 1994). Recently Osei-Bryson (2004) proposed a multi-criteria approach for comparing multiple DTs.

2.2 Multiple Criteria Decision-Making (MCDM)

In formal terms, multiple criteria decision-making (MCDM) problems are said to involve the prioritization of a set of alternatives in situations that involve multiple, sometimes conflicting criteria. MCDM problems often have no single alternative that provides the best value for each criterion. Rather for each problem there is a set of alternatives that are said to be **non-dominated**. An alternative is **non-dominated** if there is no other alternative that outscores it with regard to each criterion. Given that MCDM problems do not in general have an objectively unique `best' alternative, then procedures for addressing MCDMs aim to aid the decision-maker(s) in analyzing the given decision-making problem, and facilitate the identification of a ranking of the alternatives that is consistent with the decision maker's beliefs in the importance of the various criteria. Various formal techniques have been proposed for addressing MCDMs including the weighing model and outranking methods. Our solution approach will involve the use of the weighing model formulation for MCDMs. For alternative i, given performance vector $\boldsymbol{v}_i = (v_{i1}, v_{i2},.., v_{i|J|})$ where v_{ij} is alternative i's score with regards to performance measure j, and J is the index set of the performance measures, then with the weighting model formulation the composite score is $s_i = \sum_{j\in J} v_{ij}w_j$, where w_j (> 0) is the weight of performance measure "j", and $\sum_{j\in J} w_j = 1$.

Various approaches are available for generating weights w_j from the subjective inputs of evaluators, both for individual and group decision-making contexts, and for situations when the inputs are precise or imprecise (e.g. Saaty, 1980; Saaty, 1989; Bryson, 1995; Bryson et. al., 1995; Bryson and Joseph, 2000). The application of those techniques requires estimates of the relative importance of pairs of performance measures, and result in a weight vector that is a synthesis of the input pairwise comparison information. Given the nature of our evaluation problem, we will assume that initially the evaluator is not certain about the numeric estimate of the pairwise comparisons and as such we will provide for the evaluator to make imprecise numeric estimates in the form of numeric intervals. For situations involving an individual evaluator, techniques described in Bryson et al. (1995) can be used to generate the corresponding interval weight vector, while for situations involving a group of evaluators, techniques described in Bryson and Joseph (2000) can be used to generate a set of consistent weights the corresponding normalized interval weight vector.

3. OVERVIEW ON PERFORMANCE MEASURES

A major limitation of many clustering algorithms is that they do not provide a detailed description of their output that is easily interpretable. Although Liu et al. (2000) claim that their DT-based clustering method provides interpretable detailed descriptions of the output clusters, interpretability is just one of a set of performance measures that are relevant for comparing the outputs of various clustering algorithms. In their work, Liu et al. (2000) used accuracy and execution time as performance measures, but only the former is relevant for comparing cluster quality. Given the fact that we will be comparing clusters indirectly by doing a comparison of associated DTs, we will also use other DT-oriented performance measures that are also relevant to clustering.

Various measures have been proposed for evaluating the performance of DTs. If a DT is being used to describe a given set of clusters then it is important that given DT should be accurate, simple (so as not to overburden the evaluators who have to compare multiple sets of clusters), and stable. Further if all clusters are relevant then we would expect that each cluster should appear in at least one leaf with strong discriminating power. We will be using the set of measures recently used by Osei-Bryson (2004) which includes equivalent measures. The reader should note that we do not claim that this set of performance measures is exhaustive or that each performance measure of this set is relevant in every situation.

3.1 Accuracy (ACC)

The most commonly used performance criterion for a DT is the predictive *accuracy rate*. Let ACC_T, ACC_V, and ACC_G be the accuracy rates of the training, validation, and test datasets respectively. ACC_G is used to estimate the generalization of the accuracy rate of the DT.

3.2 Stability (STAB)

The *stability* performance criterion concerns our interest that there should not be much variation in this *predictive accuracy rate* when a DT is applied to different datasets. Thus at a minimum one might expect that there should not be much variation in predictive accuracy of the DT on the validation dataset when compared to that for the training dataset. Typically there is no numeric performance measure for the stability property, but Osei-Bryson (2004) proposed a numeric measure of stability. Let acc_{Tf} and acc_{Vf} be the accuracy rates of leaf f based on the training and validation datasets respectively, and ρ_{Vf} be the proportion of validation cases associated with leaf f. The stability of leaf f based on the training and validation datasets can be defined as $stab_f = Min\ \{acc_{Tf}/acc_{Vf}, acc_{Vf}/acc_{Tf}\}$, where $stab_f \in (0, 1]$, with higher values indicating higher stability. Given this measure the stability of the DT with regards to its performance on the training and validation datasets can be defined as $STAB = \sum_{f\in F} \rho_{Vf} stab_f$, where $STAB \in (0, 1]$, with higher values of STAB indicating higher stability. The reader may note STAB is just the weighted sum of the stability of the individual leaves.

3.3 Simplicity (SIMPL)

Tree simplicity has also been considered by many researchers. For some, a measure of tree simplicity has been limited to the number of leaves in the DT (e.g. Shafer, Agrawal and Mehta, 1996) while others have also suggested that the sizes of the corresponding rules (i.e. number of predictor variables) are also relevant, particularly when the rules are to be applied by human beings rather than computers (e.g. Han and Kamber, 2001).

3.3.1 Simplicity based on Number of Leaves ($SIMPL_{Leaf}$)

It is often assumed with regards to tree simplicity, the fewer the leaves the better. However, usually we are often not interested in a DT with only a single leaf and for other situations even a DT with two leaves might not be useful. In other words for different problem instances there may be different value functions that map the number of leaves to the simplicity measure. Let us assume that we have such a function (e.g. trapezoidal function) such that the complexity $SIMPL_{LEAF} = v_{Leaf}(|F|)$, where $|F|$ is the number of the leaves in the DT, and $v_{Leaf}(|F|)$, is a concave piece-wise linear function such that

$SIMPL_{Leaf} \in (0, 1]$, with higher values of $SIMPL_{Leaf}$ indicating higher simplicity.

3.3.2 Simplicity based on Average Chain Length ($SIMPL_{Rule}$)

For a given rule, its length (i.e. number of predictor variables) provides a measure of the simplicity of the rule then another simplicity measure for the DT could be based on the average rule length of the rules in the DT. Let len_f be the rule length for rule $f \in F$. The mean rule length of the DT could be defined as $Len_{Mean} = \sum_{f \in F} \rho_{Vf} len_f$, which is just the weighted sum of the length of each rule based on the validation dataset. The corresponding rule length based simplicity measure is defined as $SIMPL_{Rule} = v_{Rule}(Len_{Mean})$, where $v_{Rule}(Len_{Mean})$ is a concave piece-wise linear function such that $SIMPL_{Rule} \in (0, 1]$, with higher values of $SIMPL_{Rule}$ indicating higher simplicity. We will provide an example of such a function in our illustrative example.

3.4 Discriminatory Power (DSCPR)

Ideally one would like to have leaves that are totally pure but that is unlikely to occur. However, for a human being a given rule might not be considered to be particularly useful if the training posterior probability of the assigned class is less than some specified cut-off value τ.(> 0.50). Thus for some situations a *Discriminatory Power* performance measure might also be appropriate for evaluating the performance of the DT. For leaf f, let $dscpr_f = 1$ if $acc_{Tf} \geq \tau$; and $dscpr_f = 0$ if $acc_{Tf} < \tau$. A measure of discriminatory power could be defined as: $DSCPR = \sum_f \rho_{Vf} dscpr_f$, where $DSCPR \in [0, 1]$, with higher values of DSCPR indicating higher discriminatory power. The rationale here is that predictions of those leaves whose maximum posterior probability is less than the user defined cut-off values are questionable.

4. FORMULATING THE EVALUATION PROBLEM

As stated earlier, our approach assessing cluster quality will be based on the multi-criteria approach for evaluating DTs that was recently presented by Osei-Bryson (2004). For each set of clusters we will attempt to select the DT that is the most appropriate in terms of the performance measures that were described in the last section.

Let $\Omega_{ClusAlg}$ be the set of clustering algorithms and parameter setting combinations (e.g. Least Squares with Ward option for combining clusters), $\Omega_{SpltMthd}$ be the set of selected splitting methods (e.g. Entropy, Chi-Square,

Gini), $\Omega_{MinLeaf}$ be the set of selected values for the Minimum Number of Cases per Leaf, $\Omega_{MinSplit}$ be the set of selected values for the Minimum Number of Cases for a Split Search, $\Omega_{MaxBran}$ be the set of selected values for the Maximum Number of Branches from a Node, $\Omega_{MaxDpth}$ be the set of selected values for the Maximum Depth of Tree.

Step 1: Preparation

a) Specify $\Omega_{ClusAlg}$, the set of clustering algorithms and parameter setting combinations that are to be considered.
b) Specify the rule for partitioning the clustering labeled dataset into training, validation and test datasets.
c) Specify the set of performance measures J.
d) Specify the DT parameter sets: $\Omega_{SpltMthd}$, $\Omega_{MinLeaf}$, $\Omega_{MinSplit}$, $\Omega_{MaxBran}$, $\Omega_{MaxDpth}$.
e) Specify the cut-off value τ for rule discriminatory power.
f) Specify threshold values for accuracy ξ_{ACC_G}, stability ξ_{STAB}, and any other performance measure.
g) Specify the value function for Simplicity based on the Number of Leaves, and the value function for Simplicity based on the Chain Lengths of the Rules.

Step 2: Generate Weights for the Performance Measures

a) The evaluator(s) from the DM project team specify numeric pairwise comparison data on relevant importance of pairs of performance measures. It is not necessary that a pairwise comparison entry be made for each pair of performance measures but each performance measure must be included in at least one pairwise comparison.
b) Generate the corresponding output consistent pairwise comparison matrix $C = \{c_{jk} = [c_{Ljk}, c_{Ujk}]\}$, and consistency indicator using a weight vector generation technique (e.g. Bryson et. al., 1995; Bryson and Joseph, 2000).
c) If the consistency indicator value is acceptable then go to step 3, otherwise repeat step 2.

Step 3: Generate Clusters & Decision Trees

For each $a \in \Omega_{ClusAlg}$:

a) Generate the set of clusters
b) Using the labeled output data from 3a, for each combination of parameter values from the DT parameter sets (i.e. $\Omega_{SpltMthd}$, $\Omega_{MinLeaf}$, $\Omega_{MinSplit}$, $\Omega_{MaxBran}$, $\Omega_{MaxDpth}$), generate the corresponding DT_i's and calculate the performance measures.

   Let Φ be the set of all DTs that were generated in this step. It should be noted that each DT_i is associated with a single $a \in \Omega_{ClusAlg}$, but for each $a \in \Omega_{ClusAlg}$ there could be multiple DT_i's.

Step 4: Determine Set of Relevant Decision Trees

a) Exclude from Φ those DTs which violate any of the threshold values for the performance criteria.

b) Identify and exclude *dominated* DTs from Φ. At this step Φ now only contains those non-dominated DTs that satisfy all threshold constraints. Let Φ_{ND} be the set of *non-dominated* DTs that satisfy all threshold constraints.

Step 5: Determine 'Most Appropriate' Set of Clusters

a) Formulate and solve problem P_{DTh} for each $h \in \Phi_{ND}$.

P_{DTh}: Max s_h

1) $\sum_{j \in J} v_{ij} w_j - s_i = 0 \quad \forall i \in \Phi_{ND}$
2) $w_j - c_{Ljk} w_k \geq 0 \quad \forall j, k \in J, j \neq k$
3) $w_j - c_{Ujk} w_k \leq 0 \quad \forall j, k \in J, j \neq k$
4) $\sum_{j \in J} w_j = 1$
5) $w_j \geq 0 \quad \forall j \in J$

b) Order the DTs in Φ_{ND} in descending sequence based on their values of s_h. Let DT_{h*} be the top-ranked DT, then the set of clusters associated with its corresponding $a_{h*} \in \Omega_{ClusAlg}$ is the most appropriate one.

5. ILLUSTRATIVE EXAMPLE

Our illustrative example involves clustering the Abalone dataset that was obtained from the UCI Irvine machine library (Murphy and Aha, 1994). We assume that the decision makers are of the belief that the appropriate set of clusters would consist of between 3 to 6 clusters.

Step 1: Preparation

- For this illustrative example there are two possible choice of clustering algorithms, and we will label them by their distance measure: a) Least Squares distance function with the Ward combination option (LS-Ward); and b) Mean Absolute Deviation (MAD).
- Minimum Number of Clusters = 3, Maximum Number of Clusters = 6
- For DT induction, the splitting method options are: ChiSquare, Entropy, Gini.
- For each DT, the Minimum Number of Cases per Leaf parameter is to be set to 20, and Minimum Number of Cases for a Split parameter is to be set to 40.
- The discriminatory power threshold $\tau = 0.75$.

Specify the Simplicity value functions

For each of our simplicity measures we will be using a trapezoidal value function that has parameters CutBot, IdealBot, IdealTop, and CutTop where with regards to the given simplicity measure z (i.e. Number of Leaves,

Average Chain Length): the DT would be considered to be ideal if z ∈ [IdealBot, IdealTop]; the DT would be considered to be unacceptable if z ≤ CutBot or z ≥ CutTop; the value of other acceptable DTs would be based on how well they compared with an ideal DT with regard to the v(z). Given these assumptions, the value function v(z) can be defined as follows:

$$v(z) = 0 \quad \text{if } z \leq \text{CutBot or } z \geq \text{CutTop}$$
$$v(z) = (z - \text{CutBot})/(\text{IdealBot} - \text{CutBot}) \quad \text{if } z \in (\text{CutBot, IdealBot})$$
$$v(z) = 1 \quad \text{if } z \in [\text{IdealBot, IdealTop}]$$
$$v(z) = (\text{CutTop} - z)/(\text{CutTop} - \text{IdealTop}) \quad \text{if } z \in (\text{IdealTop, CutTop})$$

We assume that the following values have been specified for the relevant value functions:

Table 1 Parameter Value for Simplicity Measures

Simplicity Measures	Cut Bot	IdealBot	IdealTop	CutTop
Number of Leaves	2	3	8	9
Average Chain Length	1.5	2.0	4.0	5.0

Step 2: Generate Weights for the Performance Measures

For generating our set of consistent weights, we will use the pairwise comparison data of the illustrative example of Osei-Bryson (2004).

*Table 2.*Pairwise Comparisons of Relative Importance of Performance Measures

	ACC	DSCPR	STAB	$Simpl_{Rule}$	$Simpl_{Leaf}$
ACC			I: [0.90, 1.00] O: [0.80,1.00}		
DSCPR	I: [0.80, 1:00] O: [0.80, 1.00]		I: O: [0.77, 0.83]		
STAB	I: [0.50, 0.80] O: [0.52, 0.77]	I: [0.50, 0.80] O: [0.63, 0.80]	I: [0.50, 0.80] O: [0.50, 0.58]		I: [0.90, 1.11] O: [0.90, 1.11]
$Simpl_{Rule}$	I: O: [0.58, 0.69]	I: O: [0.69, 0.72]	I: O: [0.56, 0.58]	I: [0.90, 1.11] O: [0.90, 1.11]	

I: Input Entries O: Consistent Output Entries

Step 3: Generate Clusters & Decision Trees

Clustering was done using the SAS Enterprise Miner software. The LS-Ward approach allows the user to specify a range (i.e. Minimum Number of Clusters = 3, Maximum Number of Clusters = 6) for the Number of Clusters parameter, after which the software automatically selects the most appropriate number of clusters using a heuristic. The MAD approach does not allow the user to specify a range but rather a specific value for the Number of Clusters parameter (e.g. 3, 4, 5).

For each cluster output, 3 DTs are generated, one for each splitting method option (i.e. ChiSquare, Entropy, Gini). In some cases the DT that was selected by Enterprise Miner had more than 8 leaves. In those cases we selected the DT that had between 3 and 8 leaves and which provided the highest test accuracy rates, which usually resulted in a DT with 8 leaves.

Steps 4-5: Determine Set of Relevant DTs & the Most Appropriate Set of Clusters

First we identified the DTs that were dominated & eliminated them. Four DTs remained (see Tables 3a and 3b).We then calculated the composite scores of these DTs. The top two DTs were associated with the output of LS-Ward approach, with the top DT being associated with the Chi-Square splitting method. Thus given the decision-makers preference and value functions, the 'best' set of clusters is the one that was generated using the LS-Ward algorithm with the specified parameter values.

Table 3a. Results of Clustering, DT Induction and DT Evaluation

Clust. Alg.	Number of Clusters	Splitting Method	ACC	STAB	DSCPR	$Simpl_{Rule}$	$Simpl_{Leaf}$
LS-Ward	3 (3-6)	Chi	0.962	**0.986**	0.975	**1.000**	**1.000**
		Ent	0.957	**0.986**	0.974	**1.000**	**1.000**
		Gini	**0.966**	0.980	0.979	**1.000**	**1.000**
MAD	3	Chi	0.951	0.975	0.932	**1.000**	0.257
		Ent	0.952	0.956	**0.985**	**1.000**	**1.000**
		Gini	0.947	0.975	0.934	**1.000**	**1.000**
	4	Chi	0.905	0.981	0.857	**1.000**	**1.000**
		Ent	0.905	0.974	**0.985**	**1.000**	**1.000**
		Gini	0.915	0.973	0.947	**1.000**	**1.000**
	5	Chi	0.872	0.858	0.891	**1.000**	**1.000**
		Ent	0.805	0.955	0.899	**1.000**	0.626
		Gini	0.867	0.854	0.891	**1.000**	**1.000**

For each performance measure, highest data values are in bold print

Table 3b. Results of Clustering, DT Induction and DT Evaluation

Clust. Alg.	Number of Clusters	Splitting Method	Dominated	Score
LS-Ward	3 (3-6)	Chi	No	0.983
		Ent	Yes	N/A
		Gini	No	0.982
MAD	3	Chi	Yes	N/A
		Ent	No	0.975
		Gini	Yes	N/A
	4	Chi	Yes	N/A
		Ent	No	0.969
		Gini	Yes	N/A
	5	Chi	Yes	N/A
		Ent	Yes	N/A
		Gini	Yes	N/A

N/A: Not Applicable since DT is Dominated

6. CONCLUSION

In this paper we have presented a formal approach for evaluating cluster output that involves the use of decision tree induction and multi-criteria decision analysis. This research problem is an important one that has not been adequately addressed in the clustering literature (e.g. Ankerst et al., 1999; Jain et al., 1999). Jain et al. (1999) describe cluster validity as the assessment of the set of clusters that are generated by the given clustering algorithm. They note that there are three approaches for assessing validity: 1) External assessment which involves comparing the generated set of clusters with an a priori structure, typically provided by some experts; 2) Internal assessment which attempts to determine if the generated set of clusters is "intrinsically appropriate" for the data; and 3) Relative assessment which involves comparing two sets of clusters based on some measures (e.g. Jain and Dunes, 1988; Dubes, 1993) and measure their relative performance. Our multi-criteria DT-based approach could be considered to have some relationship to these three types of approaches:

- External: Our approach does not require the decision-makers to provide an a priori clustering structure, but does require them to provide preference and value structures which are used to indirectly evaluate and compare the sets of clusters.
- Internal: We assess each set of clusters by assessing the associated DTs. At least two of our performance measures (i.e. accuracy, and stability) that would appear to provide some indication as to whether the given set of clusters is "intrinsically appropriate" for the data.

- Relative: Our objective is to select the most appropriate set of clusters. Since we are never sure which algorithm/set of parameter values combination is the most appropriate, then we experimented with multiple combinations. However, it is almost impossible to experiment with all possible combinations, and so the set of clusters that we select as the best one is relative to our set of experimental combinations.

Acknowledgements

This research was supported in part by a grant from the 2004 Summer Research Program of the School of Business of Virginia Commonwealth University. I also wish to thank the anonymous referees for their valuable comments that enabled me to improve the quality of this paper.

REFERENCES

Ankerst, M., Breunig, M., Kriegel, H.-P., and Sander, J. (1999) "OPTICS: Ordering Points To Identify the Clustering Structure", ***Proceedings of ACM SIGMOD'99 International Conference on the Management of Data***, pp. 49-60. Philadelphia, PA.

Banfield, J. and Raftery, A. (1992) "Identifying Ice Floes in Satellite Images", ***Naval Research Reviews 43***, pp. 2-18.

Ben-Dor, A. and Yakhini, Z. (1999) "Clustering Gene Expression Patterns", ***Proceedings of the 3rd Annual International Conference on Computational Molecular Biology*** *(RECOMB 99)*, pp. 11-14, Lyon, France.

Bohanec, M. and Bratko, I. (1994) "Trading Accuracy for Simplicity in Decision Trees", ***Machine Learning 15***, pp. 223-250.

Bryson, N. (1995) "A Goal Programming for Generating Priority Vectors", ***Journal of the Operational Research Society 46***, pp. 641-648.

Bryson, N., Mobolurin, A., and Ngwenyama, O. (1995) "Modelling Pairwise Comparisons on Ratio Scales", ***European Journal of Operational Research 83***, pp. 639-654.

Bryson, N. (K-M), and Joseph, A. (2000) "Generating Consensus Priority Interval Vectors For Group Decision Making In The AHP", ***Journal of Multi-Criteria Decision Analysis 9:4***, pp. 127-137.

Bezdek, J. (1981) ***Pattern Recognition with Fuzzy Objective Function Algorithms***. Plenum Press, New York, NY.

Bock, H. (1996) "Probability Models in Partitional Cluster Analysis", ***Computational Statistics and Data Analysis 23***, pp. 5-28.

Cristofor, D. and Simovici, D. (2002) "An Information-Theoretical Approach to Clustering Categorical Databases using Genetic Algorithms", ***Proceedings of the SIAM DM Workshop on Clustering High Dimensional Data***, pp. 37-46. Arlington, VA.

Dave, R. (1992) "Generalized Fuzzy C-Shells Clustering and Detection of Circular and Elliptic Boundaries", ***Pattern Recognition 25***, pp. 713–722.

Dhillon, I. (2001) "Co-Clustering Documents and Words Using Bipartite Spectral Graph Partitioning", ***Proceedings of the 7th ACM SIGKDD***, pp. 269-274, San Francisco, CA.

Dubes, R. (1993). "Cluster Analysis and Related Issues", in ***Handbook of Pattern Recognition & Computer Vision***, C. Chen, L. Pau, and P. Wang, Eds. World Scientific Publishing Co., Inc., River Edge, NJ, pp. 3–32.

Fisher, D. (1987) "Knowledge Acquisition via Incremental Conceptual Clustering", ***Machine Learning 2***, pp. 139–172.

Jain, A. and Dubes, R. (1988) ***Algorithms for Clustering Data***. Prentice-Hall Advanced Reference Series. Prentice-Hall, Inc., Upper Saddle River, NJ.

Jain, A. and Flynn, P. (1993) ***Three Dimensional Object Recognition Systems***. Elsevier Science Inc., New York, NY.

Jain, A., Murty, M. and Flynn, P. (1999) "Data Clustering: A Review", ***ACM Computing Surveys 31:3***, pp. 264-323.

Han, J. and Kamber, M. (2001) ***Data Mining: Concepts and Techniques***, Morgan Kaufman, New York, NY.

Huang, Z. (1997) "*A Fast Clustering Algorithm to Cluster Very Large Categorical Data Sets in Data Mining*", ***Proceedings SIGMOD Workshop on Research Issues on Data Mining and Knowledge Discovery***, Tech. Report 97-07, UBC, Dept. of CS.

Kim, H. and Koehler, G. (1995) "Theory and Practice of Decision Tree Induction", ***Omega 23:6***, pp. pp. 637-652.

Liu, B., Yiyuan, X., and Yu, P. (2000) "Clustering through Decision Tree Construction", ***Proceedings of the Ninth International Conference on Information and Knowledge Management (CIKM'00)***, pp. 20-29.

Murphy, P., and Aha, D. (1994) ***UCI Repository of Machine Learning Databases***. University of California, Department of Information and Computer Science:

Murtagh, F. (1983) "A Survey of Recent Advances in Hierarchical Clustering Algorithms which Use Cluster Centers", ***Computer Journal 26***, pp. 354–359.

Osei-Bryson, K.-M. (2004) "Evaluation of Decision Trees: A Multi-Criteria Approach", ***Computers & Operations Research 31:11***, pp. 1933-1945.

Saaty, T. (1980) *The Analytic Hierarchy Process: Planning, Priority Setting, Resource Allocation*, McGraw-Hill, New York

Saaty, T. (1989) "Group Decision Making and the AHP", in B. Golden, E. Wasil, and P. Harker (Editors), ***The Analytic Hierarchy Process: Application and Studies***, pp. 59-67.

Ward, J. (1963) "Hierarchical Grouping to Optimize An Objective Function", ***J. Am. Stat. Assoc. 58***, pp. 236–244.

DISPERSION OF GROUP JUDGMENTS

The Geometric Expected Value Operator

THOMAS L. SAATY and LUIS G. VARGAS

Joseph M. Katz Graduate School of Business, University of Pittsburgh

Abstract: To achieve a decision with which the group is satisfied, the group members must accept the judgments, and ultimately the priorities. This requires that (a) the judgments be homogeneous, and (b) the priorities of the individual group members be compatible with the group priorities. There are three levels in which the homogeneity of group preference needs to be considered: (1) for a single paired comparison (monogeneity), (2) for an entire matrix of paired comparisons (multigeneity), and (3) for a hierarchy or network (omnigeneity). In this paper we study monogeneity and the impact it has on group priorities.

Keywords: reciprocal uniform distribution, geometric mean, geometric dispersion, group cohesiveness, group liaison, principal right eigenvector, beta distribution.

1. INTRODUCTION

In all facets of life groups of people get together to make decisions. The group members may or may not be in agreement about some issues and that is reflected in how homogeneous the group is in its thinking. In the AHP groups make decisions by building a hierarchy together and providing judgments expressed on a 1 to 9 discrete scale having the reciprocal property. Condon et al. (2003) mentioned that there are four different ways in which groups estimate weights in the AHP: "...consensus, vote or compromise, geometric mean of the individual judgments, and weighted arithmetic mean." The first three deal with judgments of individuals while the last deals with the priorities derived from the judgments.

To achieve a decision with which the group is satisfied, the judgments, and ultimately the priorities, must be accepted by the group members. This requires that (a) the judgments be homogeneous, and (b) the priorities of the individual group members be compatible with the group priorities.

There are three levels in which the homogeneity of group preference needs to be considered: (1) for a single paired comparison (monogeneity), (2) for an entire matrix of paired comparisons (multigeneity), and (3) for a hierarchy or network (omnigeneity). Monogeneity relates to the dispersion of the judgments around their geometric mean. The geometric mean of group judgments is the mathematical equivalent of consensus if all the members are considered equal. Otherwise one would use the weighted geometric mean. Aczel and Saaty (1983) showed that the only mathematically valid way to synthesize reciprocal judgments preserving the reciprocal condition is the geometric mean. If the group judgments for a single paired comparison are too dispersed, i.e., they are not close to their geometric mean, the resulting geometric mean may not be used as the representative judgment for the group.

Multigeneity relates to the compatibility index of the priority vectors. The closeness of two priority vectors $v=(v_1,\ldots,v_n)^T$ and $w=(w_1,\ldots,w_n)^T$ can be tested through their compatibility index (Saaty, 1994) given by $\frac{1}{n^2}e^T V \circ W^T e$, where $\circ$ is the Hadamard or elementwise product, $V=\left(v_i/v_j\right)$ and $W=\left(w_i/w_j\right)$. Note that for a reciprocal matrix $A=(a_{ij})$ with principal eigenvalue $\lambda_{\max}$ and corresponding right eigenvector $w=(w_1,\ldots,w_n)$, $\frac{1}{n^2}e^T A \circ W^T e = \lambda_{\max}/n^2$. Thus, one can test the compatibility of each individual vector with that derived from the group judgments. A homogeneous group should have compatible individuals. It is clear that homogeneity at the paired comparisons level implies compatibility at the group level, but the converse is not always true. At the hierarchy or network level, it appears that it is more meaningful to speak of compatibility than of homogeneity.

The main thrust of this paper is to study monogeneity.

Dispersion in judgments leads to violations of Pareto Optimality at both the pairwise comparison level and/or the entire matrix from which priorities are derived. Ramanathan and Ganesh (1994) explored two methods of combining judgments in hierarchies but they violated the Pareto Optimality Principle for pairwise comparisons (Saaty and Vargas, 2003), and hence, they incorrectly concluded that the geometric mean violates Pareto Optimality. Pareto Optimality at the pairwise level is not sufficient to ensure Pareto Optimality at the priority level. Fundamentally, Pareto Optimality means that if all individuals prefer A to B then so should the group. The group may be homogeneous in some paired comparisons and heterogeneous in others thus violating Pareto Optimality. The degree of violation of Pareto Optimality can be measured by computing compatibility along the rows, which yields a vector of compatibility values. What does one do when a group is not homogeneous in all its comparisons? Lack of homogeneity (heterogeneity) on some issues may lead to breaking up the group into smaller homogeneous groups. How should one separate the group into homogeneous subgroups? Since homogeneity relates to dispersion around the geometric mean, and dispersion itself involves uncertainties, how much of the dispersion is innate and how much is noise that when filtered one can speak of true homogeneity? In other words, how does one separate random considerations from committed beliefs?

Dispersion at the single paired comparison level affects the priorities obtained by each group member individually and could lead to violating Pareto Optimality. Should one combine or synthesize the priorities of the individuals to obtain the group priority or should one combine their judgments?

Here we develop a way to test monogeneity, i.e., how homogeneous the judgments of the members of a group are for each judgment they give in response to paired comparisons. This is done by deriving a measure of the dispersion of the judgments based on the geometric mean. Computing the dispersion around the geometric mean requires a multiplicative approach rather than the usual additive expected value used to calculate moments around the arithmetic mean. This leads to a new multiplicative or geometric expected value used to define the concept of geometric dispersion. The geometric dispersion of a finite set of values is given by the geometric mean of the ratios of the values to their geometric mean, if the ratio is greater than 1, or the reciprocal, if the ratio is less than or equal to 1. This measure of variability or dispersion of the judgments around the geometric mean allows us to (a) determine if the geometric mean of the judgments of a group can be used as the synthesized group judgment, (b) if the geometric mean cannot be used, divide the group into subgroups according to their geometric dispersion, and (c) measure the variability of the priorities corresponding to the matrix of judgments synthesized for the group.

In general, unless a group decides through consensus which judgments to assign in response to a paired comparison, the individual members may give different judgments. We need to find if the dispersion of this set of judgments is a normal occurrence in the group behavior. To do this, we compare the dispersion of the group with the dispersion of a group providing random responses to the paired comparison. Thus, we assume that an individual's pairwise comparison judgments about homogeneous elements is considered random, and expressed on a discrete 1/9, ..., 1/2, 1, 2,..., 9 scale of seventeen equally likely values. A sample consists of a set of values selected at random from the set of seventeen values, one for each member of the group. It is the dispersion of this sample of numbers around its geometric mean that concerns us. This dispersion can be considered a random variable with a distribution. Because treating the judgments as discrete variables becomes an intractable computational problem as the group size increases, we assume that judgments belong to a continuous random distribution. For example, if there are five people each choosing one of 17 numbers in the scale 1/9, ...,1, ..., 9, there are $17^5 = 1{,}419{,}857$ possible combinations of which 20,417 are different. Thus, the dispersion of each sample from its geometric mean has a large number of values for which one needs to determine the frequency and thus the probability distribution. To deal with this complexity, we use the continuous generalization instead. This allows us to fit probability distributions to the geometric dispersion for groups of arbitrary size. Once we have the continuous distribution of the geometric dispersion, the parameters that characterize this distribution are a function of the number of individuals n in the group.

To use the geometric mean to synthesize a set of judgments given by several individuals in response to a single pairwise comparison, as the representative judgment for the entire group, the dispersion of the set of judgments from the geometric mean must be within some prescribed bounds. To determine these bounds, we use the probability distribution of the sample geometric dispersion mentioned above. We can then find how likely the observed value of the sample geometric dispersion is. This is done by computing the cumulative probability below the observed value of the sample dispersion in the theoretical distribution of the dispersion. If it is small then the observed value is less likely to be random, and we can then infer that the geometric dispersion of the group is "small" and the judgments can be considered homogeneous or α-cohesive at that specified α level. On the other hand, if the dispersion is unacceptable, then we could divide the group of individuals into subgroups representing similarity in judgment.

The remainder of the paper is structured as follows. In section 2 we give a summary of the geometric expected value concept and its generalization to the continuous case that leads to the concept of product integral. In section 3 we define the geometric dispersion of a positive random variable and apply it to the judgments of groups. In section 4 we approximate the distribution of the group geometric dispersion. In section 5 we sketch how groups could be divided into subgroups if the geometric dispersion is large, and in section 6 we show the impact of the dispersion of a group's judgments on the priorities associated with their judgments.

2. GENERALIZATION OF THE GEOMETRIC MEAN TO THE CONTINUOUS CASE

Let X be a random variable. Given a sample from this random variable $\tilde{x} = (x_1, ..., x_n)$, the sample geometric mean is given by $\overline{x}_G \equiv \prod_{i=1}^{n} x_i^{1/n}$. Let us assume that not all the values are equally likely, and their absolute frequencies are equal to $m_1,\ldots,m_k$ with $\sum_{i=1}^{k} m_i = n$. Then, the sample geometric mean is given by: $\bar{x}_G \equiv \left[\prod_{i=1}^{k} x_i^{m_i}\right]^{1/n} = \prod_{i=1}^{k} x_i^{m_i/n}$. An estimate of the probabilities $p_i = P[X = x_i]$ is given by $\hat{p}_i = \frac{m_i}{n}$. Thus the geometric expected value of a discrete random variable X is given by:

$$E_G[X] = \prod_{\forall x_i} x_i^{P[X=x_i]} = e^{\left\{\sum_{\forall x_i} P[X=x_i]\ln x_i\right\}} = e^{E[\ln X]} \tag{1}$$

In the continuous case, because $P[X = x] = 0$ for all x, we need to use intervals rather than points, and hence, we obtain:

$$E_G[X] = \lim_{\Delta x \to 0} \prod_{\forall x} x^{P[x<X\leq x+\Delta x]} = \prod_{\forall x} x^{f(x)dx} \tag{2}$$

Equation (2) is known as the product integral (Gill and Johansen, 1990). If X is defined in the interval $(s,t]$, we have

$$\ln E_G[X] = \lim_{\Delta x \to 0} \sum_{s\leq x\leq t} P[x < X \leq x+\Delta x]\ln x_i = \int_{(s,t]} f(x)\ln x dx .$$

In general, we have

$$E_G[X] = \prod_{\mathfrak{D}(X)} x^{f(x)dx} = e^{\int_{\mathfrak{D}(X)} f(x)\ln x dx} = e^{\{E[\ln X]\}} \tag{3}$$

where $\mathfrak{D}(X)$ is the domain of the variable X and $\int_{\mathfrak{D}(X)} f(x)dx = 1$.

3. THE GEOMETRIC DISPERSION OF A POSITIVE RANDOM VARIABLE

Using the geometric expected value, we define a measure of dispersion similar to the standard deviation. Let σ_G be the geometric dispersion of a positive random variable X given by $\sigma_G(X) = E_G\left[\left|\frac{X}{\mu_G}\right|_G\right]$, where

$|x|_G = \begin{cases} x & \text{if } x > 1 \\ \frac{1}{x} & \text{if } x \le 1 \end{cases}$. For $|\ln x| = \begin{cases} \ln x, & x > 1 \\ \ln \frac{1}{x}, & x \le 1 \end{cases}$, then $e^{|\ln x|} = \begin{cases} x, & x > 1 \\ \frac{1}{x}, & x \le 1 \end{cases}$

and $\sigma_G(X) = \exp\left\{E[|\ln \frac{x}{\mu_G}|\right\} = \mu_G^{2F(\mu_G)} \exp\left\{-2\int_0^{\mu_G} (\ln x) f(x) dx\right\}$. It is possible now to write $x = \mu_G \omega^{\sigma_G}$, where the variable ω has a geometric mean equal to 1 and a geometric dispersion equal to $e^{-2\int_0^1 \ln x f(x) dx}$.

3.1 Geometric Dispersion of Group Judgments

Let X_k, $k = 1, 2, \ldots, n$ be the independent identically distributed random variables associated with the judgments. Let $\{X_k, k = 1, 2, \ldots, n\}$ be continuous random variables distributed according to a reciprocal uniform $RU[\frac{1}{9}, 9]$, i.e., the variable $Y_k = \ln X_k$ is a uniform random variable defined in the interval $[-\ln 9, \ln 9]$. The probability density function (pdf) of Y_k is given by $g(y) = \dfrac{1}{2\ln 9} I_{[-\ln 9, \ln 9]}(y)$, and hence, the pdf of X_k is given by $f(x) = \dfrac{1}{2\ln 9} \dfrac{1}{x} I_{[\frac{1}{9}, 9]}(x)$.

The sample geometric dispersion is given by:

$$s_G(x_1, \ldots, x_n) = \left[\prod_{k=1}^{n} \left|\frac{x_k}{\bar{x}_G}\right|_G\right]^{1/n} = \left[\prod_{k=1}^{n} e^{\left|\ln \frac{x_k}{\bar{x}_G}\right|}\right]^{1/n} \qquad (4).$$

Let $(x_{[1:n]}, \ldots, x_{[n:n]})$ be the order statistics corresponding to the sample $\{x_k, k = 1, 2, \ldots, n\}$, i.e., $x_{[h:n]} \le x_{[k:n]}$ if $h \le k$. Let n_1 be a value for which $x_{[k:n]} \le \bar{x}_G$ for $k = 1, 2, \ldots, n_1$. We have

$$\ln s_G(x_{[1:n]}, \ldots, x_{[n:n]}) = \frac{1}{n}\sum_{k=1}^{n} \left|\ln \frac{x_{[k:n]}}{\bar{x}_G}\right| = \frac{2n_1}{n}\left[\ln \bar{x}_G - \ln \bar{x}^G_{[n_1:n]}\right]$$

and hence, we obtain

$$s_G(x_1, \ldots, x_n) = s_G(x_{[1:n]}, \ldots, x_{[n:n]}) = \left(\bar{x}_G / \bar{x}^G_{[n_1:n]}\right)^{2n_1/n}.$$

For a group consisting of n individuals, the distribution of $S_G(X_1, \ldots, X_n)$ is given by

$$P[S_G \le s] = \sum_{n_1=1}^{n} P\left[\left[\frac{\overline{X}_G}{\overline{X}^G_{[n_1:n]}}\right]^{2n_1/n} \le s \mid \upsilon_n = n_1\right] P[\upsilon_n = n_1]$$

where $\upsilon_n = \upsilon_n(A, \overline{x}_G)$ represents the number of occurrences of the event $A \equiv \{X_k \le \overline{x}_G\}$, and it is also equal to the index of the largest order statistic less than or equal to the sample geometric mean (Galambos, 1978). Let $S_{k,n}(\overline{x}_G) = \sum_{1 \le i_1 < i_2 < \cdots < i_k \le n} P[X_{i_1} \ge \overline{x}_G, X_{i_2} \ge \overline{x}_G, \ldots, X_{i_k} \ge \overline{x}_G]$. Since $S_{k,n}(\overline{x}_G) = \sum_{r=k}^{n} \binom{r}{k} P[\upsilon_n = r]$, and $P[\upsilon_n = r] = \sum_{k=0}^{n-r} (-1)^k \binom{k+r}{r} S_{k+r,n}$ we have $P[S_G \le s] = \sum_{t=1}^{n} P\left[\left[\frac{\overline{X}_G}{\overline{X}^G_{[t:n]}}\right]^{2t/n} \le s \mid \upsilon_n = t\right] \sum_{k=0}^{n-t} (-1)^k \binom{k+t}{t} S_{k+t,n}$.

Thus, the density function is given by:

$$f_{GD}(s) = \sum_{t=1}^{n} f_{GD}(s \mid t) \sum_{k=0}^{n-t} (-1)^k \binom{k+t}{t} S_{k+t,n} \qquad (5)$$

that is a convex combination of density functions of variables of the form $\left(\prod_{k=1}^{n} (X_k)^{1/n} \Big/ \prod_{h=1}^{n_1} (X_h)^{1/n_1}\right)^{\frac{2n_1}{n}}$, i.e., the ratio of products of reciprocal uniform variates. These density functions are of the form $\frac{1}{z}(a_0 + a_1 \ln[z] + \cdots + a_{n-1} \ln[z]^{n-1})$.

There are closed form expressions for the density function of the geometric dispersion for a group consisting of three or less individuals, but for groups larger than three, it is cumbersome and not much precision is gained from it. Instead, we approximate them using simulation.

4. APPROXIMATIONS OF THE GEOMETRIC DISPERSION OF GROUP JUDGMENTS

We computed the geometric dispersion of randomly generated samples of size 20,000 under the assumption that the judgments are distributed according to a continuous reciprocal uniform distribution $RU[\frac{1}{9}, 9]$. We did this for groups consisting of 4, 5,..., 15, 20, 25, 30, 35, 40, 45, and 50 individuals. We found that as the group size increases, the geometric dispersion tends to become gamma distributed with density function given by $Gamma(\alpha,\beta,\gamma) = \beta^{\alpha}\Gamma(\alpha)^{-1}(x-\gamma)^{\alpha-1}e^{-\beta(x-\gamma)}$. The parameters α and β of these gamma distributions with location parameter (γ) equal to 1 are given in Table 1.

Table 1. Gamma Distribution Parameters ($\gamma = 1$)

n	Shape (α)	Scale (β)
4	2.80051	1.27561
5	4.03976	1.76548
6	5.40204	2.27523
7	6.55616	2.69154
8	7.67909	3.1141
9	9.29459	3.68852
10	10.4217	4.08574
11	11.8255	4.59905
12	13.0628	5.04772
13	14.4586	5.55345
14	16.0157	6.10734
15	17.4963	6.65405
20	24.2381	9.02191
25	31.4048	11.6058
30	38.5573	14.1547
35	45.6409	16.6991
40	53.1646	19.3885
45	60.1011	21.8493
50	67.254	24.429

To extend these models to groups of any size, we fit regression models to the parameters of the gamma distributions. Regression models of the shape (α) and the scale (β) parameters versus n appear to be surprisingly robust:

α(shape) = -3.48226 + 1.40829*n (R-squared = 99.9741)
β(scale) = 0.897865 + 0.504361*n (R-squared = 99.981)

In addition, the average and variance of the geometric dispersion can also be estimated from the parameters of these models:

mean = exp(1.03505 – 1.01298/n) (R-squared = 99.8463)
variance = $7.23275 * n^{-1.0664}$ (R-squared = 99.9706)

Note that as n tends to infinity, the average geometric dispersion tends to 2.81524 (99% C.I. (2.79228,2.8384)) and the variance tends to zero (99% C.I. (1.44E-9, 2.31E-9)).

We now have the basis for a statistical test to decide if the dispersion of a group can be considered larger than usual, i.e., that the probability of obtaining the value of the sample geometric dispersion of the group is greater than a pre-specified significance level (e.g., 5 percent) in the distribution of the group geometric dispersion. For example, for a group of size 6, whose judgments on a given issue are equal to {2, 3, 7, 9, 1, 2}, the geometric dispersion of the group is equal to 1.9052169. The average geometric dispersion is estimated to be equal to exp(1.03505 – 1.01298/6) = 2.378. Taking the usual significance level of 5 percent, we observe that $P[S_G(6) < 1.9052169] = 0.0376176 < 0.05$. Thus, the p-value corresponding to the sample geometric dispersion indicates that it seems rare to observe values of the geometric dispersion smaller than the sample geometric dispersion, and hence, the geometric dispersion of the group is not unusually large, which in turn implies that the geometric mean can be used as the representative preference judgment for the entire group.

5. GROUP MEMBER CLASSIFICATION BY THE GEOMETRIC DISPERSION

Let us assume that $\{x_k,\ k = 1,2,...,n\}$ is a group of judgments and let $\{x_{[k:n]},\ k = 1,2,...,n\}$ be their order statistics. If $F_{GD}[s_G(x_1,...,x_n)] \equiv P[S_G(X_1,...,X_n) \le s_G(x_1,...,x_n)] < \alpha$ (where α is usually taken to be equal to 0.05) then the geometric mean can be used as a representative of the group judgment. On the other hand, if $F_{GD}[s_G(x_1,...,x_n)] \equiv P[S_G(X_1,...,X_n) \le s_G(x_1,...,x_n)] > \alpha$ then the group needs to discuss the paired comparisons further in an attempt to reach consensus. To determine which members of a group disagree the most and hence make the geometric dispersion large, we find the p-values corresponding to the geometric dispersions of the groups of judgments given by: $\{x_{[1:n]}, x_{[2:n]}\},...,\{x_{[1:n]}, x_{[2:n]},..., x_{[k:n]}\},...,\{x_{[1:n]}, x_{[2:n]},..., x_{[n:n]}\}$.

Let $s_G(k) = s_G(x_{[1:n]},...,x_{[k:n]})$, k = 2, …, n. We give without proof because of space limitations the following results.

Lemma 1: $s_G(k) = s_G(x_{[1:n]},...,x_{[k:n]})$ is a non-decreasing function of k, i.e., $s_G(k) \ge s_G(k-1)$.

Theorem 3: Given a set of judgments $\{x_{[k:n]},\ k = 1,2,...,n\}$ with corresponding ordered geometric dispersions $\{s_G(k),\ k = 1,2,...,n\}$, if for any k, $P[S_G(k) \le s_G(k)] \le \alpha$ then $P[S_G(k-1) \le s_G(k-1)] \le \alpha$.

Definition: A group of judgments $\{x_k,\ k=1,2,...,n\}$ is said to be *α-cohesive* if $P[S_G(n) \le s_G(n)] \le \alpha$.

Definition: A member of a group of *α-cohesive* judgments is said to be *a liaison* of the group if the group is not *α-cohesive* after the elimination of the corresponding judgment from the set of judgments.

The Liaison Theorem: Given a group of *n α-cohesive* judgments, a liaison does not exist if and only if all subgroups of cardinality (*n*-1) are *α-cohesive.*

The existence of a liaison means that we may be able to divide a group into two subgroups whose preferences differ, and for which the geometric mean cannot be used as the representative group judgment. This is the subject of further study.

6. GEOMETRIC DISPERSION AND PRIORITY VARIATION

To study the relationship that exists between the geometric dispersion of a group and the dispersion of the corresponding eigenvectors, we find the range of variability of each component of the eigenvector for given sets of group judgments. This is done by first finding the distribution of the eigenvector components for random reciprocal matrices whose entries are distributed according to reciprocal uniform distributions $\mathrm{RU}[l_{ij}, u_{ij}]$.

Theorem 4: For a random reciprocal matrix $X=(x_{ij})$ with entries distributed according to a reciprocal uniform distribution, $x_{ij} \sim RU[l_{ij}, u_{ij}]$, the components of the random variable $w=(w_1,...,w_n)^T$ corresponding to the principal right eigenvector are distributed according to a beta, $\frac{w_i - \underline{w}_i}{\overline{w}_i - \underline{w}_i} \sim Beta(\alpha_i, \beta_i)$, where $\underline{w}_i = \min\{w_i\}$ and $\overline{w}_i = \max\{w_i\}$, and the principal right eigenvector of the reciprocal matrix whose entries are given by the geometric mean of its entries, $E_G[x_{ij}]$, is given by:

$$(E[w_1],...,E[w_n])^T = \left(\tfrac{\alpha_1}{\alpha_1+\beta_1}(\overline{w}_1 - \underline{w}_1) + \underline{w}_1, \cdots, \tfrac{\alpha_n}{\alpha_n+\beta_n}(\overline{w}_n - \underline{w}_n) + \underline{w}_n\right)^T.$$

Let $x_{ij} = \mu_{ij} w_{ij}^{\sigma_{ij}}$ where $\mu_{ij} = \sqrt{l_{ij} u_{ij}}$ is the geometric mean and σ_{ij} is the geometric dispersion of $x_{ij} \sim RU[l_{ij}, u_{ij}]$. By definition, $\mu_{ji} = 1/\mu_{ij}$ and $\sigma_{ji} = \sigma_{ij}$. Thus, we have $w_{ji} = 1/w_{ij}$. Let us assume that the reciprocal matrix of geometric means is consistent, i.e., $\mu_{ij}\mu_{jk} = \mu_{ik}$. Then the principal right (pr-) eigenvector of the matrix $(x_{ij} = \mu_{ij} w_{ij}^{\sigma_{ij}})$ is given by the Hadamard product of the pr-eigenvector of the matrix (μ_{ij}), μ_w, and the pr-eigenvector of the matrix $(w_{ij}^{\sigma_{ij}})$. The entries of this matrix are random reciprocal uniform variables $RU[l_{ij}/\mu_{ij}, u_{ij}/\mu_{ij}]$ whose geometric dispersion is given by $\left(u_{ij}/l_{ij}\right)^{1/4}$. Since the geometric dispersion of the variables x_{ij} and that of the variables $w_{ij}^{\sigma_{ij}}$ is the same, because $x_{ij}/\mu_{ij} = w_{ij}^{\sigma_{ij}}$, we have $\sigma_{ij} = \left(u_{ij}/l_{ij}\right)^{1/4}$. Thus, bounding the dispersion of the entries of the matrix $(w_{ij}^{\sigma_{ij}})$ bounds the dispersion of the entries of the matrix $(x_{ij} = \mu_{ij} w_{ij}^{\sigma_{ij}})$. For example, consider a group of five people who provide the judgments given in the following matrix:

$$\begin{pmatrix} 1 & (2,3,4,5,6) & (1/2,2,1,1/3,4) & (3,4,1/2,2,8) \\ & 1 & (1,2,3,4,5) & (5,4,3,2,1) \\ & & 1 & (1/4,1/3,1,2,5) \\ & & & 1 \end{pmatrix}$$

The geometric dispersion of each group and their corresponding p-values (see Table 2) show that the judgments (1,3), (1,4) and (3,4) have large geometric dispersion. This leads to large dispersion on the values of the eigenvector components (See Table 3a) and a violation of Pareto Optimality.

Table 2. Geometric Dispersions and p-values

GD				
	1	1.39930601	2.194046345	2.586241042
	1.39930601	1	1.630260481	1.630260481
	2.194046345	1.630260481	1	2.624239686
	2.586241042	1.630260481	2.624239686	1
p-value	0	0.005	**0.157**	**0.3**
		0	0.025	0.025
			0	**0.315**
				0

Reducing the dispersion of the judgments as in the following matrix

$$\begin{pmatrix} 1 & (2,3,4,5,6) & (2,2,1,1,2) & (3,4,3,2,8) \\ & 1 & (1,2,3,4,5) & (5,4,3,2,1) \\ & & 1 & (1,2,1,2,5) \\ & & & 1 \end{pmatrix}$$

leads to less dispersed eigenvectors that satisfy Pareto Optimality (See Table 3b).

Table 3. Individual Eigenvectors and Eigenvector of the Geometric Mean

(a)	P1	P2	P3	P4	P5	GM
w_1	0.2882933	0.460725	0.2671401	0.351129	0.581191	0.41676
w_2	0.3012872	0.274057	0.2835793	0.281388	0.207615	0.266142
w_3	0.2016469	0.116421	0.1354193	0.273203	0.136396	0.178548
w_4	0.2087724	0.148798	0.3138611	0.09428	0.074798	0.138551
(b)	P1	P2	P3	P4	P5	GM
w_1	0.401242	0.473463	0.4394576	0.438378	0.535109	0.471061
w_2	0.295662	0.268317	0.2699693	0.265162	0.227089	0.253814
w_3	0.187044	0.172147	0.1768845	0.184266	0.161707	0.176787
w_4	0.116052	0.086073	0.1136883	0.112194	0.076095	0.098339

7. CONCLUSIONS

In this paper we put forth a framework to study group decision-making in the context of the AHP. A principal component of this framework is the study of the homogeneity of judgments provided by the group. We developed a new measure of the dispersion of a set of judgments from a group for a single paired comparison, and illustrated the impact that this dispersion has on the group priorities.

References

Aczel, J. and T. L. Saaty, 1983, Procedures for synthesizing ratio judgments. *Journal of Mathematical Psychology* **27**: 93-102.

Condon, E., B. Golden and E. Wasil, 2003, Visualizing group decisions in the analytic hierarchy process. *Computers & Operations Research* **30**: 1435-1445.

Galambos, J., 1978, *The asymptotic theory of extreme order statistics*. New York, J. Wiley.

Gill, R. D. and S. Johansen, 1990, A Survey of product-integration with a view toward application in survival analysis. *The Annals of Statistics* **18**(4): 1501-1555.

Ramanathan, R. and L. S. Ganesh, 1994, Group preference aggregation methods employed in the AHP: An evaluation and an Intrinsic process for deriving member's weightages. *European Journal of Operational Research* **79**: 249-269.

Saaty, T. L., 1994, *Fundamentals of decision making*. Pittsburgh, PA, RWS Publications.

Saaty, T. L. and L. G. Vargas, 2003, The Possibility of group choice: Pairwise comparisons and merging functions. Working Paper, The Joseph M. Katz Graduate School of Business, University of Pittsburgh, Pittsburgh, PA.